# BSAVA
# Small Animal Reproduction and Neonatology

**General Editor**

**Gillian M. Simpson**
BVM&S MRCVS
Edinburgh's Telford College
Crewe Road, Edinburgh EH4 2NZ, UK

**Scientific Editors**

**Gary C.W. England**
BVetMed PhD DVetMed CertVA DVR DVRep DipACT FRCVS
Royal Veterinary College, University of London
Hawkshead Lane, North Mymms,
Hatfield, Herts AL9 7TA, UK

**and**

**Mike Harvey**
BVMS PhD MRCVS
Department of Veterinary Clinical Studies
University of Glasgow Veterinary School
Bearsden Road, Bearsden, Glasgow G61 1QH, UK

Published by:

**British Small Animal Veterinary Association**
Kingsley House, Church Lane
Shurdington, Cheltenham
GL51 5TQ, United Kingdom

A Company Limited by Guarantee in England.
Registered Company No. 2837793.
Registered as a Charity.



The line illustrations in this book were drawn by Ian Lennox and are printed with his permission.

A catalogue record for this book is available from the British Library.

ISBN 0 905214 36 6

The publishers and contributors cannot take responsibility for information provided on dosages and methods of application of drugs mentioned in this publication. Details of this kind must be verified by individual users from the appropriate literature.

Typeset and printed by: Fusion Design, Fordingbridge, Hampshire, UK

# Other Manuals

**Other titles in the BSAVA Manuals series:**

*Manual of Anaesthesia for Small Animal Practice*
*Manual of Canine and Feline Gastroenterology*
*Manual of Canine and Feline Nephrology and Urology*
*Manual of Companion Animal Nutrition and Feeding*
*Manual of Canine Behaviour*
*Manual of Exotic Pets*
*Manual of Feline Behaviour*
*Manual of Ornamental Fish*
*Manual of Psittacine Birds*
*Manual of Raptors, Pigeons and Waterfowl*
*Manual of Reptiles*
*Manual of Small Animal Arthrology*
*Manual of Small Animal Cardiorespiratory Medicine and Surgery*
*Manual of Small Animal Clinical Pathology*
*Manual of Small Animal Dentistry, 2nd edition*
*Manual of Small Animal Dermatology*
*Manual of Small Animal Diagnostic Imaging*
*Manual of Small Animal Endocrinology, 2nd edition*
*Manual of Small Animal Fracture Repair and Management*
*Manual of Small Animal Neurology, 2nd edition*
*Manual of Small Animal Oncology*
*Manual of Small Animal Ophthalmology*

# Contents

# Contributors

**David J. Argyle** BVMS MRCVS
Department of Veterinary Clinical Studies, University of Glasgow Veterinary School, Bearsden Road, Bearsden, Glasgow G61 1QH

**Eva Axner** DVM
Department of Obstetrics & Gynaecology, Faculty of Veterinary Medicine, PO Box 7039, S-750 07, Uppsala, Sweden

**Tony S. Blunden** BVetMed PhD MRCVS
Animal Health Trust, Balaton Lodge, PO Box 5, Newmarket, Suffolk CB8 7DW, UK

**Annelie Eneroth** DVM
Department of Obstetrics & Gynaecology, Faculty of Veterinary Medicine, PO Box 7039, S-750 07, Uppsala, Sweden

**Gary C.W. England** BVetMed PhD DVetMed CertVA DVR DVRep DipACT FRCVS
Royal Veterinary College, University of London, Hawkshead Lane, North Mymms, Hatfield, Herts AL9 7TA, UK

**Wenche Farstad** MDNV PhD
Department of Reproduction and Forensic Medicine, Norwegian College of Veterinary Medicine, PO Box 8146, N- 0033, Oslo, Norway

**Mike Harvey** BVMS PhD MRCVS
Department of Veterinary Clinical Studies, University of Glasgow Veterinary School, Bearsden Road, Bearsden, Glasgow G61 1QH

**Denise Hewitt** BSc
Royal Veterinary College, University of London, Hawkshead Lane, North Mymms, Hatfield, Herts AL9 7TA, UK

**Paula Hotston Moore** VN
Veterinary Nursing Department, University of Bristol School of Veterinary Science, Langford House, Langford, Bristol BS18 7DU, UK

**Ian A. Jeffcoate** BSc PhD
Department of Veterinary Physiology, University of Glasgow Veterinary School, Bearsden Road, Bearsden, Glasgow G61 1QH

**Laurence R.J. Keenan** MVB MVM PhD MRCVS
Department of Veterinary Clinical Studies, Faculty of Veterinary Medicine, University College Dublin, Ballsbridge, Dublin 4, Republic of Ireland

**Catharina Linde-Forsberg** DVM PhD
Department of Obstetrics & Gynaecology, Faculty of Veterinary Medicine, PO Box 7039, S-750 07, Uppsala, Sweden

**Kit Sturgess** MA VetMB PhD CertVR MRCVS
Royal Veterinary College, Hawkshead Lane, North Mymms, Hatfield, Herts AL9 7TA, UK

**John P. Verstegen** DVM MSc PhD
Bd Colonster 20, B44, Small Animal Reproduction Department, Veterinary College, University of Liège, B 4000 Liège, Belgium

**Robert N. White** BSc BVetMed CertVA MRCVS
Davies White, Manor Farm Business Park, Higham Gobion, Hitchin, Herts SG5 3HR

**John R. Watts** BVSc
Department of Veterinary Sciences, University of Melbourne, Princes Highway, Werribee, 3030 Victoria, Australia

**Patrick J. Wright** BVSc MVSc PhD
Department of Veterinary Sciences, University of Melbourne, Princes Highway, Werribee, 3030 Victoria, Australia

# Foreword

An appreciation of the complexity of the reproductive process and the parts played by the male and female is pivotal to the successful management of the problems encountered by the owners and breeders of companion animals. The *Manual of Small Animal Reproduction and Neonatology* is a new addition to the BSAVA Manual Series and comprehensively describes the requirements for normal reproduction and the problems likely to be recognized in clinical practice.

The Manual reviews the normal physiology and endocrinology of the dog and cat, looking first at female infertility, the function and dysfunction of the mammary gland and problems of the non-pregnant animal. It then goes on to consider the male dog and cat in similar detail. Mating and artificial insemination in each species are discussed, followed by a consideration of pregnancy and parturition. The care of the neonate, including a review of congenital problems and the role of infectious diseases and their prevention, is also described. The Manual concludes with chapters on the medical and surgical therapies applicable to reproductive problems.

The BSAVA is fortunate that the Editors have invited such an array of internationally recognized authors to contribute to this project and the use of full colour illustrations once more complements the text admirably. I have no doubt that this Manual will be widely used, not only by veterinary nurses and veterinary surgeons in practice and those in training but also by our breeder clients who will appreciate the wealth of information contained herein.

H. Simon Orr BVSc DVR MRCVS
BSAVA Senior Vice-President 1997-98

# Preface

The first edition of this manual has been designed for veterinary practitioners and nurses who are presented with dogs and cats with reproductive disorders. It is also written for undergraduates who have both a need and hopefully a desire to learn more about the area in order to pass their examinations and, equally importantly, to become better clinicians. We have attempted to cover, in addition to all aspects of reproduction, the related areas of the mammary gland and the neonate.

We accepted at the outset that there would be some duplication of coverage of certain conditions, but felt that that would be no bad thing. There may even be minor differences of opinion in some areas. We have tried to restrict these, but it would be a peculiar world if total agreement reigned in all matters and, to say the least, veterinary science is no exception.

We have to thank a large number of people, although not necessarily by name. Over the last two decades, the field of canine and feline reproduction has developed and progressed very rapidly. We would like to acknowledge the many people who contributed to this explosion of this knowledge. This manual is a tribute to them all. We much appreciate the cooperation of all the authors, who gave much time to the writing of their chapters while at the same time coping with extremely heavy clinical and teaching loads. Without fail, they accepted the suggested alterations with good nature and made the requested changes extremely rapidly. We would like to thank the BSAVA for commissioning this manual on Reproduction and Neonatology, and finally to thank Marion Jowett, the Publishing Manager, who, with unfailing good manners, managed to push the whole thing through to fruition while still allowing the Editors to feel that they were in charge.

Gillian M. Simpson
Gary C.W. England
Mike Harvey
February 1998

CHAPTER ONE

# Physiology and Endocrinology of the Bitch

*Ian Jeffcoate*

## INTRODUCTION TO THE OESTROUS CYCLE

The oestrous cycle is a co-ordinated sequence of ovarian, utero-vaginal and behavioural changes which has evolved in mammals to ensure production and fertilization of female gametes and intrauterine development of the conceptus. Most non-pregnant domestic animals show continuous oestrous cycles, each comprising the following stages: *oestrus*, when mature ova are shed into the tubal reproductive tract and sexual receptivity optimizes chances of internal fertilization occurring; *dioestrus*, when preparations for pregnancy occur; and finally, if impregnation fails, a return to *pro-oestrus*, when renewed follicle development leads up to oestrus again and further opportunities for breeding to occur. Animals such as pigs, cattle and horses return to oestrus about every 3 weeks but if pregnancy arises, there is substantial prolongation of dioestrus due to the requirements of pregnancy.

The cycle of the domestic bitch differs from this general scheme in several respects. First, each cycle is at least 5 months in duration; secondly, pregnancy occurs within the normal dioestrous phase rather than prolonging it and, thirdly, a long period of relative ovarian inactivity termed *anoestrus* arises between cycles whether the bitch is pregnant or not. This has given rise to a change in cycle terminology in bitches which requires some explanation. Since in most species pro-oestrus is generally so brief that its duration can be ignored, dioestrus (literally meaning 'between oestrus') is often, but incorrectly, used to describe the inter-oestrus interval. This contains both the luteal and the pro-oestral phases. However, as the bitch's inter-oestrous interval consists not only of the luteal phase but also the extremely long anoestrus phase, the term 'metoestrus' is often used to describe the luteal phase in preference to 'dioestrus'.

Despite these differences, the oestrous cycle of the bitch has been shown to be controlled by the same hormonal interactions as those demonstrated in other animals and acting between:

- The hypothalamus and pituitary gland, where primary control resides
- The ovary, the source of the steroid hormones oestrogen and progesterone
- The tubular reproductive tract, accessory sex glands and brain behavioural centres; the ultimate target tissues of the ovarian hormones.

This chapter will describe the canine oestrous cycle and review the endocrine features and developmental processes, pointing out when endocrine measurements or interventions might be appropriate.

## PRO-OESTRUS

Various clinical signs can be taken as evidence of pro-oestrus. These include vulval reddening and enlargement and, later, the appearance of a serosanguineous vulval discharge. This discharge is usually taken as marking the first day of pro-oestrus. Behavioural changes can be observed which include urine marking, restlessness and disobedience, with a tendency to roam and increased attractiveness to dogs. These changes are attributed to increasing plasma oestradiol concentration (Figure 1.1) due to the activity of developing ovarian follicles. Oestradiol promotes various processes in the reproductive tract, stimulating growth and activity of the glandular epithelium and promoting mucosal vascularity and oedema. The latter effects cause the mucosal capillaries to become fragile, and leaky endothelial cell junctions permit passage of blood plasma and cells into the uterine lumen, thus accounting for the bloodstained vulval discharge. The general oedema of the reproductive tract during pro-oestrus can be visualized as a decrease in ultrasonographic echogenicity. Additionally, mucosal development results in impressive ballooning and reddening of the vaginal mucosal folds, which can be viewed directly by endoscopy. Dramatic epithelial cell proliferation also occurs in the vaginal mucosa under oestradiol stimulation. These changes in vaginal morphology will be discussed below in relation to their usefulness in determining the time of ovulation and optimal fertility.

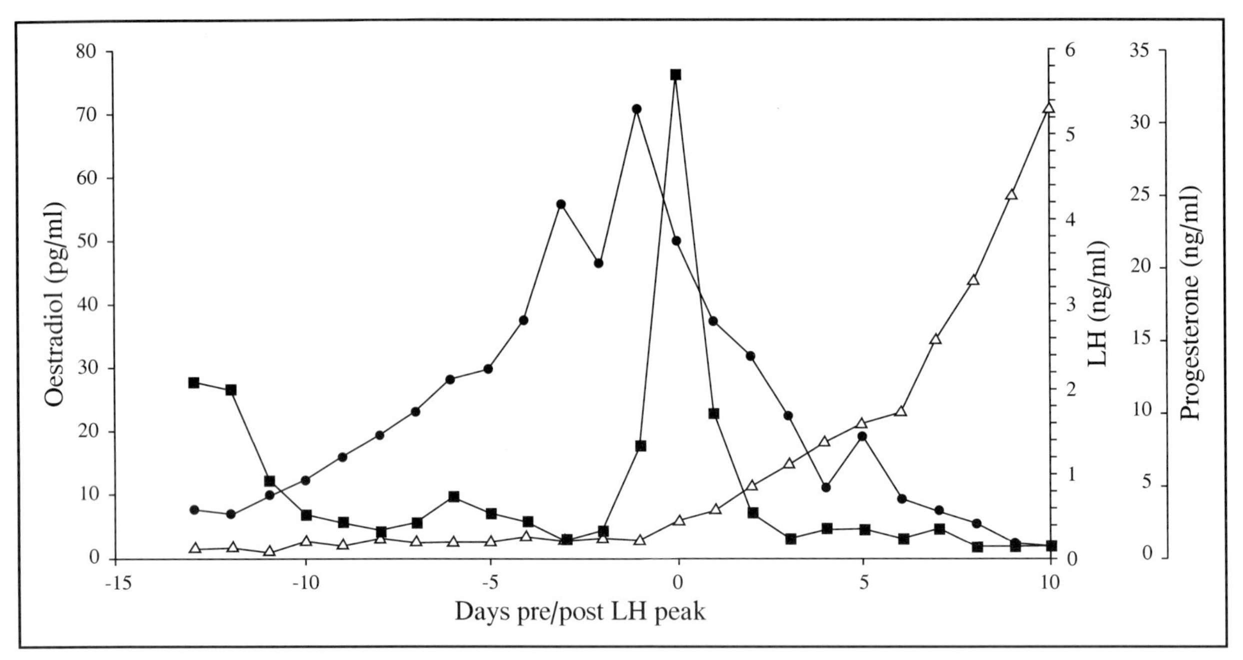


*Figure 1.1: Changes in plasma concentration of oestradiol (•), luteinizing hormone (LH, ■) and progesterone (Δ) in the bitch. The graphs have been aligned to the day of the LH peak. Note, peak oestradiol on day –1 and a detectable increase in progesterone concentrations on day 0. (Conversion factors: 1 pg/ml oestradiol = 3.6 pmol/l; 1 ng/ml progesterone ≈ 3 nmol/l.)*

Pro-oestrus is associated with follicle development stimulated by the gonadotrophic hormones luteinizing hormone (LH) and follicle stimulating hormone (FSH), which are secreted by the anterior pituitary gland under the influence of hypothalamic gonadotrophin-releasing hormone (GnRH). In the bitch, just as in other animals, the peak in the plasma concentration of oestradiol results, a day or so later, in the peak concentrations of LH (Figure 1.1). The surge in plasma LH concentration will, therefore, always occur automatically after a certain level of follicular activity and oestradiol output has been attained and, since ovulation follows the LH peak, it is said to be *spontaneous*. This is in contrast to the situation in cats and mustelids, which are described as *induced ovulators*, because copulation is required to induce LH release and ovulation. Such animals appear to lack the ability for oestradiol to augment LH secretion.

A facilitatory level of FSH is also required to stimulate follicle development and oestradiol secretion, but circulating FSH does not increase as noticeably as LH because the follicles secrete 'inhibin', a selective inhibitor of FSH secretion. Nevertheless, FSH plays an important role in the maturation of the follicle and in equipping its cells for conversion to corpora lutea after ovulation. This process is of key importance in the bitch's cycle because there is a rise in progesterone secretion by the pre-ovulatory follicles (see Figure 1.1), which appears to play a central role in triggering ovulation and standing oestrus. Increased plasma progesterone concentration before ovulation in bitches is unusual in comparison with other domestic animals, particularly ruminants, where progesterone concentrations must be minimal at this time to permit oestrus and ovulation. However, the amount of progesterone circulating in pre-ovulatory bitches is still relatively low, and it is to be remembered that oestrus and ovulation can be reliably postponed by administration of large doses of progestogens in pro-oestrus.

The pre-ovulatory LH surge is often depicted as the central event of the cycle because of its role in stimulating ovulation and transition to the progesterone-dominated luteal phase, metoestrus. For this reason, and because the onset of standing oestrus varies considerably between different bitches, pregnancy stages are best timed with respect to the day of the LH peak, rather than to the first day of oestrus as is the normal practice in other species.

## OESTRUS

### Ovarian steroids and behaviour

Oestrus is derived from the Latin for gadfly, and literally refers to frenzied behaviour, i.e. the transition from the attractive but unreceptive behavioural characteristic of pro-oestrus, to posturally inviting and receptive behaviour (lordosis). Pheromones are chemical signalling compounds important in delineating this transition. They are secreted by the bitch under the influence of oestradiol and are detected by the dogs' olfactory or vomeronasal organs. They are produced in the kidneys and reproductive tract and are voided with urine or in the vulval discharge, specifically for the purpose of advertising sexual status. When combined with the behavioural signs, pheromones increase sexual attractiveness and stimulate male reproductive activity. One bitch pheromone is methyl-*p*-hydroxybenzoate, and if this compound is applied to the vulva of anoestrous or even spayed bitches, it will stimulate excitement and mounting attempts by dogs. In addition, bitch pheromone effects are not restricted to dogs, since they can also have significant oestrus-advancement or cycle-synchronizing effects on other bitches, particularly in a kennel environment. This shows that pheromones can influence the activity of GnRH centres in the hypothalamus, ultimately leading to increased ovarian activity.

Brain behavioural centres in bitches appear to be primed by rising oestradiol levels but full expression of oestrus requires withdrawal of oestrogen in the presence of progesterone. Bitch cycles in which follicles fail to ovulate (and enter the progesterone secreting luteal phase) are therefore likely to be characterized by a period of pro-oestrous behaviour which wanes and then recurs later if a new cohort of oestrogen-secreting follicles subsequently develop without an intervening luteal phase. This is commonly referred to as a split oestrus.

## Ovulation

The requirement for circulating progesterone concentration to increase before either full expression of oestrus will occur or ovulation is triggered, implies that the onset of the initial rise in plasma progesterone should occur just before ovulation and that this might be a useful predictor of the same.

Bitches normally have multiple ovulations, and histological and laparoscopic examinations show that although most ovulations occur in a range of 30–48 hours after the LH peak, some follicles may not ovulate until as late as 96 hours after the LH peak. Canine oocytes must also undergo a further meiotic maturational stage after ovulation before fertilization. This process is difficult to study, but by using carefully timed inseminations with short-lived frozen semen from various dogs to yield identifiable puppies, it is possible to show that oocyte maturation takes from 2 to 3 days. When this figure is added to the interval between the LH peak and ovulation, it can be seen that fertilization of bitch oocytes cannot occur until 4 days after the LH peak, and this might be termed the start of the *fertile period* (see Figure 1.2). Confusion as to the relevance of the fertile period can arise because bitches are quite likely to be in standing oestrus up to 5 days before its onset, and may continue to stand for as long as 5 days after this time, with pregnancies still ensuing in some instances. However several factors are responsible for an *apparently* long fertile period in the bitch. These include prolonged sperm survival time in the bitch's genital tract, thought to be several days, and the likely time-spread of ovulation and oocyte maturation (2–3 days) and oocyte survival (1–2 days). When all these time factors are considered, the fertile period in bitches probably ranges from days 4 to 7 after the LH peak. As a consequence of the variability, however, gestation length cannot be related to mating date. If it were, pregnancy would appear to vary from 58 to 72 days, but if it is related to the day of the pre-ovulatory plasma LH peak, pregnancy duration is 65 days with very little variability.

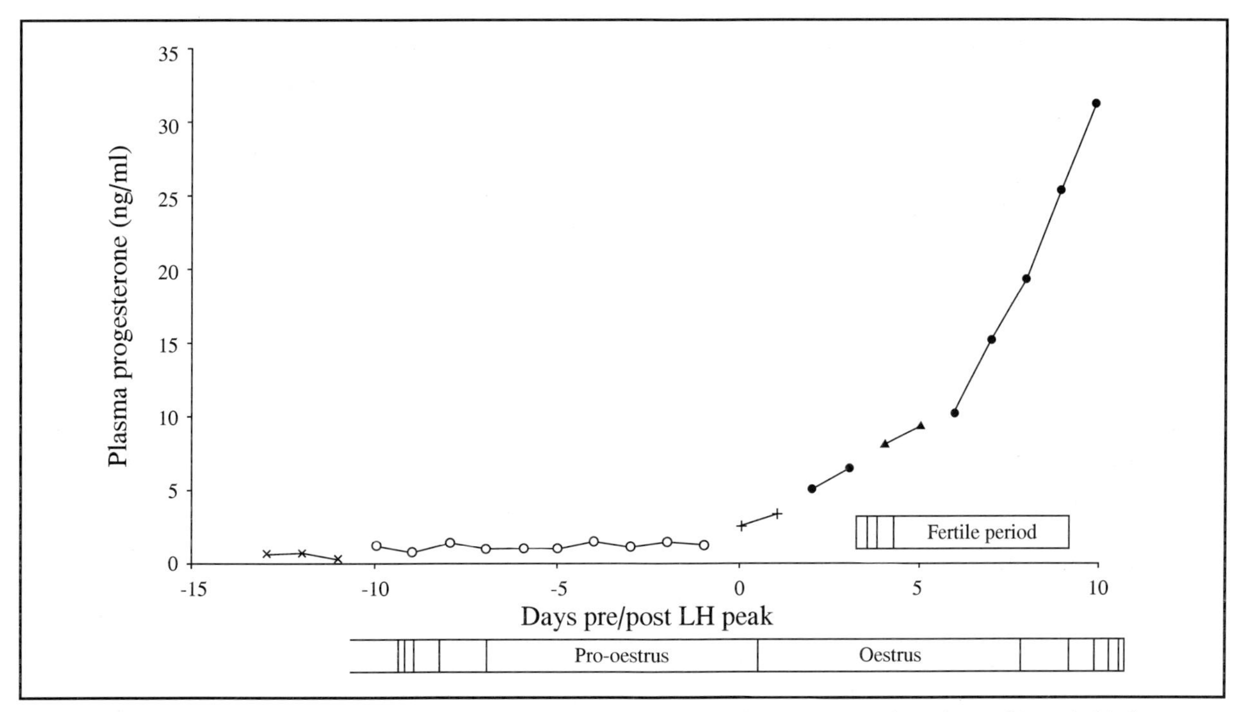


***Figure 1.2**: The typical plasma progesterone profile and cycle stages drawn from average values obtained in eight bitches relative to the day of the LH peak. Key features which may be identified and used to predict the fertile period (for definition see text) are: progesterone concentration is basal (maybe undetectable) during late anoestrus (x–x) and rises to low but detectable concentrations during pro-oestrus (o–o). There is a pre-ovulatory increase in plasma progesterone concentration at the same time as the pre-ovulatory LH surge (+–+). Because there are relatively consistent intervals between the LH surge and ovulation, the progesterone concentration has risen to quite characteristic values (▲–▲) at the onset of the fertile period and should then continue to rise (●–●) until about day 30. (The conversion factor from ng/ml to nmol/l is approximately 3.)*

| Bitch number | Blood on vulva | Onset of vaginal mucosal oedema | First day of oestrus smear | First day of oestrus (acceptance) | Onset of vaginal mucosal shrinkage | Onset of metoestrus | Last day of oestrus |
|---|---|---|---|---|---|---|---|
| 1 | -5 | -7 | 0 | +3 | -1 | +10 | +9 |
| 2 | -5 | -8 | +1 | +3 | -1 | +11 | +10 |
| 3 | -5 | -6 | +1 | -4 | -3 | +13 | +8 |
| 4 | -12 | -12 | -1 | +1 | -1 | +6 | +5 |
| 5 | -8 | -8 | +1 | +1 | -1 | +9 | +9 |
| 6 | -3 | -6 | 0 | +2 | -2 | +10 | +8 |
| 7 | -5 | -11 | -3 | +1 | -4 | +7 | +6 |
| 8 | -11 | -11 | -7 | -1 | -7 | +11 | +11 |
| **Average** | **-6.8** | **-8.6** | **-1** | **+1** | **-2.5** | **+9.6** | **+8.3** |

Points to note are:
- The oestrous smear may be obtained many days before oestrus.
- First day of acceptance is often close to the start of the fertile period, but this cannot be guaranteed.
- The last day of acceptance can be several days after the estimated end of the fertile period, i.e. closer to the onset of metoestrus.
- The onset of mucosal shrinkage usually occurs just before the LH peak, but subsequent more marked stages of mucosal angulation (see Figure 1.3b) arise at the onset of the fertile period.

***Table 1.1**: Timing of certain key clinical signs relative to the day of the LH peak in a group of eight bitches. The presumed fertile period is from days +4 to +7 for reasons stated in the text.*

### Fertilization

The oestrogen priming of the mucosa is in most animals credited with stimulating secretions which favour gamete survival and transport. Oestrogen is also known to stimulate smooth muscle excitability, probably indirectly via adrenergic receptors. This and the stimulation of tubal ciliary action promote sperm transport through the utero-tubal junctions and the delivery of ovulated oocytes to the site of fertilization in the uterine tubes. Later, as follicle luteinization proceeds, the increasing plasma concentration of progesterone is responsible for altering the character of mucosal secretions, thereby impairing further transport and survival of spermatozoa in the tract and in particular the migration of spermatozoa through the cervix (see below). Progesterone also decreases smooth muscle excitability and is thought to slow the passage of ova or embryos in the uterine tubes, thereby delaying entry of embryos into the uterus until conditions are more suitable.

These well documented effects of the two main ovarian steroids are considered equally valid in the bitch, despite the comparatively early decrease in the oestrogen:progesterone ratio as a result of pre-ovulatory luteinization (Figure 1.1). We must therefore assume that in the bitch, progesterone is facilitatory to gamete survival, transport and fertilization but that these actions may well require a long (pro-oestrous) period of oestradiol dominance. In the bitch, as in other species, unbalancing the oestrogen: progesterone ratio by pharmacological administration of oestrogen during the post-ovulatory period increases tubal motility, and embryos will arrive in the uterine horns too soon, before proper progesterone conditioning has occurred. This is the rationale behind post-coital oestrogen therapy.

## TIMING THE FERTILE PERIOD

With the long period of sexual receptivity in bitches, the spread in ovulation times and longer gamete maturation and viability than usual, the fertile period has been set somewhere between 4 and 7 days after the LH peak. However, fertility is probably poor towards the end of the fertile period, and there may be specific instances when good timing of insemination becomes crucial, for example when using frozen semen or a dog of low fertility. Various signs can be used to help pinpoint the fertile period and the optimum mating time in bitches. These include exfoliative vaginal cytology (vaginal smears), vaginal endoscopy and plasma hormone measurements.

### Vaginal cytology

Increasing plasma oestradiol concentration during pro-oestrus stimulates cell division in the basal layers of the vaginal epithelium, but then as plasma oestradiol concentration falls, endocrine support of this new highly stratified epithelium diminishes and there is a marked increase in number of dead cornified cells, some of which are easily sloughed off. As pro-oestrus progresses, therefore, there is a decrease in the number of nucleated epithelial cells. Additionally, there are usually large numbers of red blood corpuscles in the early to mid pro-oestrous smear. Peak cornification occurs as plasma progesterone concentration begins to rise, roughly at the onset of oestrus, but there are no distinctive features of vaginal cytology which can be related to the time of peak fertility which usually occurs some days later (Table 1.1). Nevertheless, vaginal cytology remains a popular tool for assessing sexual status of a bitch, largely because it is cheap, quick and simple to perform. Cells can be obtained using a cotton bud inserted into the vagina, preferably using a small speculum, to scrape the mucosa gently; or by flushing with saline. Adherent cells can be transferred to a microscope slide by rolling and then stained using a differential (trichrome) or non-differential (e.g. Diff-Quik®) stain. By using differential staining, dead keratinized cells appear as orange flakes, whereas the active nuclei of less superficial basal or parabasal epithelial cells stain various shades of blue to green.

Towards the end of oestrus, the vaginal smear reveals characteristic changes which include the return of nucleated epithelial cells and the appearance of large numbers of leucocytes. These changes usually commence 7–9 days after the LH peak, and when seen are referred to as the 'metoestrous vaginal smear'. There is a transitional period between 'oestrous' and 'metoestrous' smears, when increasing numbers of nucleated cells can be observed, and this often starts at the end of the fertilization period, about 7–9 days after the LH peak. However, by the time such changes are obvious, the fertile period and possibly also oestrus may have passed (Table 1.1).

Vaginal cytology can undoubtedly provide useful information on the breeding status of the bitch, but considerable variability in the timing of the main features relative to the likely time of peak fertility limits the usefulness of the technique in the absence of corroboratory evidence.

## Vaginal endoscopy

The switch from oestradiol to progesterone secretion by the pre-ovulatory follicles, which was discussed above, reduces mucosal vascularity and oedema, and this is manifested by a marked change in the appearance of the vaginal mucosa. Vaginal endoscopy offers the opportunity to observe these changes in character and appearance of the vaginal mucosa, as its folds. It first becomes more convoluted and oedematous during pro-oestrus and then shows progressive shrinkage and becomes very pale, even white in appearance, close to the time of the LH peak, followed by abrupt angulation as progesterone concentrations rise and oestradiol concentrations fall (Figure 1.1). There follows a stage of progressively increasing angulation towards the time of the ovulation and subsequently the onset of the fertile period. These changes can be easily recognized with a little practice (Figure 1.3) and are sufficiently repeatable that each stage can be ascribed a score. Vaginal endoscopy is proving very useful in forecasting the optimal mating time in individual bitches. After the end of the fertile period, i.e. at the onset of metoestrus, the vaginal mucosa appears pale and thin with increasingly rounded folds, and perhaps most characteristically, the mucosal folds in the anterior vagina appear irritable and may be provoked to close rapidly, forming a rosette-like pattern.

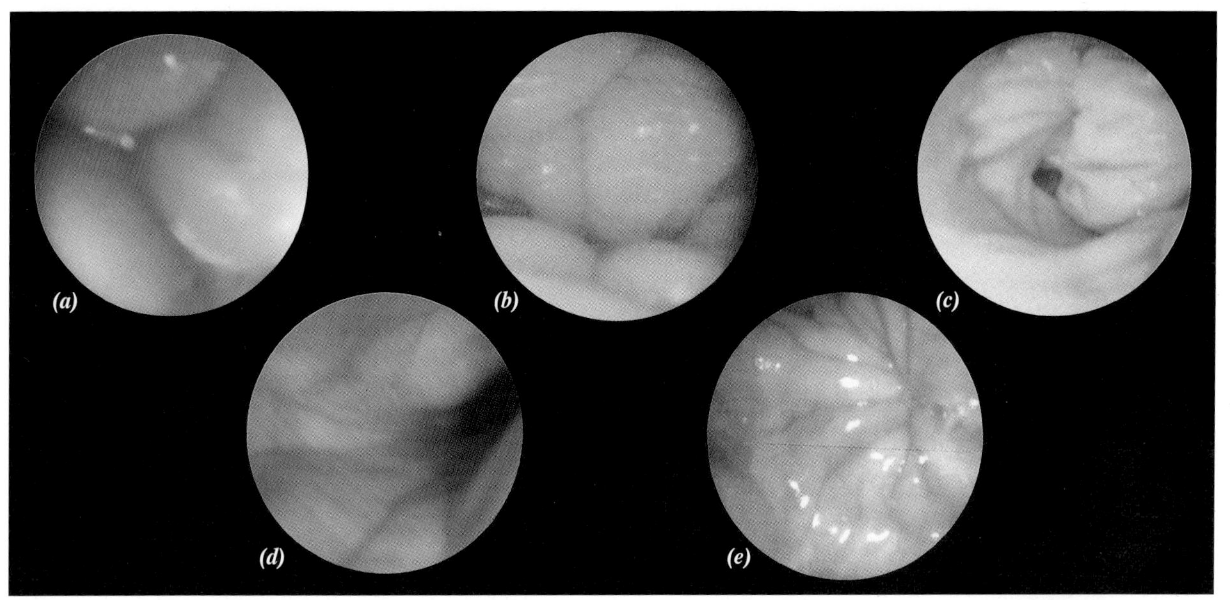


***Figure 1.3.*** *Endoscopic appearance of the vaginal mucosa at key stages in its development during the bitch cycle. (a) Pro-oestrus; note pink fluid and oedematous glossy folds. (b) Onset of oestrus; note pale colour and onset of wrinkling of the mucosa. This stage generally occurs just before the LH peak. (c) Mid-oestrus, note pale creamy colour, marked wrinkling (shrinkage) and angular mucosal profiles. This corresponds to the middle of the fertile period. (d) Early metoestrus; mucosal folds are rounded with pale patches, and irritable, sometimes closing to form a rosette pattern as in (e).*

*Courtesy of F. E. F. Lindsay.*

### Hormone measurements

Measurements of plasma progesterone concentration can be very useful in predicting the fertile period of the bitch. The background behind the changes in the concentrations of the hormones has been more fully explained in preceding sections (see Figure 1.2).

Progesterone concentration should be basal (quite probably undetectable) during late anoestrus and will rise to low but detectable concentrations (around 3 nmol/l (1 ng/ml)) by late pro-oestrus. As explained above, plasma progesterone concentration begins to rise at the time of the pre-ovulatory LH surge, reaching 3-6 nmol/l (1-2 ng/ml); it will probably have reached 6-12 nmol/l (2-4 ng/ml) 2 days later (the day of ovulation), and by day 4, the onset of the fertile period, circulating progesterone will probably have risen to 18-30 nmol/l (6-10 ng/ml) (Figure 1.1). Blood samples are necessary to determine progesterone, as assays to measure progesterone in other body fluids such as saliva and urine are not sufficiently robust at present. Sampling can safely be postponed until well after the first signs of pro-oestrus, but samples should be collected at least every 2-3 days during late pro-oestrus or after oestrous behaviour has commenced. This will allow the rapid periovulatory increase in progesterone to be detected to forecast possible mating dates. In cases in which a very short pro-oestrus is anticipated, the start of blood sampling should be earlier to avoid being too late to detect ovulation.

Some differences arise in the values quoted for plasma progesterone concentration, depending whether radioimmunoassay or enzyme immunoassays have been used, the latter giving higher but well correlated results. Much of the published data (including those in the present article) have been obtained using radioimmunoassay. If using a commercial enzyme immunoassay, plasma progesterone concentration during the bitch's oestrous cycle will be nearly doubled but the relative changes with cycle stage will be the same. Absolute progesterone concentrations cannot be obtained with qualitative tests, but the relative changes in plasma progesterone concentration can be assessed after the start of pro-oestrus as the cycle progresses, and this should be useful for corroborating other clinical signs.

Progesterone is the only hormone currently of value in predicting the fertile period. LH and oestradiol would be equally or more predictive but sample collection and assay of these hormones is not currently practical.

### Other aspects defining the fertile period

Recent work using infusion of radio-opaque contrast dye into the anterior vagina has revealed interesting links between circulating oestradiol and progesterone concentrations and cervical opening. These findings are relevant to the fertile period in bitches because the cervix remains closed (impenetrable to dye and presumably also spermatozoa) until plasma oestradiol declines and the ratio of progesterone to oestradiol increases during the pre-ovulatory period. The cervix then remains open until about 6 days after the LH peak, i.e. about 2 days before full cytological metoestrus, at which time plasma progesterone is high and oestradiol virtually undetectable. These results show that spermatozoa may only be able to gain access to the uterus during a certain proportion of oestrus, despite the common observation that mating can occur before and after this 'open-cervix' period. Changes in cervical patency may arise from the early oedema-inducing effects of oestrogen during pro-oestrus, which close the cervix, followed by an open period due to the mucosal shrinkage that occurs as the progesterone ratio increases. Interestingly, the 'open-cervix' period correlates very well with the optimal mating time predicted by the appearance of the vaginal mucosa at endoscopy. Despite the convenient interpretation, these radiological contrast study observations do not account for the ability of blood and uterine secretions to pass through the oedematous cervix during pro-oestrus. Other cervical activators may be involved, e.g. oestrogens via their effect on smooth muscle function and prostaglandins contained in semen.

The importance of increasing progesterone concentrations for full expression of oestrus and for induction of the pre-ovulatory LH surge suggests that onset of standing heat should not occur until close to the start of the fertile period. Unfortunately, when the onset of standing heat is related to the timing of the LH surge there is considerable disparity in certain animals (Table 1.1), but signs of standing oestrus may nonetheless be useful as a rough guide to the beginning of the fertile period, as would be entirely expected from an evolutionary standpoint.

## METOESTRUS

This phase follows oestrus, and is defined as commencing when a bitch refuses to stand to be mated, usually 6-8 days after the onset of oestrus, or 8-10 days after the pre-ovulatory LH peak. Changes in both the vaginal smear, such as the influx of leucocytes and increased number of nucleated cells and in the endoscopic view, with rounding and thinning of mucosal profiles, may be useful in judging the end of the fertile period and transition into metoestrus. On the other hand, it is difficult to find any discrete endocrine changes that are particularly characteristic of the onset of metoestrus, except plasma progesterone concentration, which should have increased progressively from the onset of oestrus and be in the range 30-90 nmol/l (10-30 ng/ml) (see Figure 1.2). As described earlier, the entire gestation period fits into metoestrus, which is therefore dominated by the effects of progesterone. In both non-pregnant and pregnant bitches, this is secreted exclusively by the corpora lutea, since there is no placental contribution to circulating progesterone.

Metoestrus includes the early stages of embryo development, and it is interesting to review the timing of certain important pre-implantation events in the bitch and to show them in relation to plasma progesterone and oestradiol concentration (see Figure 1.1). Times are best stated using the day of the pre-ovulatory LH peak as a reference point. As already discussed, ovulation can be estimated as occurring on day 2, and after a period of maturation in the oviduct, these ova would be ready for fertilization between days 4 and 7. Resultant embryos appear to be retained in the uterine tubes until the utero-tubal junction permits their entry into the uterine horns around day 10. An increasing ratio of progesterone to oestradiol (see Figure 1.1) is likely to be of vital importance in the timing of this event. Blastocysts can then be located in the ipsilateral uterine horns until about day 13, after which time the blastocysts are free to migrate between horns until around day 16, when endometrial swellings indicate the presence of embryonic attachment sites.

Plasma progesterone concentration continues to increase after the onset of metoestrus reaching peak values (range 90–270 nmol/l; 30–90 ng/ml) about 30 days after the LH peak. Thereafter, progesterone declines slowly to about 30 nmol/l (10 ng/ml) by day 60 after the LH peak. Progesterone concentration in blood should lie within the range 30–90 nmol/l (10–30 ng/ml) for approximately 2 months after oestrus, and failure to do so is indicative of an anovulatory cycle or an abnormality in luteal function. Until day 60, the timing and magnitude of the plasma progesterone profile does not essentially differ in pregnant and non-pregnant bitches. Thereafter, differences arise as plasma progesterone concentration falls abruptly before term (see later) whereas in non-pregnant bitches levels fall gradually to 3–9 nmol/l (1–3 ng/ml) over the next 30–60 days because of the absence of a luteolytic mechanism. It is not known whether luteal function proceeds for a programmed time period or declines due to diminishing luteotrophic support.

Raised plasma prolactin concentrations, which were evident during the period from 30–65 days after the LH peak in pregnant and non-pregnant bitches, continue to be a feature from days 60–90, especially in suckling bitches, as prolactin is required for milk secretion (note: not let-down). False lactation *pseudocyesis* is a situation which arises spontaneously in some cycling bitches, or it may be induced in bitches which are spayed during metoestrus. Possible causes include raised plasma prolactin concentration in late (non-pregnant) metoestrus as a consequence of less effective feedback inhibition from the declining plasma progesterone concentration. It is also likely that progesterone normally reduces mammary prolactin sensitivity, thus allowing prolactin to stimulate mammary function in late metoestrus as luteal progesterone output falls. This condition probably has ancestral origins in wild dog packs, in which dominant bitches would have synchronized their cycles with the non-breeding females, who would then have been left to raise the puppies.

## Termination of pregnancy by interfering with progesterone secretion

Experimental studies show that the corpora lutea are initially quite autonomous and can secrete progesterone without pituitary luteotrophic support for about the first 20 days of metoestrus. After this, it seems likely that both LH and prolactin are necessary luteotrophins, with prolactin a particular requirement after day 30. The similarities between pregnant and non-pregnant metoestrus reveal the unlikelihood of there being an important embryonic signal in bitches to ensure continuation of progesterone secretion. This is quite unlike the situation known to exist in many other domestic species, where embryonically derived factors are required to ensure survival of the new corpus luteum and thus continuation of progesterone secretion for at least the first part of pregnancy. Many of these embryonically initiated mechanisms for corpus luteum survival act through alterations in uterine prostaglandin synthesis, and termination of pregnancy can often be achieved simply by one injection of suitable prostaglandin. That such a luteolytic mechanism does not exist in bitch metoestrus is shown by the finding that hysterectomy has no effect on length of metoestrus. Also, the bitch corpus luteum is very refractory to injection of prostaglandin analogues. In bitches, single prostaglandin injections only cause transient reductions in circulating progesterone, and repeated doses are required if circulating progesterone is to be lowered sufficiently and for long enough to cause abortion. Such treatments are, however, very badly tolerated and cause sickness and diarrhoea.

Prolactin's role in supporting the corpora lutea after about day 30, has led to various attempts to shorten metoestrus and induce oestrus using dopamine agonist drugs such as bromocriptine, cabergoline and metergoline. This will be described in a later chapter.

## Parturition

A pre-partum decrease in progesterone occurs in most domestic species, as a consequence of the rapid fetal acquisition of adrenal steroid secretion leading to oestrogen and prostaglandin secretion from the endometrium. In bitches, parturition is also preceded by a precipitous drop in plasma progesterone concentration from around 30 nmol/l (10 ng/ml) to <6 nmol/l (<2 ng/ml), and this is now also widely accepted to be the final trigger for the initiation of parturition. This is further confirmed by evidence that parturition can be postponed by progesterone administration and induced by progesterone-blocking drugs such as epostane. It has been suggested that the timing of the fall in circulating progesterone may be reliable enough to predict the time of whelping. If, after analysis of a blood sample, plasma progesterone concentration is still >6 nmol/l (>2 ng/ml), whelping will not commence within 14 hours.

### Endocrine pregnancy tests

As reviewed above, plasma progesterone increases to peak values around day 30 and then declines gradually, reaching around 30 nmol/l (10 ng/ml) towards the end of gestation. A shallower gradient to the declining plasma progesterone profile after day 30 in pregnant bitches has been suggested but the difference between this and the profile in non-pregnant bitches is quite subtle, implying that progesterone measurements are not useful for pregnancy testing in bitches.

Oestrogens are also essential hormones of pregnancy. Embryo attachment involves oestrogen synthesized by the embryonic trophoblast in many species and probably also in bitches. Measurement of urinary oestrone sulphate (the major oestrogen metabolite) at 3 weeks after mating, a time coinciding with embryo attachment, has been suggested as a possible pregnancy test in bitches but the idea has not been tested further. Plasma oestradiol is also detectable later in pregnancy, but its concentration is no higher than during metoestrus, thus ruling out the use of an oestradiol assay at this stage.

Plasma proteins such as fibrinogen, one of the so-called acute phase proteins, increase in pregnant bitches around day 30–40 but not at equivalent stages of metoestrus in non-pregnant bitches. Increased plasma fibrinogen concentration can be detected immunologically, and form the basis of a commercial pregnancy test, although to some extent the popular advent of ultrasonography has replaced the need for pregnancy tests at this stage of gestation.

## ANOESTRUS

In pregnant cycles, this phase would be defined as the interval between parturition and pro-oestrus and would normally incorporate lactation. In non-pregnant cycles, anoestrus is the interval between the end of the luteal phase and the onset of pro-oestrus. Determining the end of the luteal phase is difficult in practice because of extremely low plasma progesterone concentration at this time, usually a maximum of 3 nmol/l (1 ng/ml) by 90 days after the LH peak (i.e. at least 60 days before the onset of the next pro-oestrus). Plasma oestradiol and LH concentrations do appear to alter throughout anoestrus, but these changes are subtle and require frequent collection of blood. Of interest is recent research which shows that despite the presence of small ovarian follicles during anoestrus, they may be unresponsive to both endogenous and exogenous hormone stimulation. This may be a receptor problem, but whatever the cause it will certainly be one factor responsible for the difficulty encountered trying to induce fertile oestrus in bitches.

Relatively high plasma oestradiol concentrations can be measured for 10–20 days before any external manifestations of pro-oestrus are present in the bitch, so there may be quite a long transitional period out of anoestrus. It is interesting to speculate on mechanisms which might be involved in triggering this transition. Possibilities include increased LH secretion, suggested by some research and seen in other species at comparable stages in the reproductive cycle, for example just prior to the transition from anoestrus to oestrus in seasonal breeders like sheep. However, onset of the breeding season in sheep is driven by seasonal day length changes which indirectly increases pituitary and then ovarian activity, but breeding patterns in domestic bitches cannot be reliably and uniformly linked to day length changes.

Alterations in prolactin secretion might offer another explanation for anoestrus in bitches, since prolactin concentration is raised in bitches during metoestrus and early anoestrus, particularly in pregnant cycles, and in other species (e.g. sow, mare) prolactin is responsible for lactational anoestrus. Prolactin reduces ovarian sensitivity to gonadotrophins in other species, but when plasma hormone levels are studied in bitches, there does not appear to be any consistent temporal correlation between decreasing prolactin and increasing LH secretion at the onset of pro-oestrus. If dopamine agonists (prolactin suppressants) are given during metoestrus or anoestrus however, plasma prolactin levels are reduced and anoestrus is shorter.

## SUMMARY

In conclusion, the ovarian cycle in domestic bitches has many similarities in terms of hormonal control to the regulatory pathways that have been closely elaborated in more economically important domestic species. The main differences arise in the rate of progression from one cycle stage to the next. While this aspect of bitch reproduction might cause confusion, various simple procedures have been described which will help the veterinarian to predict dates to optimize breeding success. Additionally, there is much more information available on procedures to modify the reproductive patterns in bitches with normal and abnormal cycles. With the physiology of the bitch's oestrous cycle now much more clearly elucidated, the outlook for methods to induce fertile oestrus and shorten metoestrus looks good, and many potential drugs and treatment protocols are already in various stages of development.

## REFERENCES AND FURTHER READING

Concannon PW, Morton DB and Weir BJ (1989) Dog and cat reproduction, contraception and artificial insemination. *Journal of Reproduction and Fertility*, Supplement **39**

Concannon PW, England GCW, Verstegen JP and Russell HA (1993) Fertility and infertility in dogs cats and other carnivores. *Journal of Reproduction and Fertility*, Supplement **47**

England GCW, Allen WE and Porter DJ (1989) A comparison of radioimmunoassay with qualitative and quantitative enzyme-linked immunoassay for plasma progesterone detection in bitches. *Veterinary Record* **125**, 107–108

Jeffcoate IA and Lindsay FEF (1989) Ovulation detection and timing of insemination based on hormone concentrations, vaginal cytology and the endoscopic appearance of the vagina in domestic bitches. *Journal of Reproduction and Fertility*, Suppl. **39**, 277–287

Lindsay FEF (1983) The normal endoscopic appearance of the caudal reproductive tract of the cyclic and non-cyclic bitch: post-uterine endoscopy. *Journal of Small Animal Practice* **24**, 1–15

CHAPTER TWO

# Physiology and Endocrinology of Reproduction in Female Cats

*John P. Verstegen*

## INTRODUCTION

Over the past decade, a considerable database has been generated on behavioural-ovarian-endocrine interrelationships in cats during anoestrus, oestrus, the luteal phase after sterile or fertile matings, during pregnancy and in the post-partum period. However, many aspects of the cat's reproductive physiology and endocrine cycle are still poorly understood and controversial. The study of reproductive endocrinology in this species needs further investigation.

The cat is a seasonally polyoestrous species (but sometimes non-seasonal) and shares with other mammalian species such as the rabbit, mink and ferret, the peculiarity of being an induced ovulator.

## PUBERTY

The average age of puberty in the domestic cat is variable. Cats usually start to show oestrus when they have reached about 2.3–2.5 kg body weight. The normal queen may enter puberty as early as 4 months of age or as late as 18 months, with the majority showing the first signs of sexual activity around 6–9 months.

To a certain degree, sexual maturity may be hereditary, with some breeds always showing early or delayed puberty. Short-haired breeds such as Siamese or Burmese appear to be more precocious and reach puberty at a lower body weight than long-haired breeds such as Persians, which may not commence their first oestrous cycle until older than 18 months. However, the main factor responsible for both sexual maturity and cyclicity appears to be the amount and duration of light received in relation to day length in outdoor cats, or of artificial light in breeding colonies.

The onset of puberty is seasonally influenced, and usually occurs when the hours of daylight increase. Queens reaching development compatible with sexual maturity at the end of summer will, due to longer exposure to daylight, show signs of oestrus earlier than queens who reach this stage during autumn or winter, the latter showing oestrus only when day length increases in the following spring.

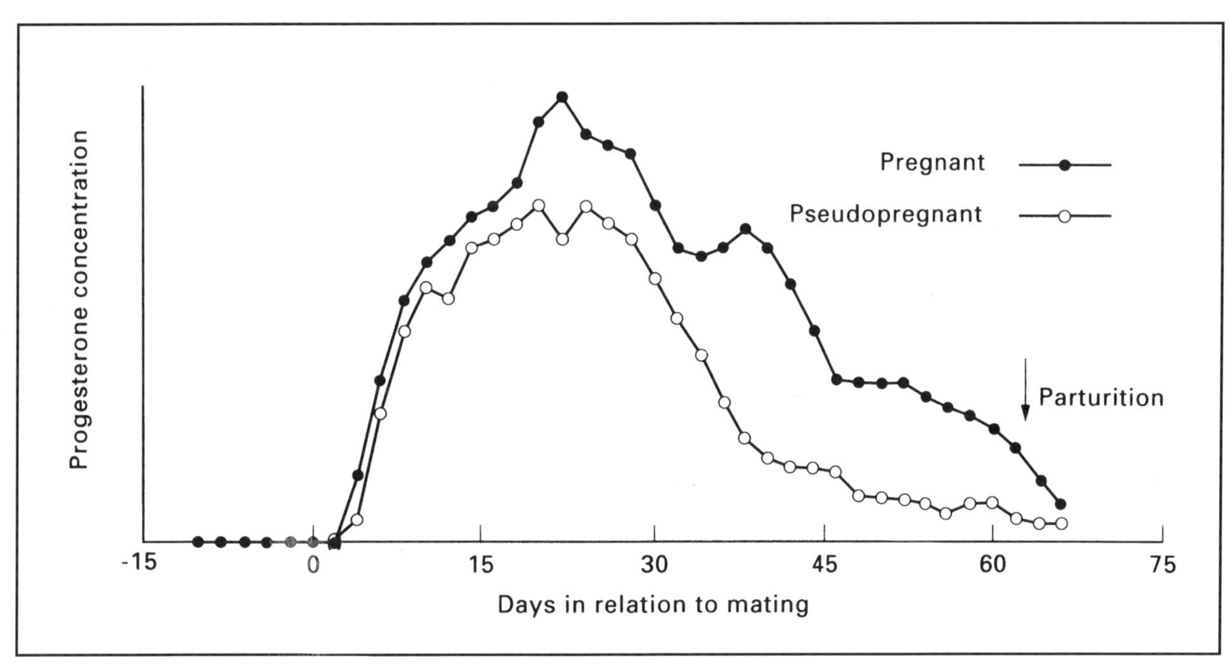


*Figure 2.1: Changes in plasma progesterone concentration in the pregnant and pseudopregnant queen. Reproduced from Lane and Cooper (1994)* Veterinary Nursing, *with the permission of Butterworth Heinemann.*

## SEASONALITY AND THE OESTROUS CYCLE

In the absence of pregnancy or in false pregnancy, the cat will show repeated oestrous cycles every 2 or 3 weeks in spring, summer and autumn. In breeding colonies with controlled light and in households with light at night, cats can become non-seasonal breeders and also show oestrus during winter. Photoperiods influence the reproductive processes via the pineal gland and its main hormone, melatonin, in a manner believed to be similar to that seen in the mare.

### Behavioural cycle

In terms of behaviour, the cycle of the cat may be divided into the heat and the non-heat period.

#### Pro-oestrus and oestrus

The heat period can be divided into pro-oestrus and oestrus. The behavioural pro-oestrus and oestrus period together last between 3 and 10 days. It is, in contrast to the bitch, rather difficult in cats to separate pro-oestrus and oestrus, because no significant external signs are visible. Oestrus may be defined as male acceptance, and the first days (from 1 to 4) of behavioural heat, which are associated with calling but non-acceptance of the male, can be termed pro-oestrus. Pro-oestrus *per se* is clinically difficult to detect, especially as some cats express oestrous behaviour and accept mating without the preliminary transitory period of pro-oestrus. In pro-oestrus, follicular growth begins with plasma concentrations of oestradiol still low and insufficient to allow maximum behavioural expression of heat.

Oestrus begins with the queen allowing mating, and ends when this behaviour ceases. Oestrus is characterized by maximum synthesis and effects of follicular oestrogens.

During heat, the cat rubs against animate and inanimate objects, rolls, and often becomes extremely affectionate. She is vociferous, constantly vocalizing, with much howling and calling. If stroked, especially along the back, the cat will adopt the mating posture, i.e. fore-end crouched, hind-end raised, tail deflected, with paddling movements of the limbs. Male cats are attracted to her, and when the queen is fully in heat she will attempt to escape and allow mating. Male thrusting causes the female to vocalize violently, which is followed immediately by the female forcefully and sometime aggressively freeing herself from the male's neck grip. The queen will then show a temporary lack of interest in mating; however, this is soon followed by multiple copulation typically observed throughout oestrus. Although controversial and certainly difficult to assess in conditions other than controlled breeding colonies, it appears that oestrus is shortened by mating.

External genitalia show few if any typical signs, sometimes a little reddening, softening or oedema of the vulva, but no discharge is present. The vulva may present a small aperture not observed in anoestrous queens.

#### Interoestrus and anoestrus

In the absence of mating and/or spontaneous ovulation, heat periods are observed every 10–14 days during the reproductive season. The phase observed between two periods of intense sexual activity and vocalization during which the cat shows no specific physical or behavioural signs is called interoestrus. Interoestrus corresponds to an apparent quiescence at the level of ovaries and uterus. The ovary may, however, already be preparing for the next follicular growth of the next oestrus. During interoestrus, oestrogens usually decline to basal values.

In some cats, however, sequential follicular growth waves may overlap, oestrogen concentrations may not decline and the queen may appear in constant oestrus. This phenomenon is often falsely called 'nymphomania' or prolonged oestrus.

True ovarian quiescence is only found during anoestrus which, in contrast to interoestrus, is a longer period without sexual activity and reproductive behaviour. Anoestrus occurs when animals are living under very short natural daylight (winter) or are exposed to short light intervals of 4–6 h/day.

Anoestrus occurs during late autumn and early winter in queens exposed to short hours of natural daylight in the northern hemisphere but may be absent in animals submitted to constant long light days.

## MATING AND INDUCED OVULATION

The cat is generally considered to be a reflex mediated induced ovulator, i.e. a coital stimulus (natural during copulation or artificial during vaginal manipulation) is required for ovulation to take place, although evidence is accumulating that some cats may ovulate spontaneously. Vaginal stimulation during mating induces increased neural firings at the medioventral hypothalamus, triggering an increased release of gonadotrophin releasing hormone (GnRH). This GnRH release causes the luteinizing hormone (LH) surge to occur. Similarly, administration of gonadotrophin or GnRH can induce ovulation of several ova within 24–48 hours of administration. Interestingly, ovulation rate seems to be directly related to the amplitude of the LH surge, which is itself correlated to the number, interval and the quality of the matings. The interval between vaginal stimulation and ovulation is indirectly proportional to the number of matings and the endocrine status at copulation. Cats that have been in oestrus for several days ovulate sooner after mating than cats that have only just entered oestrus. Ovulation can therefore occur as early as 24 hours or as late as 52 hours after the induced LH surge. After ovulation, the cat goes out of heat within

24–48 hours. However, it is possible for an LH surge to occur spontaneously without mating or vaginal stimulation and to be followed by ovulation. Increases in plasma progesterone concentration typical of those following ovulation have been observed in queens which have not been mated. Spontaneous ovulation seems to occur more frequently in old cats and in breeding colonies where the males are housed in the same room as females, although there is no visual or physical contact. This observation may be attributable to the effects of pheromones as in other species.

The behavioural aspects of matings and possible problems encountered are covered in Chapter 10.

## HORMONAL CONTROL OF OVULATION, PSEUDOPREGNANCY AND PREGNANCY

It is believed that follicular development in the cat is under the influence of follicle stimulating hormone (FSH) as in other species, and oestrogen production causes the clinical and behavioural signs of pro-oestrus and oestrus. However, at present, there are no convincing data concerning plasma FSH concentrations during the oestrous cycle. If ovulation does not occur, the follicles become atretic and oestrogen concentrations decline. Luteinization does not occur, and newly recruited follicles may start to grow after a few days of interoestrus.

In the case of an induced or spontaneous effective LH surge, ovulation will ensue and corpora lutea develop. Although queens may accept the male during the second or third day of follicular growth (first and second day of oestrus), some cats may show a sufficient release of LH in response to copulation only by the fourth or fifth day. Cats seem to require oestrogen priming for several days before copulation can induce an LH surge which is sufficient to cause ovulation. In contrast to the bitch, in this species, events after ovulation will depend on whether the mating was fertile or not.

### Ovulation without fertilization

When oocytes are not fertilized after ovulation, corpora lutea develop and produce progesterone for a period of about 25–45 days. This luteal phase is shorter than that of pregnancy and is therefore often called pseudopregnancy. In the cat, pseudopregnancy is therefore the luteal phase following an ovulation without fertilization or implantation, but is not, as in dogs, associated with behavioural changes or lactation. During this time, high concentrations of progesterone are maintained by a centrally mediated blockage of GnRH release which does not allow the cat to come into heat. At the end of the luteal phase or pseudopregnancy, a brief period of interoestrus will precede the next return of oestrus provided it is still the breeding season, otherwise anoestrus occurs.

### Ovulation and fertilization

If the cat is mated with a fertile male or is successfully inseminated, the embryos develop for 4–5 days after fertilization in the oviducts and then migrate into the uterine horns where implantation occurs around day 12–16 post-mating (first mating is considered as the ovulatory stimulus). Before implantation, the blastocysts move freely into the uterine horns, migrating from one horn to the other. These intra- and inter-horn migrations seem to ensure an even distribution of fetuses in both horns. After implantation, pregnancy will continue for another 50 days, so that the total length of gestation is around 64 days (timed from the first mating). As with pseudopregnancy, progesterone secretion during pregnancy generally prevents oestrus occurring.

When the cat ceases lactation after kittening, if it is still the breeding season, a brief period of interoestrus will follow until the next return of oestrus; otherwise anoestrus occurs. In lactating queens, this oestrous period usually commences about 10–15 days after weaning. However, in some cases oestrus may be seen in nursing females 10–15 days after kittening. In this case, if she is not mated or if ovulation does not occur, the queen will return to oestrus (normally) every 10–20 days. The first mating after kittening is often not fertile due to incomplete uterine involution, but mating during the next oestrus may be fertile and it is not uncommon to see a nursing queen which is pregnant again before weaning. Spontaneous or induced follicular growth with oestrous behaviour and mating have been described during pregnancy, suggesting that the ovaries are responsive to gonadotrophins during gestation in this species.

## ENDOCRINE EVENTS DURING THE REPRODUCTIVE CYCLE

### Oestrogens

Follicular phases are associated with production of 17-β-oestradiol secreted from the developing follicles with oestrus itself associated with maximal production. Oestradiol concentrations are basal in anoestrus or interoestrus (less than 60–70 pmol/l) and rise over a few days from the end of anoestrus or interoestrus to more than 150–300 pmol/l during oestrus itself. If no ovulation occurs, oestradiol decreases to basal concentrations within 5–10 days. If ovulation occurs, however, oestradiol concentrations will decline within 2–3 days. The increase in oestradiol during follicular development and oestrus is essential for oestrous behaviour and for priming the gonadotrophin surge associated with mating and ovulation.

In some cats, follicular waves follow each other so closely that the follicular phase and therefore oestrous behaviour seem continuous. This phenomenon is often observed in specific breeds such as Siamese cats.

During the first part of the luteal phase, in pregnant and non-pregnant animals, oestradiol concentrations are generally below 70 nmol/l but they may increase occasionally in individual animals. Increased basal levels of 17-β-oestradiol may also be observed after day 35–45 of pregnancy.

## FSH and LH

Up to now, no convincing studies concerning plasma FSH concentration in the cat have been reported. It is, however, believed that similar to other species, dynamic changes in FSH secretion at the end of anoestrus and interoestrus are responsible for the induction of the follicular growth.

During pro-oestrus and oestrus, before the stimulus triggering the LH surge, plasma LH concentration is basal. An ovulatory LH surge begins within minutes of vaginal stimulation or in some instances spontaneously, peaks within 2 hours and returns to basal concentrations within 12–24 hours. The amplitude and duration of the LH surge is highly variable and depends on the intensity, duration and frequency of the coital stimulus. The LH surge is higher and lasts longer when multiple matings occur, compared with the LH values measured after a single coitus. The amplitude varies from less than 10 ng/ml before mating to more than 100 ng/ml after maximal stimulation. The duration of the LH surge varies from a few minutes after the coital stimulus to a maximum of 24 hours. The optimal LH surge is observed when a maximum of four matings occur within a 2–4 hour interval. More coital stimulus during this time-frame or over a longer time interval may not significantly increase the LH response. This is apparently due to exhaustion of the pituitary LH content.

## Progesterone

Plasma progesterone concentrations are basal during anoestrus, interoestrus, pro-oestrus and oestrus prior to ovulation.

In both pregnant and pseudopregnant cats plasma progesterone begins to increase after ovulation, beginning between 24 hours and 50 hours after the LH surge. Concentrations may reach a maximum of 100–200 nmol/l around day 20–25 after the first mating.

In pseudopregnant animals plasma concentrations start to decrease on about day 25 and reach basal values around days 30–40. This slow decline in progesterone is characteristic of the pseudopregnant cat, in contrast to the pregnant cat, in which progesterone concentrations remain high before declining rapidly at the end of gestation. The slower decline in progesterone at the end of pseudopregnancy is similar to the dog, and is probably due to a missing luteolytic factor that is produced in pregnant cats at the end of pregnancy. In pseudopregnant animals, the corpus luteum appears to be pre-programmed to undergo atrophy after 25–35 days when there is no luteotrophic support from the embryo or placenta. Hysterectomies performed during the luteal phase in pseudopregnant animals do not change the lifespan of the corpus luteum, demonstrating that the uterus is not involved in luteolysis.

In the pregnant cat, around day 25–35 an initial decrease in progesterone occurs, but values then remain stable at approximately 15–30 nmol/l (5–10 ng/ml) and do not decrease below 3–5 nmol/l (1–1.6 ng/ml) until day 60. In the cat, progesterone is necessary throughout gestation to maintain pregnancy, although queens can still remain pregnant for a few days after plasma concentrations have decreased below 3 nmol/l (<1 ng/ml), the apparent minimal required concentration of progesterone to maintain pregnancy. Corpora lutea remain the main source of progesterone. The placenta either produces a very small amount insufficient to maintain pregnancy, or does not secrete any progesterone at all. In fact, ovariectomy performed during any stage of pregnancy results in a decline of plasma progesterone below 3 nmol/l (<1 ng/ml) within 48 hours of surgery.

What causes the difference in corpus luteum activity of pregnant *versus* pseudopregnant cats is still not known. Pregnancy may, as in other species, involve some specific luteotropic factors originating from the fetus, the placenta and/or the pituitary, and acting as a signal to maintain the corpus luteum and prevent its regression. In fact, the initial signal is most likely to originate from the uterus, the fetus and/or the placenta, since hysterectomies performed in early pregnancy (before day 20–25) are associated with a corpus luteum lifespan of 25–35 days, similar to that observed in pseudopregnant animals. This indirectly demonstrates that the fetus and/or the placenta is/are involved, directly or indirectly, in the prolongation of progesterone secretion by the corpus luteum. Possible factors involved in sustaining the corpus luteum during pregnancy are prolactin and relaxin, which have been identified as possible luteotrophins in dogs.

## Relaxin

Relaxin is the only pregnancy-specific hormone in both cats and dogs. Relaxin is essentially secreted by the placenta in carnivores, although some production by the ovaries cannot be excluded. Plasma relaxin concentrations increase from days 20–30 post-mating exclusively in pregnant animals, suggesting its role as a luteotrophic factor specific for pregnancy. Relaxin may act directly on the corpus luteum or may indirectly stimulate prolactin secretion at the level of the pituitary, or by stimulating other still unknown luteotropic factors. It was shown that relaxin can sustain and increase prolactin secretion both *in vivo* and *in vitro*. Relaxin is measurable throughout pregnancy and during the first few days after kittening.

## Prolactin

Prolactin secretion, which is basal during the oestrous cycle, increases around 25–35 days after the first mating in the pregnant queen and reaches a maximum a few days before parturition, after which it increases dramatically during lactation. The suppression of prolactin by the administration of a dopamine agonist such as bromocryptine or cabergoline during pregnancy causes a decline in progesterone secretion in cats, and will result in abortion when concentrations of progesterone fall below 3 nmol/l (1 ng/ml). These observations demonstrate the central luteotrophic role played by prolactin in the cat, as in dogs.

### Melatonin

In cats, melatonin secretion is associated with the photoperiod, i.e. varying according to the length of the day. Plasma melatonin concentrations are in synchrony with concentrations of plasma prolactin. Melatonin and prolactin concentrations appear to be high during the period of darkness and low during the period of high light intensity. The exact relationship between the secretion of melatonin, prolactin and the follicular growth associated with oestrogen production is not well understood, requiring more studies in future. Melatonin and prolactin secretions are likely to play a role in ovarian function in the cat, since concentrations of both hormones are lower during periods of ovarian activity (oestrus) than during periods of ovarian quiescence (anoestrus and interoestrus).

## VAGINAL CYTOLOGY

Vaginal cytology in queens reflects hormonal changes (Figure 2.2), particularly during the period of heat. In breeding management vaginal cytology is less useful in cats than in dogs, particularly as the cycle may be affected by carrying out the procedure.

During the follicular phase, superficial and keratinized cells appear on the smear, reflecting the increased effects of 17-β-oestradiol. The proportion of anuclear superficial cells to nuclear superficial cells does not change significantly. A main characteristic of follicular phase vaginal smears is the slow reduction in nuclear intermediate and basal cells. However, these changes are less impressive and characteristic than they are in the bitch. The most significant indication of oestrogen activity is the clearing of the background on the slide associated with a dramatic reduction in the cellular debris or mucus. Eosinophilic staining is less consistent and marked than in dogs, and red blood cells are not observed.

During interoestrus, the majority of cells are intermediate with a few basal or parabasal and keratinized nuclear cells. Generally, there is debris which gives the background on the vaginal smear a dirty appearance.

During anoestrus, cells are rare, mucus is obvious and abundant and the majority of cells are basal or parabasal with some intermediate cells. Leucocytes can sometimes be observed.

Vaginal cytology in cats can be used to detect or confirm the follicular phase, namely pro-oestrus and oestrus. It allows the confirmation of mating by the identification of spermatozoa when carried out soon after copulation.

## REFERENCES AND FURTHER READING

Banks DH and Stabenfeldt GH (1982) Luteinizing hormone release in the cat in response to coitus on consecutive days of estrus. *Biology of Reproduction* **26**, 603–611

Burke TJ (1976) Feline reproduction. *Veterinary Clinics of North America* **6**, 317–321

Concannon P, Hodgson B and Lein D (1980) Reflex LH release in

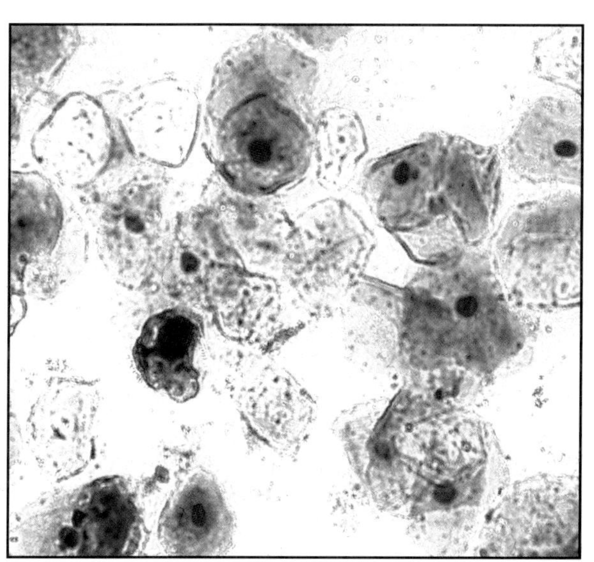

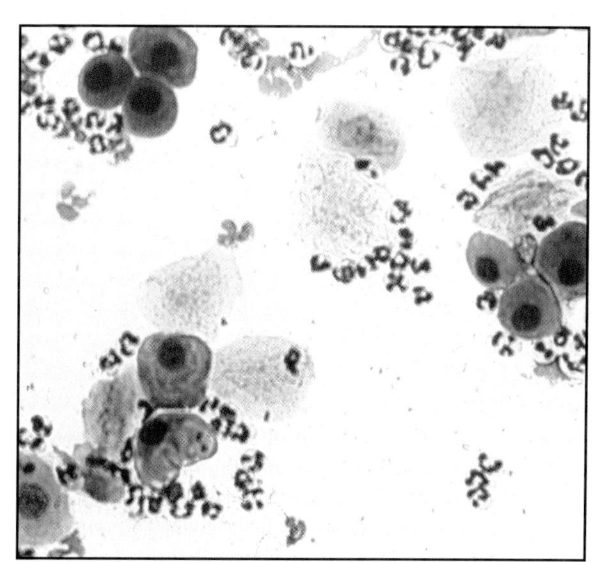

***Figure 2.2:*** *Feline vaginal smears. (a) Smear typical of the oestrous phase of the cycle. The majority of cells are superficial anucleated keratinized cells or cells with pyknotic nuclei. Intermediate cells can still be observed. (b) Smear typical of a metoestrus 'rush' sometimes observed in queens at the end of the oestrous phase. Superficial and intermediate cells are observed, together with numerous white blood cells. This metoestrus rush is of short duration (24–48 hours) in the queen.*

estrous cats following single and multiple copulations. *Biology of Reproduction* **23**, 111–117
Goodrowe KL, Howard JG, Schmidt PM and Wildt DE (1989) Reproductive biology of the domestic cat with special reference to endocrinology, sperm function and in-vitro fertilization. *Journal of Reproduction and Fertility* Supplement 39, 73–90
Hurni H (1981) Daylength and breeding in the domestic cat. *Laboratory Animals* **15**, 229–233
Olson PN, Husted PW, Allen TA and Nett TM (1984) Reproductive endocrinology and physiology of the bitch and queen. *Veterinary Clinics of North America: Small Animal Practice* **14**, 927–946
Schmidt PM, Chakraborty PK and Wildt DE (1983) Ovarian activity, circulating hormones and sexual behaviour in the cat II: relationships during pregancy, parturition, lactation and the postpartum estrus. *Biology of Reproduction* **28**, 657–671
Shille VM, Lundstrom KE and Stabenfeldt GH (1979) Follicular function in the domestic cat as determined by estradiol 17 beta concentrations in plasma: relation to estrous behavior and cornification of exfoliated vaginal epithelium. *Biology of Reproduction* **21**, 953–963
Stewart DR and Stabenfeldt GH (1983) Relaxin activity in the pregnant cat. *Biology of Reproduction* **32**, 848–854
Tsutsui T and Stabenfeldt GH (1993) Biology of ovarian cycles, pregnancy and pseudopregnancy in the domestic cat. *Journal of Reproduction and Fertility* Supplement 47, 29–35
Verstegen JP, Onclin K, Silva LDM, Wouters-Ballman P, Delahaut F and Ectors F (1993) Regulation of progesterone during pregnancy in the cat: studies on the roles of corpora lutea, placenta and prolactin secretion. *Journal of Reproduction and Fertility* Supplement 47, 165–173

## CHAPTER THREE

# The Infertile Female

*Patrick J. Wright and John R. Watts*

## INTRODUCTION

Infertility is the reduction in the ability to produce young. Older (>6–7 years) bitches and queens may cycle less frequently, have reduced rates of pregnancy and litters with lower numbers than normal. Infertility may reflect a problem with the male and/or the female, and can be associated with normal or abnormal oestrous cycles, or failure to mate. Aetiologies include improper management, as well as disorders of behaviour, development, anatomy or function of the reproductive system, and may involve infection, neoplasia or be iatrogenic.

The approach to the bitch or queen presented for investigation of infertility needs to be ordered, sensible and cognizant of the likelihood of the various disorders. The investigation involves obtaining a history, performing an initial clinical examination, and then further examination and tests as indicated.

The considerations in this section relate first to the bitch free from *Brucella canis*. In bitches, *B. canis* can cause infertility, embryonic loss, abortion, stillbirths and death of puppies; in male dogs, it can cause epididymitis. No cases of *B. canis* infection have been reported in Australia or in the UK, except in imported animals. Nonetheless, *B. canis* is an important cause of infertility in other countries and disease caused by this organism is described in the final section of this chapter. Infertility resulting from failure to mate and from abortion is considered in other chapters.

## CASE MANAGEMENT

### History

The aims of obtaining a history are: to determine the general health and management of the bitch or queen; to define the problem; to assess the competency and understanding of the owner; and to obtain information relevant to the investigation. The taking of the history should be thorough, painstaking and concerned with the owner's observations and not interpretations. Important aspects of the general history include vaccination and anthelmintic programmes, past and current illnesses and treatments.

Important aspects of reproductive history include: the fertility of siblings; the number of oestrous episodes; the intervals between the occurrence of successive periods of pro-oestrus; the duration of stages (pro-oestrus, oestrus, metoestrus/anoestrus); the presence and nature of vulval discharges; and the nature and timing of sexual behaviour. Information on the number of heats at which the female was mated, the days on which mating occurred, whether pregnancy occurred, the number of pregnancies, the number of puppies or kittens born, normality or otherwise of parturition, and any losses of puppies or kittens should be obtained. The fertility of past sires around the time of infertile matings should be determined. Early cycles in the young bitch or queen may be irregular and associated with infertility. Owners should be assured that this pattern may reflect the young age of the animal and will most probably be self-limiting.

After the history has been obtained it should be possible to define the presenting problem as:

- Failure to show oestrus
- Infertility associated with abnormal cycling activity (short inter-pro-oestrous interval, short pro-oestrus, prolonged pro-oestrus/oestrus)
- Infertility associated with apparently normal oestrous cycles
- Failure to mate or abortion.

### Initial examination

The general clinical examination of the bitch or queen should include assessment of her disposition and demeanour, body weight and condition, and cardinal signs.

The examination of the reproductive system should include the following:

- Palpation of the mammary glands (for evidence of mastitis, tumours, nodules, lactation, hypertrophy)
- Inspection of the vulva (to determine its size, relative turgidity/flaccidity and presence of any vulvar discharge)
- A vaginal smear, to determine the stage of the oestrous cycle and to evaluate the numbers of leucocytes

- If oestrus is suspected, the assessment of positive postural reflexes (flagging of tail, lifting of vulva) may be assessed by stroking the perineal region
- The vagina of the bitch should be examined using an endoscope and by digital palpation (the vagina of the queen is more difficult to examine; see Diagnostic aids, vaginoscopy)
- Abdominal palpation, which may reveal fetuses or abnormalities of the uterus such as tumours or uterine enlargement due to pyometra.

The stage of the cycle of the normal, non-pregnant, post-pubertal bitch can be determined from vaginal cytology and plasma progesterone concentrations (Table 3.1). It should be noted that the findings for the prepubertal bitch are similar to those for the anoestrous bitch. The findings for the metoestrous and pregnant bitch may also be similar, and procedures for the diagnosis of pregnancy are needed to determine the reproductive status.

The extent of further routine tests such as haematological and biochemical examinations, and urinalysis may reflect cultural expectations, commercial imperatives, the maintenance of mystique, a litigious environment and diagnostic incompetence rather than what is appropriate for diagnosis of the problem. It is unethical and unprofessional to perform unnecessary tests without discussion with the owners as to the real value and cost of these investigations.

## Disorders characterized by failure to observe oestrus

Reported anoestrus may reflect the failure of commencement of ovarian cycling, an unexpected prolongation of the inter-pro-oestrous interval (e.g. >12 months for bitches, >20 days for cats) or the cessation of cyclic activity. The condition may be due to a failure to observe overt signs of pro-oestrus and oestrus, or to structural or functional disorders within the hypothalamus (or higher centres)—pituitary-ovarian axis.

The inter-pro-oestrous interval in bitches is 4–12 months, with considerable consistency within an individual. In older bitches, the normal inter-pro-oestrous interval may be longer (e.g. 10–12 months) than in younger bitches.

### Differential diagnoses

- *Pre-pubertal.* Puberty normally occurs at 6–24 months in bitches and at around 10 months in queens. It is associated with the attainment of mature body weight, occurring approximately 2 months after reaching 70–80% of mature body weight. However, up to 2–3 years can be allowed for normal puberty to occur in the bitch, and in some Greyhounds it may not occur until their fourth or fifth year.
- *Ovarian aplasia/hypoplasia.* The congenital absence or lack of development of one or both ovaries is very rare. Plasma gonadotrophin concentrations are elevated in these cases, indicating a lack of inhibitory effect of ovarian hormones on the hypothalamic-pituitary axis. A GnRH stimulation test would show increase in plasma oestradiol concentrations in cases of unilateral but not bilateral aplasia. Inspection, via a laparotomy, reveals very small or absent ovaries.
- *Seasonal.* Domestic dogs (with the exception of African breeds, e.g. Basenjis) do not have cyclicity linked to season. Wild dogs (dingoes, wolves) mate in autumn. Cats under natural

| Diagnosis | | Oestrogenic activity | Plasma progesterone concentrations |
|---|---|---|---|
| Anoestrus | | – | <1.5 nmol/l (<0.5 ng/ml) |
| Pro-oestrus | | + | <6 nmol/l (<2.0 ng/ml) |
| Oestrus | | + | >6 nmol/l (>2.0 ng/ml) |
| Metoestrus<br>early<br>late | | <br>+ or –<br>– | <br>>32 nmol/l (>10 ng/ml)<br>>1.5 nmol/l (>0.5 ng/ml) |
| Pregnancy | | – | >6–9 nmol/l (>2–3 ng/ml) |
| Cystic follicles | (oestrogen-secreting)<br>(progesterone-secreting) | +<br>– | <1.5 nmol/l (<0.5 ng/ml)<br>>6 nmol/l (>2.0 ng/ml) |
| Tumour | (oestrogen-secreting)<br>(progesterone-secreting) | +<br>– | <1.5 nmol/l (<0.5 ng/ml)<br>>6 nmol/l (>2.0 ng/ml) |

***Table 3.1:*** *Determination of the stage of the reproductive cycle of the bitch and some conditions of the reproductive tract from evidence of oestrogenic activity (indicated from vaginal cytology) and plasma progesterone concentrations. + present; – absent*

photoperiod cycle in spring, summer and autumn and are commonly anoestrous during winter. Exposure to artificial patterns of photoperiod associated with indoor housing may influence their cyclic behaviour.

- *Inappropriate photoperiod.* Seasonal breeders (cats, wild dogs) require appropriate patterns of light/dark for normal cyclic activity. Exposure to inhibitory photoperiod or to a continuous long duration of stimulatory photoperiod may result in acyclicity. Stimulatory photoperiod in the cat would involve 12–14 hours of light and 10–12 hours of dark, although cats are reported to cycle with a minimum of 10 hours of light (and 14 hours of dark).
- *Cats: social factors.* Queens low on the social scale may show 'silent heat'. Behavioural oestrus is not shown, but endocrine and cytological signs of oestrus are normal. Serial vaginal smears should reveal oestrogenic changes and indicate oestrus.
- *Stress, physical training, poor nutrient status.* These factors have been shown to inhibit ovarian cyclicity in a number of species. Physical training may inhibit ovarian cyclicity in Greyhounds. Stress factors for the cat include overcrowding, extremes of temperature, inadequate diet, frequent exhibition at cat shows and associated travel.
- *Inadequate observation/display of oestrous signs.* Factors contributing to a failure to observe signs of pro-oestrus/oestrus may involve the owner or the bitch or queen. Factors associated with owners include lack of frequency or care of observation, or a lack of understanding of the signs of oestrus. These factors may be exacerbated in animals housed extensively (e.g. yard, outside kennelling, in groups) compared with those housed more intimately. Factors associated with the bitch or queen include modest physical signs of pro-oestrus/oestrus (e.g. Greyhounds that may have a less marked vulval swelling than say Boxers), long hair coat (Newfoundlands, Samoyed), inadequate behavioural signs resulting in lack of male interest, or lack of enthusiasm in the female. Close observation of the bitch's behaviour and vulva twice a week, and examination of vaginal smears taken weekly may be necessary to detect pro-oestrus/oestrus. The determination of plasma progesterone concentrations is not necessary as any increase in concentrations would be preceded by oestrogenic signs in the vaginal smear. However, at the time of the initial consultation, a progesterone assay should be carried out to eliminate the possibility of an ovulation having occurred in the previous 2 months.
- *Ovarian cysts or neoplasia.* Progesterone-secreting (luteal) cysts and neoplasms (e.g. granulosa-theca cell tumours) may prevent ovarian cyclicity by inhibiting the secretion of gonadotrophins. Plasma progesterone concentrations will be >6 nmol/l (>2 ng/ml). Tumours (if large) may be detected by palpation, radiography or ultrasonography. Progesterone-secreting cysts are commonly small and may be single or multiple, involving one or both ovaries. The patterns of progesterone secretion will generally be different from those of metoestrus in that the concentration will be lower(>1.5–6 nmol/l; 0.5–2 ng/ml) and elevated for a longer period than metoestrus (60–90 days). Treatment of progesterone-secreting cysts may include the administration of prostaglandins to induce regression of luteal tissue, or surgical removal. Neoplasms and cysts (e.g. rete ovarii) that are not hormonally active, if present on both ovaries, may cause anoestrus by causing ovarian atrophy.
- *Premature ('senile') ovarian failure.* The duration of the functional life of the ovaries of the bitch or queen is not known. A stage of clear cessation of ovarian function similar to that of the menopause in women has not been defined. Ovarian function may decline with age, and in some bitches, prematurely. In older queens, oestrous activity may be less frequent or ultimately cease. Animals with ovarian failure generally have higher gonadotrophin concentrations than normal animals.
- *Immune-mediated oophoritis.* This condition is most uncommon. The diagnosis is based on histological examination of the ovary. Signs described in affected bitches include diffuse lymphocytic inflammation, follicular degeneration, degeneration and necrosis of oocytes and thickening of the zona pellucida.
- *Ovariectomy.* Previous ovariectomy may be indicated by an incision scar or from an ear tattoo where such is customary for spayed bitches. Ovariectomized bitches have concentrations of LH and FSH higher than anoestrous and metoestrous bitches. However, these assays have limited availability and further evidence of ovariectomy may be derived from inspection by laparotomy or laparoscopy. Alternatively, a gonadotrophin or GnRH stimulation test may be performed (using eCG or GnRH, see Clinical approach).
- *Cats: pseudopregnancy.* Pseudopregnancy may follow sterile mating(s), spontaneous ovulation or early embryonic death. The corpus luteum commonly persists for 35–37 days, preventing ovarian cyclicity. Cyclicity commonly resumes 7–10 days after luteal regression. The period of lack of oestrus is then 40–50 days. The cyclic

occurrence of oestrus around this interval suggests a pattern of ovulation followed by pseudopregnancy. The lack of pregnancy following mating may reflect male infertility or female infertility resulting from a failure of conception or a loss of embryos. To confirm the diagnosis of recurrent pseudopregnancy, vaginal cytology may be performed twice weekly for 4 weeks to detect oestrogenic changes and plasma progesterone concentrations measured to detect the corpora lutea of pseudopregnancy.

Pseudopregnancy may follow embryonic loss due to infections. Systemic infection insufficient to cause clinical illness in the queen may cause embryonic death, abortion and mummification . These infections include feline panleucopenia and feline infectious peritonitis. Diagnosis of infection with feline panleucopenia virus on the basis of serology is accurate only during the stage of acute viraemia. Feline leukaemia virus may specifically infect the reproductive tract and cause resorption, abortion and infertility. Elimination of feline leukaemia virus requires detection (by serological testing) of infected cats and their removal from the breeding group. Repeat testing is required to detect seronegative carriers. Testing should continue at 90 day intervals until all cats are found to be negative in two successive tests. The role of other infections in embryonic loss is not clear.

- *Cats: lactational/post-lactational anoestrus.* In the queen, oestrus is usually absent during lactation and oestrous activity recommences approximately 2–3 weeks after the end of lactation, or sooner (around 6–8 days) if the kittens are removed or die within a few days of birth. It should be noted that pregnancy and lactation have little effect on the inter-pro-oestrous interval in the bitch.
- *Abnormalities of sexual differentiation.* These animals are externally phenotypically female but the vagina and vulva are small and the clitoris may be enlarged. The gonads are either undeveloped testes (male pseudo-hermaphrodites), undeveloped ovaries, or ovotestes (true hermaphrodites). Affected animals commonly have complements of sex chromosomes different (e.g. 77,XO, 79,XXX, 79,XXY, 78,XX/78,XY, 37,XO) from the normal bitch or queen (78,XX, 38,XX). Such complements of chromosomes reflect the processes of non-disjunction during meiosis or mitosis during embryonic development. Diagnosis is confirmed by karyotypic examination and the detection of abnormal complements of chromosomes. There is no treatment for these conditions.
- *Iatrogenic.* Drugs which may prevent the occurrence of oestrus include progestogens, androgens, anabolic steroids and glucocorticoids. These preparations act by inhibiting the secretion of gonadotrophins. Bitches treated with androgens may have clitoral enlargement and a purulent vaginal discharge.
- *Intercurrent diseases.* Anoestrus can be associated with disorders of other body systems, e.g. poor body condition, debility, hypothyroidism, hyperadrenocorticism , hypoadrenocorticism. Hypothyroidism is most unlikely to be a cause of anoestrus in the clinically normal bitch showing no signs of the disease (poor coat, lethargy, poor appetite, low tolerance of cold, obesity, alopecia).
- *Pituitary insufficiency* is rare, but occurs in German Shepherd Dogs with hereditary dwarfism. Diagnosis is based on history and physical findings. Tests of endocrine function will show impaired activity of pituitary-thyroid or pituitary-adrenal function.
- *Idiopathic prolonged anoestrus.* The senior author has encountered a few bitches that do not cycle spontaneously but are responsive to treatment with gonadotrophins and have produced normal healthy litters. This condition was probably due to a failure of the normal hypothalamic-pituitary functional changes associated with the onset of pro-oestrus. Diagnosis is on the basis of response to gonadotrophin treatment.

## Clinical approach

A checklist for possible diagnoses is presented in Table 3.2.

### History

Assess whether the problem lies with the animal or with the owner's expectation or management. Determine whether the owner understands the signs of oestrus and normal patterns of reproduction and if observations have been adequate. If necessary, extend the period of anticipation for the onset of first oestrus in the bitch to 2–3 years of age, or to 12–14 months after the previous oestrus. Consider whether the patient has been treated with drugs likely to inhibit ovarian cyclicity. Consider whether the bitch or queen shows male behaviour suggestive of male pseudohermaphroditism (although it should be noted that mounting and thrusting behaviour can also occur in normal bitches).

### Initial examination

Look for evidence of ovariectomy. Determine whether the clitoris is enlarged (intersex, iatrogenic). Ensure there are no signs of disorders of other body systems.

| Aetiology | History | Examination | Additional tests |
|---|---|---|---|
| Prepubertal | + | + | |
| Ovarian aplasia/hypoplasia | | | Plasma LH/FSH, inspection, gonadotrophin/GnRH stimulation test |
| Seasonal | + | | |
| Inappropriate photoperiod | + | | |
| Social factors (cats) | + | | Serial vaginal cytology |
| Stress, training, poor nutrition | + | + | |
| Inadequate observation of display of oestrus | + | | Serial vaginal cytology, plasma progesterone determinations |
| Ovarian cysts, neoplasia | | + | Imaging, plasma progesterone |
| 'Senile' ovarian failure | + | | Plasma LH/FSH, GnRH stimulation test |
| Oophoritis | | | Inspection, biopsy |
| Ovariectomy | + | + | Plasma LH/FSH, gonadotrophin /GnRH stimulation test, inspection |
| Pseudopregnancy (cats) | + | | Plasma progesterone, serology |
| Lactation (cats) | + | + | |
| Abnormalities of sexual differentiation | + | + | Karyotyping, inspection, plasma testosterone |
| Iatrogenic | + | + | |
| Intercurrent disorders | + | + | As appropriate |
| Pituitary insufficiency | + | + | Pituitary function tests |
| Idiopathic | | | Gonadotrophin/GnRH stimulation test |

***Table 3.2:*** *Diagnosis of failure to observe oestrus – a checklist emphasizing important aids for diagnosis.*

**Further investigation**

- Confirmation of ovarian inactivity can be obtained from a vaginal smear and measurement of plasma progesterone concentrations
- In some cases, an eCG or GnRH stimulation test may be useful to determine the presence and functional capability of the pituitary-ovarian axis
- Elevated plasma progesterone concentrations indicate functioning luteal tissue, consistent with pseudopregnancy/pregnancy (cats), normal metoestrus/pregnancy (bitch), luteal cysts and ovarian neoplasia
- Elevated plasma LH/FSH concentrations in the absence of signs of oestrus indicates the lack of inhibitory effect of ovarian steroids on LH/FSH secretion. These increased plasma concentrations of LH and FSH may be present in cases of ovariectomy, ovarian hypoplasia/aplasia and premature ovarian failure.

Other diagnostic procedures which may be indicated are:

- Karotyping (intersex)
- Ultrasonographic/radiographic imaging of the reproductive tract (ovarian cysts, neoplasia)
- Inspection of the reproductive tract at laparotomy (ovarian hypoplasia/aplasia, intersex)
- An oestrous response to treatment with gonadotrophins (gonadotrophin stimulation test) indicates functional ovaries (see below).

**Treatment and management**

Owner education concerning normal reproductive patterns and the signs of oestrus may be necessary. Underlying intercurrent conditions should be treated (e.g. endocrine disorders, poor body condition) and ovarian cyclicity may resume. Many conditions are not treatable (e.g. ovarian aplasia/hypoplasia, senile ovarian failure, ovariectomy) and discussion of the poor prognosis with the owner may be necessary.

Treatments to induce oestrus (e.g. gonadotrophins, see Chapter 16) may be indicated when the animal appears normal, and may also have some value in determining whether the animal has functional ovaries (gonadotrophin stimulation test). However, there is no treatment protocol which reliably induces fertile oestrus and ovulation in all bitches. This treatment should be instituted during anoestrus and after endometrial regeneration. In the authors' experience the animal is more likely to respond to treatment if she has exhibited oestrous cycles previously.

## ABNORMAL CYCLING ACTIVITY

### Disorders characterized by a short pro-oestrus followed by anoestrus (no oestrus or ovulation)

These disorders may reflect factors which can cause a reduced intensity of the visible signs of oestrus, e.g. stress, poor nutrition, training, iatrogenic and intercurrent disease. They may also result from conditions caused by factors involved in shortened inter-pro-oestrous intervals, e.g. split heats, frequent pro-oestrus. The basis of these disorders is inadequate or inappropriate secretion of gonadotrophins, or poor ovarian responsiveness to gonadotrophins. The reader is referred to the previous and following sections for further consideration of these disorders.

## Disorders characterized by shortened inter-pro-oestrous interval

The normal inter-pro-oestrous interval in the bitch is approximately 7 months, and in the queen approximately 20 days. A shortened inter-pro-oestrous interval in the bitch can be considered as less than 4-4.5 months. In the unmated queen, the period between oestrous periods is around 8 days (range 2-18 days).

### Differential diagnoses (Table 3.3)

- *'Split heats.'* This pattern is seen more commonly in young bitches and queens. Pro-oestrous behaviour and physical signs occur then subside, and a normal pro-oestrus/oestrus, or another 'false oestrus', occurs. In the bitch, subsequent pro-oestrous activity may occur weeks or even up to 2 months later, but in the queen, it occurs within a few days. This condition reflects inadequate follicular development at the first pro-oestrus. No treatment is indicated, as the condition resolves with maturity. The condition can complicate regimens of treatment (short-term progestogens) for the suppression of oestrus, since a prompt return to pro-oestrus after treatment is interpreted by the owner as a failure of treatment - which it is. Young bitches may therefore require longer treatment periods (14 days) for the suppression of oestrus than mature bitches (8 days).
- *Frequent pro-oestrous episodes.* One or two pro-oestrous episodes not leading to oestrus and ovulation can occur in older bitches. This condition seems similar to split heats. The basis is not understood, but may reflect inadequate stimulation of the ovaries by gonadotrophins or inadequate ovarian responsiveness to gonadotrophins. Optimal regimens of treatment have not been established. Symptomatic treatment could involve the administration of gonadotrophins at the start of pro-oestrus, or a period of treatment with androgens or progestogens to postpone the episode. Standard monitoring techniques utilizing vaginal cytology and plasma progesterone measurements will reveal the ovarian activity and determine if ovulation eventually occurs, which it frequently does.
- *Follicular cysts.* These cysts are more commonly associated with prolonged oestrogenic activity (see disorders characterized by prolonged pro-oestrus/oestrus).
- *Ovulatory failure.* In the bitch, this problem may result in a shortened inter-pro-oestrous interval due to the absence of the period of metoestrus (normally lasting around 60 days) from the cycle. This shortening is similar to that observed when oestrus is suppressed using progestogens administered from the start of pro-oestrus. In the cat, ovulatory failure may reflect inadequate numbers of matings or mating early in oestrus. The queen will return to oestrus in <18 days. Diagnosis of ovulatory failure as a persistent pattern requires the detection of a low plasma progesterone concentration during (bitch only) and immediately after oestrus (bitch and queen). Ovulation may be induced by treatment with GnRH or hCG early in oestrus, although the timing of administration is critical.
- *Short anoestrus.* This syndrome is inadequately defined. Some individual bitches that present with infertility show normal oestrus and ovulation, but have shortened inter-pro-oestrous intervals. Breeds affected include German Shepherd Dogs and Rottweilers. The basis of infertility is not clear, but could involve insufficient duration of anoestrus to allow for endometrial repair after the endometrial exfoliation that occurs at the end of metoestrus/early anoestrus. If this aetiology is indeed a

| Aetiology | History | Examination | Additional tests |
|---|---|---|---|
| Split heats | + | + | Vaginal cytology, plasma progesterone |
| Frequent pro-oestrus | + | + | Plasma progesterone, vaginal cytology |
| Follicular cysts | | + | Vaginal cytology, ovarian imaging |
| Ovulatory failure | + | | Plasma progesterone after oestrus |
| Short anoestrus | + | | Endometrial cytology, biopsy |
| Cats: short inter-oestrous intervals | + | | Vaginal cytology |

***Table 3.3:*** *Diagnosis of disorders characterized by short inter-pro-oestrous interval – a checklist, emphasizing the important aids to diagnosis.*

| Aetiology | History | Examination | Tests |
|---|---|---|---|
| Young females | + | | |
| Follicular cysts | + | + | Imaging of ovaries, inspection |
| Adrenal oestrogen (cats) | + | + | Response to corticosteroid treatment |
| Prolonged oestrus (cats) | + | + | Vaginal cytology, plasma oestradiol |
| Exogenous oestrogens | + | + | |
| Non-hormonal | + | + | Vaginal cytology |
| Ovarian tumours | + | + | Imaging of ovaries, inspection |
| Hepatic disorders | | + | Hepatic function tests |

***Table 3.4:*** *Diagnosis of disorders characterized by prolonged pro-oestrus/oestrus – a checklist indicating important aids to diagnosis.*

cause of infertility, diagnosis requires assessment of endometrial cytology or uterine biopsy at the start of pro-oestrus. Anecdotal evidence indicates that extending the inter-pro-oestrous interval by postponing oestrus using progestogens or androgens can result in a fertile oestrus. However, the role of progestogens in endometrial exfoliation and repair is unclear. Treatment with progestogens to extend the inter-pro-oestrous interval may affect these processes.

Some bitches showing short inter-pro-oestrous intervals may have ovulatory failure and an anoestrus of normal duration but no metoestrus (see above). The fertility in these animals at an ovulatory oestrus should be normal.

- *Cats: shortened inter-oestrous intervals.* These intervals can have a behavioural or an endocrine basis (see disorders characterized by prolonged pro-oestrus/oestrus).

Note: In some species such as cattle and horses, acute endometritis can cause shortened inter-pro-oestrous intervals, due to the secretion of prostaglandin from the uterus resulting in the demise of the corpus luteum. This shortening has not been demonstrated in the bitch or queen. Indeed, shortened inter-oestrous intervals are not a reported feature of pyometra where plasma concentrations of prostaglandin metabolite have been shown to be elevated.

### Clinical approach

***History:*** In young bitches, split heat is the most probable diagnosis. In older bitches with a repeatable pattern, the poorly defined short anoestrus, failure of ovulation (uncommon), or frequent pro-oestrus should be suspected.

***Initial examination:*** Signs of oestrogenic activity should be assessed from vaginal cytology to distinguish from non-oestrogen producing conditions that may cause vaginal discharge. These conditions include vulvitis and vaginitis, which may also produce vulval swelling and attractiveness to dogs.

***Further investigation:*** The determination of plasma progesterone concentrations during and immediately after oestrus will assess the occurrence of ovulation. Endometrial cytology or uterine biopsy during pro-oestrus may help to determine the adequacy of endometrial repair in the bitch.

***Treatment and management:*** The pattern of split heats in young females usually resolves without treatment. Oestrous suppression in affected bitches using progestogens involves an extended regimen of treatment, but is contraindicated in bitches in their first heat. In older bitches showing repeatable patterns, accurate diagnosis is required for rational treatment as outlined under the differential diagnoses. Hormonal treatment can be instituted for conditions involving failure of ovulation, frequent pro-oestrus and short anoestrus.

## Disorders characterized by prolonged pro-oestrus/oestrus

These disorders (Table 3.4) reflect the prolonged secretion of oestrogen. The source of oestrogen may be exogenous or endogenous. Pro-oestrus can be considered to be prolonged when signs persist for more than 21 days, although pro-oestrus of up to 2 months has been reported for wild canids. Prolonged oestrus can be considered to be a period longer than 21 days in bitches and queens.

### Differential diagnoses

- *Young females.* Early cycles, particularly the pubertal one, may be abnormal or abnormally long before normal cyclic patterns develop. This condition resolves with maturity.
- *Follicular cysts.* Oestrogen-secreting follicular cysts are follicles that persist and do not ovulate. The condition is uncommon and the cause unclear, although in other species the cause is failure of the marked pre-ovulatory increase in secretion of LH (LH surge). Cysts are usually single or few, and less commonly, ovaries with many cysts (polycystic) occur. Oestrogenic activity can be determined from oestrogenic signs (such as vaginal cytology, behaviour and

the appearance of external genitalia). Diagnosis can be confirmed in some cases by abdominal ultrasonography and the cysts may be larger than normal pre-ovulatory follicles. Optimal regimens for treatment have not been determined. The condition may resolve spontaneously or the cysts may luteinize and secrete progesterone. Treatment with hCG or GnRH may induce ovulation or luteinization of the follicle. Treatment with prostaglandin may subsequently cause luteal regression. Prolonged periods of secretion of oestrogen followed by the secretion of progesterone from luteal tissue may predispose to the development of cystic endometrial hyperplasia and pyometra. Similarly, while cystic ovaries may be treated with progestogens, a similar risk of pyometra exists and such bitches are better to undergo ovarohysterectomy once the signs have subsided. Surgical removal of the cyst or ovary with the cyst can result in a return to normal cyclical activity, although in most cases, ovarohysterectomy should be undertaken, as affected bitches are usually past breeding age.

- *Cats: adrenal oestrogen.* Excess of oestrogen production may come from the adrenal gland. Treatment with prednisolone (2.2 mg/kg orally for 5 days) may result in the disappearance of signs.
- *Cats: prolonged oestrus.* This may be due to a merging of successive waves of follicular growth, or in some animals persistence of oestrous behaviour between waves of follicular growth. Vaginal cytology will distinguish these conditions. This condition does not normally warrant treatment but for convenience can be terminated by the induction of ovulation using hCG or cervical stimulation.
- *Exogenous oestrogen.* This effect may be associated with treatment after mating to prevent pregnancy (see Chapter 16).
- *Non-hormonal causes.* Signs of oestrus (attractiveness to males, swollen vulva, red vaginal discharge) may result from non-hormonal causes, e.g. vaginitis, vulvitis, vaginal foreign body or tumour, which may be mistaken for signs of pro-oestrus/oestrus. Vaginal cytology will distinguish these conditions, in which there is no oestrogenic action, from true pro-oestrus/oestrus. Vaginitis or vulvitis can be treated locally using douches containing antiseptics or antibiotics, or systemically using antibiotics.
- *Ovarian tumours.* Ovarian tumours are uncommon and occur mainly in older females. Oestrogen-secreting tumours include granulosa cell tumours, cystadenomas and adenocarcinomas. The tumours can be uni- or bilateral and small or large. They may be detected by abdominal palpation, ultrasonography, radiography or at laparotomy. Treatment is surgical removal.
- *Hepatic disorders.* Oestrogen is metabolized and cleared in the liver. Prolonged oestrogenic activity can be a result of inadequate metabolic clearance of oestrogen secondary to hepatic diseases. Such animals usually have other signs of hepatic disease.

### Clinical approach

***History:*** Determine the age of the animal, define the pattern of occurrence of the oestrogenic signs and behaviour, and whether the animal has had surgery for ovariectomy.

***Examination:*** Verify the oestrogenic basis to the signs using vaginal cytology. Abdominal palpation may reveal enlarged ovaries (tumours).

***Further investigation:*** Further investigation may involve:

- Ovarian imaging (commonly using ultrasound) to detect follicular cysts or ovarian tumours
- Measurement of basal concentrations of plasma hormones (oestrogens, progesterone) and stimulation tests can be useful to detect ovarian tissue
- Surgical exploration to view ovaries or to find ovarian tissue may be necessary in some cases.

### Treatment and management

This treatment and management can involve:

- Inspired inaction (e.g. young bitches)
- Hormonal therapy (e.g. follicular cysts, prolonged oestrus in queens)
- Surgery (e.g. ovarian cysts or tumours, intersex)
- Antibiotic/antiseptic treatment (e.g. vaginitis, vulvitis).

## Failure to achieve pregnancy despite normal oestrous cycles

Concern is commonly expressed if the female fails to get pregnant after being mated during two normal heats.

An absence of pregnancy despite normal oestrous cycles may be due to female infertility, male infertility or inadequacy of management of mating or artificial insemination.

Animals that have early embryonic loss may be included in this category. However, the incidence of embryonic loss/resorption is not known, and such loss is usually not detected and in the bitch does not affect

| Aetiology | History | Examination | Further investigations |
|---|---|---|---|
| Incorrect timing of mating | + | | Vaginal cytology, plasma progesterone determination |
| Infertile male | + | + | Examine male, semen quality assessment |
| Stress | + | | |
| Segmental aplasia of paramesonephric duct | | | Imaging, inspection |
| Uterine tumours, polyps | + | | Imaging, exploratory surgery |
| Cervical stenosis | | | Pass cannula |
| Non-patent oviducts | | | Laparotomy, tubal flushing (cats) |
| Failure of ovulation | | | Plasma progesterone determination |
| Embryonic loss | | + | Early pregnancy diagnosis (ultrasound) |
| Hypoluteodism | | | Plasma progesterone |
| CEH | | | Uterine imaging, biopsy |
| Endometritis | | + | Endometrial cytology, bacteriology |

***Table 3.5:*** *Diagnosis of infertility, oestrous cycle normal – a checklist indicating important aids to diagnosis.*

the inter-pro-oestrous interval. The cyclic pattern of the queen experiencing early embryonic loss will be that of pseudopregnancy. She will return to oestrus 40–50 days after the mated oestrus.

It should be noted that not uncommonly a diagnosis is not able to be made. The animal may become fertile or remain infertile, and the veterinarian may feel quite inadequate.

**Differential diagnosis (Table 3.5)**

- *Bitches: Incorrect timing of mating/insemination.* Natural mating or artificial insemination with fresh semen at any time during oestrus (period of acceptance of mating), except very early or in the last 2–3 days, generally results in good fertility in terms both of bitches becoming pregnant and of size of litter. This situation reflects the long life of sperm in the reproductive tract and the long period over which ova may be fertilized. Sperm are present in the endometrial glands during oestrus. Capacitation of spermatozoa takes 7 hours, and then sperm are fertile for about 4–7 days. Ovulation occurs around 2 days (36–50 hours) after the LH surge (which is commonly on the first day of oestrus). Ova require 2–3 days for maturation and are then fertilizable for up to 24–48 hours. Because these figures are estimates, and there is variation around them, fertilizable ova are considered to be present in the tract over a 3–4 day period from 2–6 days after ovulation. Clearly, the period when ova are fertilizable could be expected to be of longer duration in bitches ovulating larger numbers of ova than those ovulating fewer ova.

  The most common problem in the determination of the optimal time for mating is the variability of the day of ovulation in relation to the onset of pro-oestrus. Common breeding practice is for matings/insemination to be scheduled between days 10 and 14 after the onset of pro-oestrus, but this timing may not result in good fertility, since bitches may ovulate as early as 5 days after the onset of pro-oestrus or as late as 21 days. The timing of insemination needs to be more precise with frozen-thawed semen where the fertilizing lifespan of spermatozoa may be appreciably shorter (12 hours) than for fresh spermatozoa. It could be speculated that this precision in the timing of insemination may also be required should infertility result from 'inhospitable' uterine environment or from semen of poor quality or reduced longevity. Procedures for the determination of oestrus and the appropriate time for mating or insemination are reviewed later in this chapter.
- *Infertile male.* The fertility of males at past matings should be assessed from history of pregnancies resulting from matings to other females. However past fertility may not indicate present fertility. One such situation may be the case of the young male in which previously immature (and infertile) semen (characterized by a high incidence of spermatozoa with proximal droplets yet with good motility) has matured and is now fertile.
- *Stress* may result in infertility, but the evidence is often anecdotal and/or derived from other species. Such stress may be involved in the bitch following a long period of stressful travel to the dog for mating. In cats, stress factors that may cause failure of early pregnancy include transport, alteration of social structure, poor nutrition, parasites and infectious diseases.
- *Segmental aplasia of paramesonephric duct.* This condition, characterized by missing or rudimentary manifestations of parts of the

tubular genitalia (uterus, oviduct), is most uncommon. Diagnosis is based on radiography, ultrasonography or laparotomy. Animals which are bilaterally affected are sterile.

- *Uterine tumours, polyps,* may interfere with the transport of spermatozoa or with implantation. Diagnosis may involve palpation, radiography or ultrasonography. Treatment involves surgical removal of the lesion, and the prognosis for future fertility is guarded
- *Cervical stenosis* is uncommon, and causes infertility by obstructing entry of spermatozoa into the uterus. Affected bitches do not show pro-oestrous bleeding. Diagnosis is based on failure to pass a cannula through the cervix *per vaginam* (bitch) or from the uterus at laparotomy (bitch, cat).
- *Non-patent oviducts* are uncommon, and could reflect a congenital defect or be acquired as a sequel to infection (salpingitis). Diagnosis is difficult (see Diagnostic aids) and involves laparotomy and inspection. Attempts to render oviducts patent are unlikely to be successful.
- *Failure of ovulation* is uncommon in the bitch. However, where it does occur, oestrous behaviour is normal but there is usually a shortened inter-pro-oestrous interval. Diagnosis is based on failure of the normal increase in plasma progesterone concentration at around the start of oestrus. The cause is unclear, but may involve failure of the marked pre-ovulatory increase in plasma LH. If this pattern recurs, treatment with GnRH or hCG to induce ovulation at the start of oestrus is indicated. Mating or artificial insemination should then be performed 3–4 days later. The onset of oestrus may need to be determined from vaginal cytology. Treatment with GnRH or hCG later than the start of oestrus may induce ovulation but no fertilization due to ovulation of aged ova.

  In the queen, failure of ovulation may occur where there have been inadequate numbers of matings, or mating too early or too late in oestrus. To avoid this failure, a minimum of four matings in 1 day when in full oestrus is recommended. The occurrence of mating can be verified by the cry of the cat during mating and characteristic aggression towards the male and rolling and licking after mating. Failure of ovulation in the queen is associated with a low plasma progesterone concentration 1 week after mating, with a lack of a subsequent period of pseudopregnancy, and prompt return to oestrus.
- *Early embryonic loss* has not been well investigated because of the inaccuracy of the diagnosis of early pregnancy. Possible causes include endometritis, cystic endometrial hyperplasia, embryonic defects and possibly inbreeding. Inadequate luteal function has been postulated as a cause in some bitches but there is no evidence to support this claim (see below). Embryonic loss does not affect normal patterns of ovarian cyclicity in the bitch. In the cat, however, early embryonic death is followed by pseudopregnancy (see Disorders characterized by failure to observe oestrus).
- *Hypoluteodism* resulting in an inadequate plasma concentration of progesterone is a undocumented cause of embryonic/fetal loss in queens and bitches. Experimental studies suggest that in the bitch a plasma progesterone concentration of at least 6–9 nmol/l (2–3 ng/ml) is required to maintain pregnancy and that progesterone values below this concentration lasting around 3 days results in loss of pregnancy. If failure of pregnancy is associated with low plasma progesterone concentration it is unclear whether the failure is due to the low plasma progesterone concentration, or *vice versa.*
- *Cystic endometrial hyperplasia (CEH)* is more common in older females. Predisposing factors are age, treatment with progestogens (for the postponement of oestrus) or oestrogens (for the early termination of pregnancy). Failure of clearance of bacteria from the uterus at the end of oestrus may also be involved in the pathogenesis. Bacteria of vaginal origin are often present in the uterus during oestrus when the cervix is open. The diagnosis of CEH with macroscopic cysts can be made using ultrasonography. If the cysts are microscopic, histological examination of a section of uterine wall is required. The clinical course and treatment of CEH are not fully understood. The return of fertility after treatment with prostaglandins for pyometra (usually associated with CEH) suggests recovery is possible in the young animal. Since CEH is dependent on the action of steroids on the uterus (oestrogen, progestogens), postponement of oestrus and metoestrus using androgens may be beneficial. Antibiotic treatment to assist the removal of bacteria at the end of oestrus may also be indicated.
- *Endometritis.* The role of endometritis (apart from pyometra) as a cause of infertility is quite unclear. This fogginess reflects the difficulty in making a diagnosis of endometritis in cases other than those involving a purulent discharge. This difficulty is due to the problems of gaining access to the uterus to obtain samples for microbiology and cytology to detect inflammation. The presence of bacteria in the vagina is normal and is not diagnostic of endometritis. Diagnosis of endometritis requires examination of uterine fluids for the presence of

bacteria and leucocytes. These samples may be obtained in the bitch by pervaginal transcervical cannulation of the uterus or at laparotomy (see Diagnostic aids). Treatment can involve the local or systemic administration of antibiotics. As diagnosis is inadequate, so appropriate regimens of treatment are poorly defined.

- *Herpes virus.* The involvement of herpes virus in naturally occurring infertility in the bitch is unclear. This virus has been associated with abortions and stillbirths.
- *Anti-sperm/ova antibodies* are not documented as a naturally occurring cause of infertility in the bitch or cat. Such antibodies, naturally occurring or induced, can cause infertility in other species.

### Clinical approach

#### *History*

- Determine the likely fertility of the male(s) used at past matings
- Assess whether the management of mating (timing and frequency) has been correct
- For queens, assess the general management and programmes for the maintenance of health (feeding, vaccination, parasite control) in the cattery
- For queens, determine whether infection with viruses causing feline panleucopenia, feline leukaemia or feline infectious peritonitis are present
- For queens, determine whether mating occurred from post-coital behaviour and the interval until subsequent oestrus
- For queens, a plasma progesterone measurement can be used one week after mating to determine whether ovulation occurred
- For queens, an interval of less than 18 days indicates failure of ovulation (not mated, inadequate timing or number of matings), whilst an interval of approximately 40 days indicates ovulation has occurred but there was failure of fertilization (male infertility or female disorder, e.g. CEH, endometritis) or embryonic loss.

***Examination:*** The detection of inappropriately increased numbers of leucocytes within a vaginal smear may be indicative of infection: note that increased numbers are normal at the end of oestrus. Uterine microbiology and endometrial cytology may be performed to assess the presence of endometritis.

***Further investigation:*** Ultrasonographic imaging may be useful to detect CEH or some anatomical defects of the uterus. Laparotomy for inspection of ovaries and oviducts may be warranted. Plasma progesterone concentration should be measured after oestrus since low values may indicate failure of ovulation, and persistently low values may indicate hypoluteodism.

#### *Management*

- If the history and initial examination indicate no abnormality, a non-aggressive management approach may be used in a surveillance cycle
- Male fertility should be assessed and proper management of mating applied
- For the bitch, endometrial cytology and uterine microbiology should be performed early in pro-oestrus if the technique is available
- The uterus may be defined by contrast hysterography using radio-opaque fluid
- Plasma progesterone concentrations should be determined in the bitch to ensure the correct time of mating and that ovulation has occurred
- Ultrasonographic examination should be performed at 3–4 weeks to determine pregnancy and to provide information as to when any failure of pregnancy occurs.

Many females become pregnant during this surveillance cycle. In animals that fail to become/remain pregnant, many of the more common causes have been eliminated and more interventionist (and expensive) investigation can be considered. This investigation could include laparotomy, determination of the patency of oviducts and uterine biopsy.

## Determination of the appropriate time for insemination of bitches

It is necessary to inseminate at the appropriate time to allow fertilizable ova and sperm to be present concurrently in the reproductive tract. The time when fertilizable ova are present and, when spermatozoa need to be in the tract, is around 5 (4–8) days after the LH surge, which commonly occurs around the start of oestrus (period of acceptance of male). The time when fertilizable ova are present can be determined from the measurement of the concentrations of hormones (LH, progesterone) in the blood, or estimated from the history, behaviour and physical signs of the bitch. The measurement of hormones gives a more precise estimate than the other procedures.

The time of insemination must be determined accurately when using frozen-thawed semen due its short fertile life-span (perhaps 12–24 hours). This accuracy is not normally needed when fresh sperm of longer life-span (4–7 days) are inseminated (natural mating or insemination with fresh-collected semen). Insemination with frozen semen is recommended 5 days after the LH surge or about 2 days after the progesterone concentration first rises above 30 nmol/l (10 ng/ml). Fresh semen usually results in good fertility if inseminated any time during oestrus, except very early or over the last few days.

**History**

An indication as to the likely time for fertilization in a current oestrous period can be obtained by estimating the day of fertilization in a previous oestrous period(s). This day can be estimated by relating the day of parturition back to the day when the bitch was inseminated. The duration of pregnancy from fertilization to parturition in the bitch is approximately 60 days. Hence, if a bitch was mated on day 12 of the cycle (first day of pro-oestrus = day 1) and parturition occurred 64 days later, fertilization would have occurred on day 16 of the cycle. If there are a number of cycles available to assess this figure, it may be useful to document the variation present in an individual bitch. This information will provide an estimate of the likely time of fertilization in a current oestrus, and indicate when blood and vaginal smears should be collected to give a more precise estimate of the time of fertility.

**Behaviour**

Positive postural reflexes (flagging the tail, lifting the vulva in response to attentions of the dog or perineal stroking) commonly first occur at the time of the LH surge (day 1 of oestrus) and the commencement of the pre-ovulatory rise in the concentration of plasma progesterone. There is some variability in the time of occurrence of these signs, and they may occur sooner (ie before the LH surge) in mature bitches and later in young bitches.

**Physical signs**

The vulval discharge often changes from red (sanguineous) to straw coloured (serous) at the start of oestrus. Some bitches, however, have a sanguineous discharge throughout oestrus, whilst others have no or little discharge at all during both pro-oestrus and oestrus ('colourless heats').

The vulva often softens at the commencement of oestrus and the vaginal epithelium becomes pavement-like when observed with an endoscope.

**Laboratory aids**

Measurement of the concentration of plasma progesterone is a most useful laboratory technique for the determination of the appropriate time of mating. Some hormone assays detect an increase in plasma progesterone concentration; other procedures give a quantitative measurement of progesterone concentrations. Plasma progesterone concentrations increase from the time of the LH surge, and can be used to estimate the time of the surge (6–12 nmol/l; 2-4 ng/ml), ovulation around 2 days later (12-30 nmol/l; 4–10 ng/ml) and fertilization around 2-3 days after ovulation (30-75 nmol/l;10-25 ng/ml, or higher). The measurement of plasma progesterone concentrations in blood samples taken every 2–3 days, commencing late in pro-oestrus, will indicate the appropriate time for mating. However, in a few bitches an initial rise in the concentration of plasma progesterone is followed by plateau concentrations for some days and then values continue to increase. This pattern suggests that ovulation in these bitches may occur longer after the initial increase in plasma progesterone than in normal bitches.

The pre-ovulatory LH surge lasts 24–48 hours. The detection of the surge by measuring plasma LH concentrations therefore requires daily blood sampling, but such assays are not generally available.

| **Determinant** | **Guide** | **Inadequacy** |
|---|---|---|
| Day from onset of pro-oestrus | Day 9 | Variable duration of pro-oestrus |
| Vulval turgidity | Softens at start of oestrus | Subjective determination |
| Vulval discharge | Lessens and clears to straw colour | Some bitches have no discharge<br>Some bleed into/through oestrus |
| Vaginal appearance | Rounded folds become angular | Subjective determination |
| Behaviour | Positive postural reflexes<br>Acceptance of dog | Between-animal variability |
| Vaginal cytology | High proportion of superficial cells<br>Clearing of vaginal smear | Between-animal variability in cellular picture at start of oestrus |
| Plasma progesterone | Around 2–4 ng/ml (6–12 nmol/l) | Occasional unusual patterns |
| Glucose in vaginal discharge | Present | Unreliable |
| pH of vaginal discharge | Falls | Unreliable |
| Plasma LH | Marked increase (surge) | Daily sampling required<br>Assays not readily available |

***Table 3.6:*** *Guides to the various features commonly used to indicate the occurrence of the plasma LH surge and the first day of standing oestrus.*

Vaginal cytology is a useful technique to determine if a bitch is in oestrus, but does not indicate precisely the time of endocrine (the LH surge) or physiological events (ovulation, the presence of fertilizable ova). Indeed vaginal cytology gives its best estimate of the time of ovulation in retrospect. Ovulation occurs around 6 days before the first day of metoestrus determined as when the proportion of superficial cells in the vaginal smear falls by at least 20%.

## DIAGNOSTIC AIDS

### Vestibule, vagina, cervix

#### Vaginal discharges

Observation of the vulva may reveal a vaginal discharge. The colour and nature of the discharge are indicative of the underlying condition. The discharge may be white - indicating cells including leucocytes, red - indicating blood, or clear - indicating relatively cell-free mucus. The source of the discharge may be the reproductive or the urinary tract.

A white discharge can reflect:

- Prepubertal leucorrhoea: reflecting a low grade vaginitis in prepubertal bitches, usually resolves at the first oestrus
- Vaginitis: may be associated with a foreign body
- Early metoestrus: discharge at the end of oestrus (around day 1 of metoestrus) as there may be a purulent discharge for up to a few days
- Pyometra: open cervix
- Cystitis.

A red (blood-stained) discharge can reflect:

- Pro-oestrus: reflecting diapedesis from uterine blood vessels
- Oestrus
- Persistent ovarian follicle
- Ovarian tumour: oestrogen-secreting
- Transmissible venereal tumour
- Cystitis
- Vaginal ulceration: trauma
- Coagulopathy
- Foreign body in the vagina
- Placental separation during pregnancy
- Sub-involution of placental sites.

Other discharges include:

- Clear brownish glary mucus: normal
- Green black: normal parturition
- Clear watery: amniotic/allantoic fluid
- Brown odorous: failure of birth due to single puppy
- Thick tarry black: mummified fetus
- Brown red: metritis.

#### Vaginoscopy

The vagina of the bitch can be inspected using a rigid endoscope. An auroscope is too short to inspect the entire vagina. Visualization of the cervix and the cranial part of the vagina of the bitch is difficult, because the cranial vagina is narrow and the cervix points ventrally with no dorsal fornix. This narrowing is due to the dorsal median post-cervical fold that extends caudally for about 2 cm from the cervix on the dorsal wall of the vagina. Inspection of the cervix thus requires a narrow endoscope. Examination may be more difficult during metoestrus and anoestrus when the vagina is narrower and more thin-walled than at other stages of the cycle. During oestrus, pronounced vaginal folds may make visualization of the cervical os difficult.

Examination of the vagina of the queen is difficult because it is narrow and difficult to distend especially at the vestibulo-vaginal junction.

Vaginoscopy permits the visualization of vaginal disorders such as septae, strictures, neoplasia, and foreign bodies. Samples for biopsy can be taken. Small amounts of discharge may be seen in the cranial vagina or may be noticed on the endoscope after withdrawal.

#### Vaginal cytology

Vaginal cytology assists in the diagnosis of the stage of the reproductive cycle, inflammation of the vagina or the uterus (if the cervix is open) and transmissible venereal tumour.

Oestrogens acting on the vaginal epithelium cause hyperplasia, hypertrophy, cornification and exfoliation of the cells. Vaginal cytology reflects these changes and is an indicator of oestrogenic action. The common cell in vaginal smears of bitches at stages with minimal oestrogenic activity (anoestrus, metoestrus, prepuberty) is the parabasal cell (Figure 3.1) Parabasal cells have a round outline and a low cytoplasm to nucleus volume ratio. Under the action of oestrogen, hypertrophy of vaginal epithelial cells results in small intermediate cells (larger than parabasal cells, round outline and with a cytoplasm to nucleus ratio higher than for parabasal cells; Figures 3.1-3.3), large intermediate cells (polygonal cells with intact nuclei; Figure 3.3) and superficial cells (similar to large intermediate cells but commonly anuclear, or with pyknotic nuclei; Figure 3.4). The proportion of superficial cornified anuclear cells increases during pro-oestrus, plateaus during oestrus (Figure 3.4), and falls sharply at the end of oestrus. This sharp fall (by more than 20%) defines the first day of metoestrus (Figure 3.2). The oestrous smear is characteristically devoid of debris and stained mucus, and has a clear appearance. Increased numbers of leucocytes may suggest infectious process, and abnormal cells may indicate tumours. Increased numbers of neutrophils may be seen early during metoestrus (Figures 3.1 and 3.5). This increase is normal and does not reflect pathogenic infection.

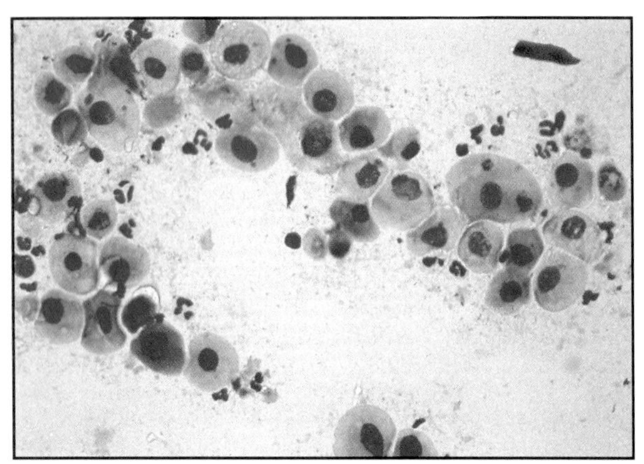

***Figure 3.1:*** *Predominantly small intermediate cells, some parabasal cells and neutrophils in a vaginal smear taken from a bitch during early metoestrus. Diff-Quik®.*

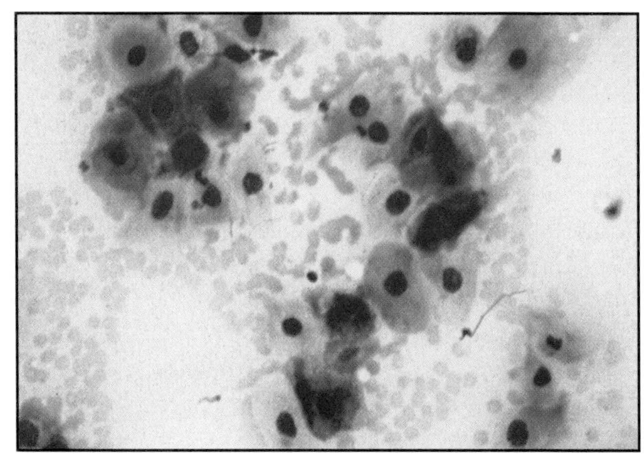

***Figure 3.3:*** *Predominantly small and some large intermediate cells and erythrocytes in a vaginal smear taken from a bitch during early pro-oestrus. Diff-Quik ®.*

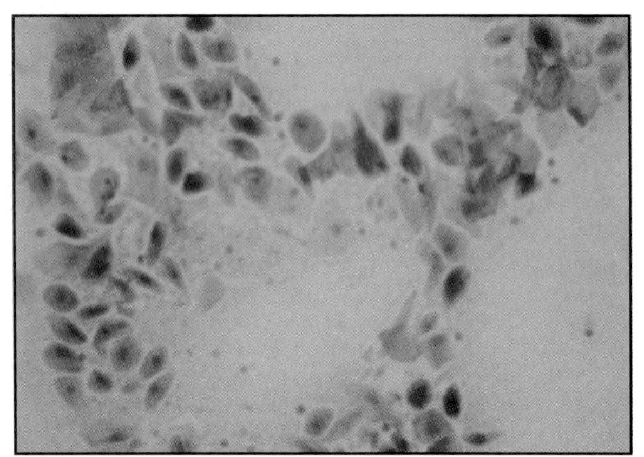

***Figure 3.2:*** *Predominantly small intermediate and some parabasal cells in a vaginal smear taken from a bitch on day 2 of metoestrus. Modified Schorr's trichrome.*

***Figure 3.4:*** *Superficial cells in a vaginal smear taken from a bitch in oestrus. Schorr's trichrome.*

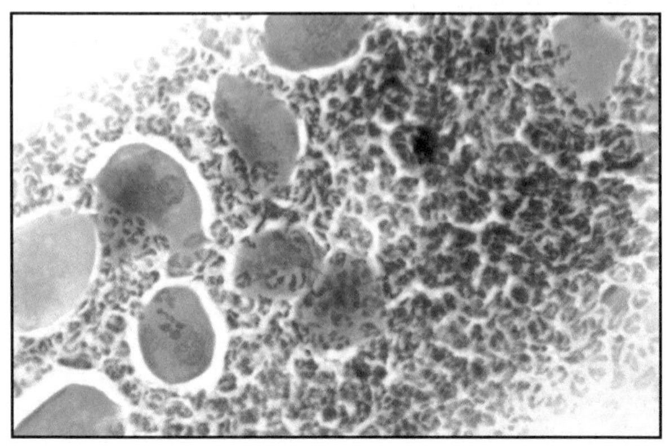

***Figure 3.5:*** *Neutrophils and small intermediate cells in a vaginal smear taken from a bitch in early metoestrus. Diff-Quik®.*

Cytological changes in cats are similar to those in the bitch but are not as pronounced, and there is no pro-oestrous bleeding.

## Vaginal microbiology

Bacteria that occasionally cause reproductive disorders can be present as part of the normal flora. During pro-oestrus and oestrus, vaginal bacteria may gain access to the uterus through a relaxed cervix. Bacteriological culture of the cranial vagina during pro-oestrus/oestrus therefore reveals bacteria that may gain entrance to the uterus rather than bacteria originating from the uterus. Findings for vaginal bacteriology therefore only have relevance in the context of other signs of reproductive tract infection, e.g. purulent discharge or excessive numbers of leucocytes in the vaginal smear. A wide variety of organisms has been isolated from the vagina, and the numbers are often increased during pro-oestrus and oestrus. The only microbe isolated from the vagina that always requires treatment is *Brucella canis*. The other bacteria are normal flora or opportunistic pathogens. The incidence and significance of endometritis around the time of mating has not been determined. The request of stud dog owners for a 'clear' vaginal swab (i.e. vaginal culture free from bacteria) before mating is allowed is nonsensical, because normal flora are commonly isolated, and the presence of this normal flora does not indicate pathogenic infection. In addition the preputial flora of male dogs is similar to the vaginal flora. A path out of the dilemma is to obtain a detailed reproductive history, examine the vagina and a vaginal smear, and then state : 'On the basis of history, clinical examination, vaginal examination and vaginal cytological data, there is no evidence of pathogenic infection of the reproductive tract in this bitch, and the findings are compatible with normal fertility'.

*Figure 3.6: An oestrous Greyhound bitch undergoing transcervical uterine cannulation with an endoscope of the uterus.*

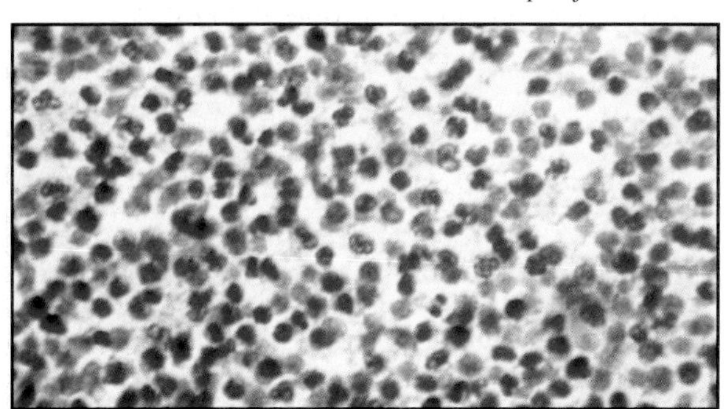

*Figure 3.7: Degenerative neutrophils in an endometrial cytological sample taken from a bitch with pyometra. Modified Schorr's trichrome.*

### Vaginal radiology

Positive contrast radiography can be used to diagnose septae, strictures and tumours. The radio-opaque medium may pass into the uterus during pro-oestrus, oestrus and *post partum*, but not at other stages of the cycle.

## Uterus

### Endometrial cytology and uterine microbiology

Procedures for the collection and examination of samples for endometrial cytology and uterine microbiology have recently been developed. These procedures involve visualization of the cervix using an endoscope (Figure 3.6), transcervical cannulation of the uterus and the injection and aspiration of 2-5 ml of sterile physiological saline. The aspirate is examined for cells and cultured for micro-organisms.

Micro-organisms are often detected during pro-oestrus and oestrus but are uncommon at other stages of the cycle. The microbes detected during pro-oestrus and oestrus are probably of vaginal origin, gaining access through the relaxed cervix. A diagnosis of endometritis requires the presence of bacteria plus increased numbers of neutrophils (Figure 3.7). Cells identified in normal smears include endometrial cells (Figure 3.8), leucocytes (Figures 3.7 and 3.9), erythrocytes, cervical cells, bacteria and sperm (Figure 3.10). The endometrial cells are commonly degenerative

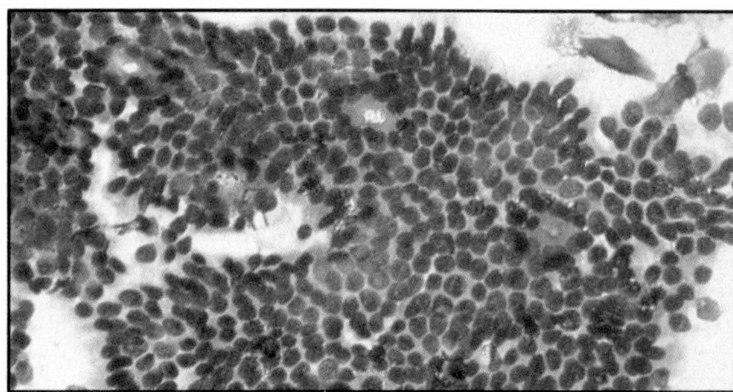

*Figure 3.8: A large group of normal endometrial epithelial cells in an endometrial cytological sample taken from a Greyhound bitch during early metoestrus. Diff-Quik®. (Reproduced with permission from* Journal of Small Animal Practice *(1998)* ***39****, 2–9).*

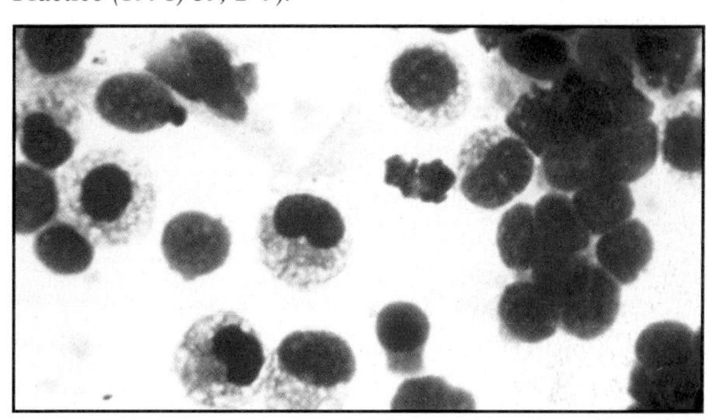

*Figure 3.9: Endometrial cells and macrophages with foamy cytoplasm in an endometrial cytological sample taken from a Greyhound bitch 46 days before the onset of pro-oestrus. The bitch had been in anoestrus for at least 114 days. Diff-Quik®. (Reproduced with permission from* Journal of Small Animal Practice *(1998)* ***39****, 2–9).*

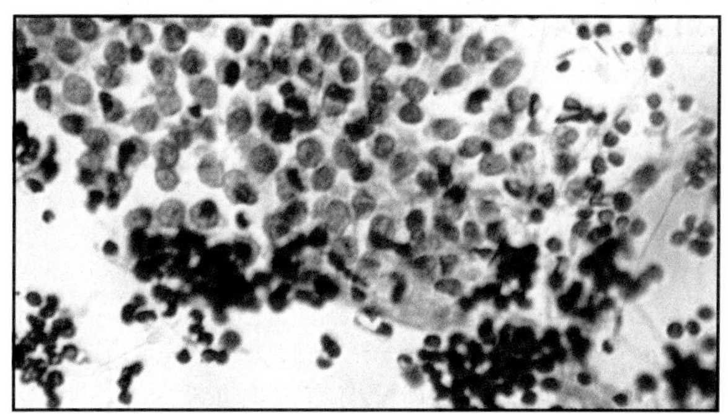

*Figure 3.10: Spermatozoa and endometrial epithelial cells in an endometrial cytological sample taken from a Greyhound bitch 24 hours after mating on day 5 of oestrus. Modified Schorr's trichrome.*

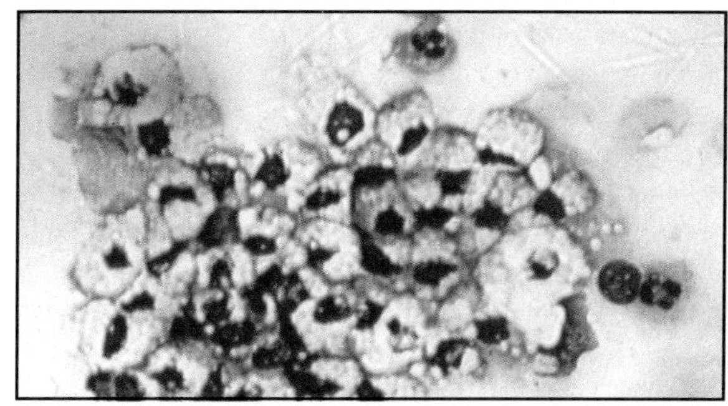

*Figure 3.11: Degenerative endometrial epithelial cells from an endometrial cytological sample taken from a Greyhound bitch during the period of endometrial exfoliation post partum. Note that many of the cells have a pyknotic nucleus and vacuolated cytoplasm. Diff-Quik®.*

(Figure 3.11) during late metoestrus and early to mid-anoestrus (reflecting exfoliation of the endometrium) but are normal at other stages of the cycle. The detection of these degenerative cells for longer than normal can be used to diagnose sub-involution of the uterus *post partum*. Neutrophils are the most common leucocytes during pro-oestrus, oestrus and metoestrus, whilst lymphocytes are predominant during anoestrus. Sperm may be present during oestrus for up to 6 days or more after mating (Figure 3.10) and their presence can be used to diagnose mismating.

### Hysteroscopy

A rigid endoscope can be inserted into the uterus for up to 3 weeks post-whelping (Figure 3.12). Such examination can facilitate the diagnosis of abortion, endometritis, uterine rupture or retained placenta.

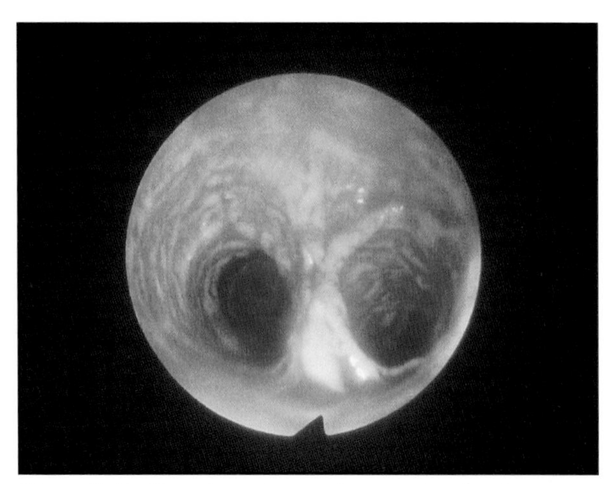

***Figure 3.12:*** *View through an endoscope showing the bifurcation of the uterine horns of a cross-bred bitch on day 23 after parturition.*

### Hysterography

The normal non-pregnant uterus is difficult to visualize radiographically.

Contrast hysterography can be used to diagnose cystic endometrial hyperplasia, pyometra, retained placenta, uterine cysts, and may be of value in the diagnosis of uterine neoplasia, torsion and rupture. The introduction of contrast medium into the uterus can be achieved by uterine cannulation at any stage of the cycle, or occasionally during pro-oestrus, oestrus and *post partum* by filling the vagina with medium.

### Ultrasonography

Ultrasonography can assist the diagnosis of pyometra, cystic endometrial hyperplasia, uterine neoplasia, retained placenta, puppies and uterine rupture. During pro-oestrus, oestrus and *post partum*, echogenic medium introduced into the vagina may pass into the uterus allowing clearer imaging. While the normal uterus cannot usually be seen during anoestrus, the infusion of echogenic fluid (such as normal saline) by uterine cannulation allows its visualization.

## Oviducts, ovaries

### Hysterosalpingography

Hysterosalpingography is an unreliable procedure for the detection of patency of oviducts. The radio-opaque medium used to outline the uterus does not always enter the oviducts of normal bitches, in contrast to women. When the dye does enter the oviducts, it may enter the ovarian bursa, outline the ovaries and enable the detection of cystic ovaries. Plain abdominal radiographs are reliable only for detecting gross ovarian enlargement.

### Ultrasonography

Ultrasonographic imaging may be useful for evaluating ovaries of bitches during pro-oestrus, oestrus, and early metoestrus and for detecting abnormal enlargements of the ovary, such as tumours. Ovaries cannot normally be seen during anoestrus. Ultrasonography is not useful for the detection of ovulation because the follicular wall does not collapse after ovulation unlike in other species.

### Determination of plasma hormone concentrations

Assays for plasma progesterone are readily available, for LH and oestradiol less so. Plasma progesterone concentrations can be used to estimate the time of the LH surge, ovulation and the fertile period of the ovum (see Determination of the appropriate time for the insemination of bitches).

Plasma oestradiol concentrations can be used to determine the presence of oestrogen-secreting structures such as follicles and granulosa cell tumours, although it should be remembered that vaginal cytology reflects such oestrogen production. An increase in the plasma concentrations of oestradiol after the administration of GnRH indicates the presence of functional ovaries (GnRH stimulation test). Lack of an increase indicates a lack of functional ovaries (ovariectomized bitch, ovarian aplasia or hypoplasia, ovarian failure).

Gonadotrophins are normally elevated during late anoestrus, pro-oestrus and early oestrus. Plasma gonadotrophin (LH, FSH) concentrations are elevated in situations of ovariectomy, premature ovarian failure and ovarian aplasia or hypoplasia due to the lack of inhibitory influence from ovarian hormones (oestradiol, progesterone). Plasma LH determinations may be used to detect the LH surge to determine the time of presence of fertilizable ova in the oviducts (see Determination of the appropriate time for the insemination of bitches).

### Laparoscopy and laparotomy

These procedures permit visualization of the reproductive organs, and enable the collection of samples for biopsy. The ovary is difficult to visualize because it is surrounded by a bursa. The patency of oviducts may be tested by injecting saline into the uterus, occluding the horn distal to the oviduct, and observing saline pass out the proximal end of the oviduct. However such passage of fluid often does not occur in the normal bitch. Possible treatment of blocked oviducts (a very rare condition ) has not been documented.

## *BRUCELLA CANIS*

### Description

*Brucella canis* may cause abortion, infertility due to embryonic death and absorption or early undetected abortions or vaginal discharge in the bitch. It is implicated in epididymitis in the dog.

### Aetiology

*B. canis* is an obligate intracellular Gram-negative coccobacillus. Natural infection is restricted to Canidae. Accidental infection of humans may be asymptomatic or may cause mild signs such as intermittent fever, headache, malaise, chills or lymphadenopathy. Infection of humans responds readily to treatment.

### Transmission

Natural transmission can be by ingestion of placental tissue, vaginal discharge or mammary secretions, venereally from an infected dog to a susceptible bitch or *vice versa*, or it may be congenital.

### Clinical signs

Fever is uncommon and most dogs show few or no systemic clinical signs. The signs of reproductive disorder are abortion (commonly between days 45 and 59 of gestation), infertility due to embryonic death and resorption, early undetected abortion, and persistent sero-purulent vaginal discharge. In male dogs, signs may include epididymitis, scrotal dermatitis (due to licking and secondary infection by organisms other than *B. canis*) and testicular degeneration (secondary to epididymitis). Orchitis is rare. Other more generalized signs include enlarged lymph nodes, anterior uveitis, discospondylitis, meningoencephalitis, protein-losing glomerulopathy, prostatitis, arthritis/ polyarthritis.

### Diagnosis

Microbiology can be carried out on specimens of blood, milk, urine, vaginal discharges, semen, placental tissue, prostate, testes, epididymides, lymph nodes and bone marrow. The organism is difficult to culture. Diagnosis is usually by serology, but titres are not detectable for some time after infection (8-12 weeks), and may fluctuate during and decrease after bacteraemia.

### Treatment

Often antimicrobial therapy does not result in the elimination of the organism from the animal. The most appropriate antibiotic therapy appears to be a combination of tetracyclines and aminoglycosides.

### Control

Procedures to eliminate the infection from kennels involve serological testing every month and the removal of positive animals. New animals to be introduced should be tested and should be negative at two tests at least 1 month apart before their introduction.

## REFERENCES AND FURTHER READING


Andersen AC (1970) *The Beagle as an Experimental Dog.* Iowa State University Press, Ames, Iowa

Burke TJ (1986) *Small Animal Reproduction and Infertility: A Clinical Approach to Diagnosis and Treatment.* Lea and Febiger, Philadelphia

Concannon PW, McCann JP and Temple M (1989) Biology and endocrinology of ovulation, pregnancy and parturition in the dog. *Journal of Reproduction and Fertility* Supplement **39**, 3-25

England GCW (1998) *Allen's Fertility and Obstetrics in the Dog, 2nd edn.* Blackwell Science, Oxford

Feldman EC and Nelson RW (1996) *Canine and Feline Endocrinology and Reproduction, 2nd edn.* WB Saunders, Philadelphia

Johnston SD and Romagnoli SE (eds) (1991) Canine reproduction. *Veterinary Clinics of North America, Small Animal Practice* **21**, No 3.

Schille VM and Sojka NJ (1995) Feline reproduction. In: *Textbook of Veterinary Internal Medicine*, ed. SJ Ettinger and EC Feldman, pp. 1690-1698. WB Saunders, Philadelphia

Watts JR and Wright PJ (1995) Investigating uterine disease in the bitch: uterine cannulation for cytology, microbiology and hysteroscopy. *Journal of Small Animal Practice* **36**, 201-206

Watts JR, Wright PJ and Lee CS (1998) Endometrial cytology of the normal bitch throughout the reproductive cycle. *Journal of Small Animal Practice* **39**, 2-9.

Watts JR, Wright PJ, Lee CS and Whithear KG (1997) New techniques using transcervical uterine cannulation for the diagnosis of uterine disorders in the bitch. *Journal of Reproduction and Fertility* Supplement **51**, 283-293

Watts JR, Wright PJ and Whithear KG (1996) The uterine, cervical and vaginal microflora of the normal bitch throughout the reproductive cycle. *Journal of Small Animal Practice* **37**, 54-60

CHAPTER FOUR

# Conditions of the Non-Pregnant Female

*Mike Harvey*

## INTRODUCTION

There are many conditions of the reproductive system involving the non-pregnant bitch and queen and the approach to the differential diagnosis uses a number of investigative techniques.

- History
- Clinical signs
- Radiography
- Ultrasonography
- Vaginal cytology
- Endocrine assays
- Vaginoscopy
- Contrast vaginal radiography
- Pregnancy diagnosis
- Haematology
- Biochemistry

One of the most important facets of such diagnoses is a full understanding of both the physiology and endocrinology of the reproductive cycle of the bitch and queen (see Chapters 1 and 2). This knowledge, taken together with the above investigative procedure, should make diagnosis relatively straightforward.

## CONDITIONS OF THE NON-PREGNANT BITCH

### Oestrous cycle

A very brief summary of the various stages of the bitch's oestrous cycle follows.

#### Pro-oestrus

Follicular development in the ovaries results in rising oestrogen concentrations which produce the typical serosanguineous discharge from the vulva, the vulval swelling and the release of pheromones that attract the male.

#### Oestrus

This is typified by the bitch standing to be mated. The vulval discharge is normally non-haemorrhagic and the vulva loses its tenseness. Plasma oestrogen concentration falls and the progesterone concentration starts to rise prior to ovulation.

From the clinical point of view, care must be taken in assessing the correct stage of the cycle. Some bitches continue to bleed although they have ovulated. In such bitches, if the owner chooses the 'correct' time to mate as being after the serosanguineous discharge stops, the result could be a mating after the ova have degenerated, leading to an infertile mating. An extreme example was a bitch seen recently at the author's clinic that had a haemorrhagic vulval discharge throughout pro-oestrus, oestrus and pregnancy but whelped normally. The bleeding then ceased, confirming that the haemorrhage was not the result of trauma, but was 'normal'.

Many bitches show a poor relationship between behaviour and hormonal events and therefore difficulties can be encountered in finding the correct time to mate without the aid of vaginal cytology, progesterone assays or other monitoring techniques (see Chapter 3).

#### Metoestrus

This stage lasts approximately 2 months and during this period, the ovary produces progesterone regardless of whether the bitch is pregnant or not. Normally, no external signs are seen, with the exception of some bitches which produce a non-purulent opaque vulval discharge. Such a discharge is also commonly seen during pseudopregnancy. If the owner is concerned, a Diff-Quik® stained smear of the fluid will eliminate the presence of polymorphonuclear leucocytes, confirming the non-purulent nature of the discharge. The most common reproductive conditions in the bitch, namely pyometra and pseudopregnancy, occur during this phase.

#### Anoestrus

From the clinical point of view, the ovaries are inactive producing neither oestrogen nor progesterone. Although small amounts of oestrogens can be measured, no clinical signs are evident. Anoestrus lasts until the next pro-oestrus, the stage varying from as short as one month up to 10 months.

A bitch that has never been in oestrus is referred to as being in primary anoestrus, whereas one which has had at least one normal, or in some cases a short and abnormal oestrus, is referred to as being in secondary anoestrus. The investigation of abnormal oestrus is covered in Chapter 3.

## Cycling abnormalities

### 'Silent heat'

Bitches with silent oestrus may show some or none of the signs of oestrus, i.e. vulval swelling, serosanguineous discharge or attractiveness to the male. They will therefore either present as apparently anoestrous or with vague signs of oestrus. Some bitches show normal pro-oestral discharge during their first few seasons, but later in life fail to show any visible signs. Vaginal cytology and plasma progesterone concentration indicate that ovarian function is normal and ovulation in these bitches can be monitored using these methods. It is debatable whether such bitches should be bred as the condition may be inherited.

In the absence of any signs indicating ovarian activity, plasma progesterone assays can be carried out on a monthly basis, a concentration of >9 mmol/l (>3 ng/ml) being evidence of ovulation. An estimate of when ovulation took place can then be made. Using this information to anticipate the next oestrus, vaginal cytology can be performed on a weekly basis beginning shortly before the expected pro-oestrus. Once the start of pro-oestrus has been detected by cytology, continued monitoring using both cytology and progesterone assays will allow the time of ovulation to be determined.

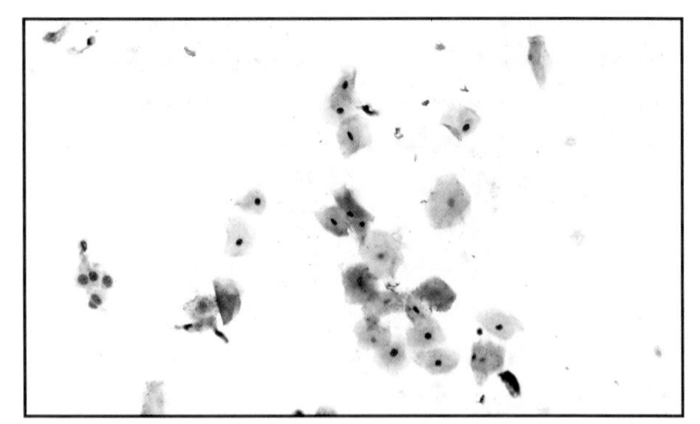

***Figure 4.1:*** *Vaginal cytology smear from a bitch with split oestrus. Day 6 following onset of pro-oestral bleeding. The cells are mainly intermediate and nucleated superficial.*

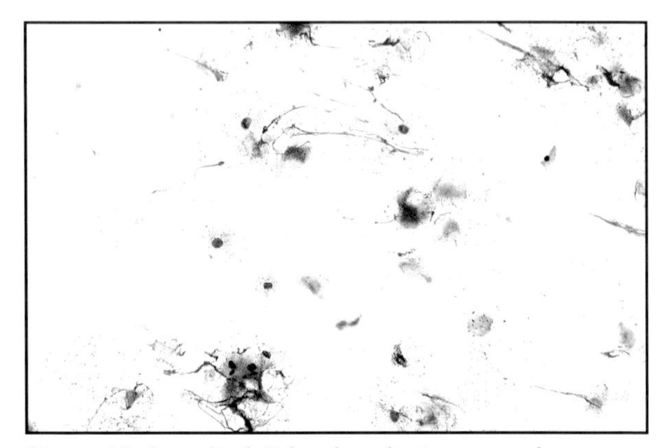

***Figure 4.2:*** *Same bitch 7 days later having reverted to anoestral smear with basal plasma progesterone concentration. The cells are now mainly parabasal and intermediate with significant amounts of mucus. Three weeks later, pro-oestral bleeding restarted followed by normal ovulation.*

### 'Split heat' or 'split oestrus'

This condition is better referred to as 'split pro-oestrus' (Figures 4.1 and 4.2). It is not uncommon for bitches to start a pro-oestral serosanguineous discharge that terminates after 4 or 5 days. The discharge then starts again within a few weeks and usually progresses to ovulation and normal fertility. Occasionally, the 'false start' occurs for a second time before a normal oestrus occurs. From the owner's point of view, the bitch is perceived as having had two heats and confusion about the stage of the cycle occurs. Greyhounds are alleged to have a higher incidence than other breeds, but there is little published detail about the condition.

The aetiology of this condition is not known. However, split oestrus is easily monitored by the use of vaginal cytology and blood progesterone concentrations. When bleeding first starts, vaginal cytology indicates follicular activity (Figure 4.1). When the serosanguineous discharge stops, the cytology indicates that a sudden decrease in oestrogen production has occurred due to the cessation of the follicular activity. This manifests as a reappearance of nucleated and parabasal cells together with a basal plasma progesterone concentration (<3 mmol/l; <1 ng/ml). When pro-oestrus re-starts, cytology is typical of oestrus. The bitch will then usually have a normal cycle in which cytology and progesterone monitoring can be used to indicate the time of ovulation (see Chapter 3).

There is no evidence that this condition results in or is associated with any uterine or ovarian pathology. A split heat is likely to recur at subsequent oestrus; however, fertility should remain the same as in a normal heat.

### Short pro-oestrus

A bitch with an extremely short pro-oestrus of less than one week has similar presenting signs to one with a split heat (Figure 4.3). Where the pro-oestral bleeding ceases because of a very short pro-oestrus, vaginal cytology and progesterone assay indicate that the bitch is either close to ovulation, with the progesterone concentration rising because of the pre-ovulation luteinization (>9 mmol/l; >3 ng/ml), or has ovulated (>32 mmol/l; >10 ng/ml). In other words, the bitch follows the normal pattern but much more rapidly than normal.

### Anovulation

Anovulation is an uncommon condition which may be confused with split oestrus. The serosanguineous vulval discharge in this condition tends to be produced for a prolonged period of time.

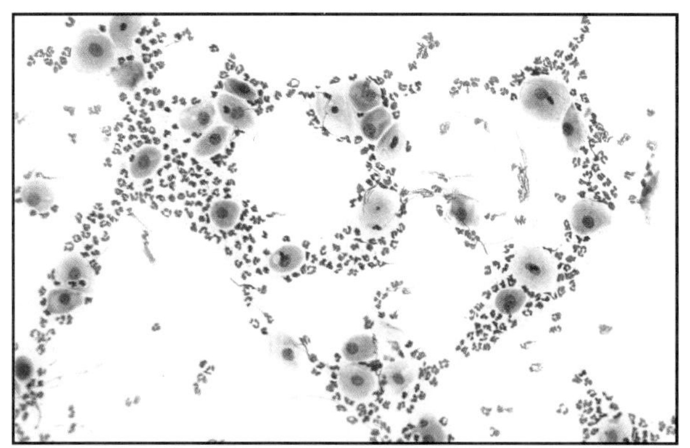

*Figure 4.3: Vaginal cytology from bitch with extremely short pro-oestrus. Smear taken day 12 after onset of pro-oestrus. An elevated plasma progesterone concentration plus the large numbers of polymorphonuclear leucocytes indicated that ovulation probably had occurred about day 6 after pro-oestrus started.*

The bitch enters pro-oestrus and appears to follow the normal pattern of cytological changes leading towards the luteinizing hormone (LH) surge, but does not progress to ovulation. Not uncommonly, the pre-ovulation rise in progesterone is seen, but values then return to basal concentrations. The use of human chorionic gonadotrophin (hCG) at a dose of 500 IU i.m. often results in ovulation. If not treated, the bitch may either eventually ovulate up to one week after the presumed initial LH surge, or she may undergo follicular atrophy and progress to anoestrus. These bitches often return to oestrus at a shorter than normal interval since there is no luteal phase.

It is difficult to differentiate between the bitch with anovulation and one with a normal but prolonged pro-oestrus. The history of previous heats may help as the bitch may repeat her previous pattern. It is reasonable to assume that normal pro-oestrus is unlikely to last longer than about 25 days; a longer time may indicate that the bitch is anovulatory.

Anovulation can be differentiated from split oestrus in two ways. First, the bitch with split oestrus tends to stop bleeding within about one week of starting pro-oestrus, unlike the anovulatory bitch, which tends to have a longer than normal bleeding phase. Secondly, the bitch with split oestrus normally starts her second oestrous phase within a few weeks of the first, whereas the bitch that has undergone anovulation will not return to heat for many months.

It is probable that these conditions are recurrent; therefore a full history indicating the nature of previous heats is essential.

**Juvenile prolonged pro-oestrus**

In the mature bitch, pro-oestrus and oestrus lasting for more than about 25 days is considered abnormal. However, in the pubertal bitch, pro-oestrus, with its accompanying serosanguineous discharge, can last for up to 40–50 days without being considered abnormal.

In attempting to understand the aetiology, it is assumed that the animal's LH surge is either insufficient or absent. In other domestic species, cyclicity problems are commonly associated with LH deficiencies and it seems reasonable to extrapolate this to the bitch.

In this condition, the bitch starts her first season in the usual fashion but continues to have a swollen vulva and attract males for much longer than normal. The bitch behaves as if in pro-oestrus and the cytology is similar to that of a bitch in late pro-oestrus. The discharge continues to have a fresh haemorrhagic nature unlike cystic ovarian disease in which the discharge tends to become tarry (see below).

Treatment is usually unnecessary, as the condition will resolve after about 6 or 7 weeks. The bitch will either ovulate, or the follicles will regress leading directly into anoestrus without an intervening luteal phase. These two possibilities can be differentiated by measuring plasma progesterone concentrations. A concentration of >9 mmol/l (>3 ng/ml) will indicate that there has been an LH surge; >30 mmol/l (>10 ng/ml) indicates that ovulation has taken place. Following ovulation, the problem will resolve.

Chorionic gonadotrophin has been used to induce ovulation, but with very limited success, and as the condition resolves naturally, any improvement with treatment is probably coincidental.

This condition does not recur at later oestruses and fertility is not affected. The main difficulty is in persuading the owner that it is not a serious problem, although the serosanguineous discharge and the persistent attentions of male dogs may become very tiresome.

While it may appear logical to use progestogens to suppress the heat, this should never be done, as there is a risk of inducing a juvenile pyometra.

**Cystic ovarian disease**

True cystic ovarian disease is occasionally found in the bitch and is normally follicular. However, para-ovarian cysts which arise from the ovarian bursa are commonly found at routine ovariohysterectomy (Figure 4.4). These are endocrinologically inactive and therefore asymptomatic. In the event of these cysts being

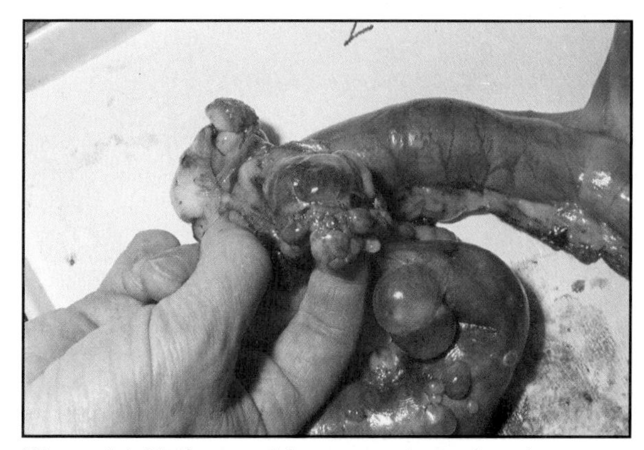

*Figure 4.4: Bitch tract with para-ovarian and uterine cysts. Note that the ovaries contain corpora lutea.*

detected ultrasonographically, unless the bitch is demonstrating an abnormality of the oestrous cycle, they should be considered to be incidental findings.

***Follicular cysts:*** These are the most common cysts found in the ovary. They are usually a condition of the older bitch which has previously had normal seasons. Pro-oestrus commences normally but continues for longer than expected. The serosanguineous vulval discharge becomes less fluid and more tarry and tacky after about 3–4 weeks. The ovarian pathology is normally that of follicular cysts, producing normal concentrations of oestrogen in the early stages which then declines with the advancing age of the cysts. It is likely that a failure of LH release has resulted in the animal's cycle being halted at late pro-oestrus.

Vaginal cytology (Figure 4.5) reveals superficial cornified cells but the degree of cornification does not alter once the bitch reaches late pro-oestrus and blood progesterone concentrations will remain basal. Ultrasound examination using a high frequency transducer will reveal larger than normal (>1 cm) non-ovulated follicles. Exploratory laparotomy can be carried out to aid diagnosis.

The treatment of choice is ovariohysterectomy, as affected bitches are normally past breeding age and the condition will probably recur at the next heat. There is a possibility of excessive bleeding during surgery if undertaken while oestrogen concentrations are high; it may therefore be preferable to postpone the procedure until after medical treatment.

Chorionic gonadotrophin has been used in this condi-tion, but its success rate is low. It may cause only partial luteinization rather than ovulation, although if the resultant progesterone suppresses the gonadotrophin output, it should cause atrophy of the cysts and thus resolution of the condition. The bitch can then be spayed shortly after the signs abate or later when she is in anoestrus.

Progestogens can be used to cause reduction of gonadotrophin production and cyst atrophy. However, there is a possibility that such treatment may result in a pyometra, particularly in an older bitch, due to the prolonged oestrogen production followed by progestogens. For this reason, surgery should be carried out very shortly after the clinical signs cease in order to limit the time available for uterine changes. In addition, because of the chance of producing a pseudopregnancy when spaying a bitch in the luteal phase (see below), this surgery is best carried out before the normal increase of prolactin occurs at about day 30 after ovulation.

***Figure 4.5:*** *Vaginal cytology from bitch with cystic ovaries. Pro-oestrus started 34 days previously. Typical late pro-oestrus smear containing superficial cells.*

It has been reported that aspirating the cysts via a laparotomy can resolve the condition, but as the bitch is unlikely to be used for breeding, this does not appear to be a viable option.

***Luteal cysts:*** A much less common type of cyst is one which produces progesterone. Instead of progesterone production ceasing 2–3 months after ovulation, it appears to continue for many months. As a result, the bitch will appear to be in prolonged anoestrus. Diagnosis is either by ultrasonography or serial progesterone assays, normally monthly. No other condition is likely to produce a prolonged luteal phase with an elevated progesterone concentration.

It is assumed that follicular cysts are caused by a deficiency of LH secretion, and that the bitch with luteal cysts has produced sufficient LH to cause some luteinization, but not ovulation. It is unknown why this luteal tissue does not undergo regression as do normal corpora lutea. As the uterus appears to play no part in luteal regression it is difficult to see why luteal cysts should have a prolonged lifespan.

While prostaglandin therapy has been used, ovariohysterectomy is the treatment of choice.

**Ovarian remnant syndrome**

In this 'syndrome' a spayed bitch is presented which is attractive to males either on a cyclical basis similar to the previous inter-oestrous intervals or, if spayed recently, at about the time the next heat would have been expected.

The bitch does not normally show a serosanguineous discharge when she cycles unless a portion of uterus has been left; however, the vulva may appear more moist than normal and most of the other signs of pro-oestrus are present. The bitch ovulates spontaneously and the corpora lutea are maintained for the usual 2–3 months. This is further evidence that in the bitch, unlike most other species, the uterus plays no part in ovarian function. Despite the variety of explanations given to owners to explain why a bitch that has been spayed should show cyclical activity, the majority are found at surgery to have one entire ovary remaining. In some cases, part of an ovary has been left and it has also been reported occasionally that extra-ovarian ovarian tissue can be found in the ovarian ligament. However, there is no chance that a bitch which becomes attractive approximately every 6 months has anything but ovarian tissue as the cause of the problem.

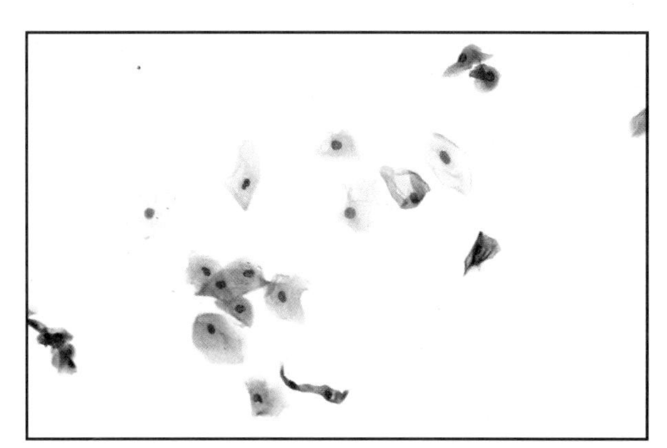

*Figure 4.6: Vaginal cytology from bitch with 'ovarian remnant syndrome', showing oestrogenic stimulation of cells during the time the bitch was attractive to male dogs. An ovary was removed following laparotomy.*

Diagnosis can be carried at two stages of the cycle. When the bitch is in pro-oestrus or oestrus, vaginal cytology will demonstrate oestrogenic stimulation. While it is feasible to carry out oestrogen assays, it is cheaper and more convenient to use cytology (Figure 4.6).

Secondly if the oestrous signs occurred in the previous 2 months, a plasma progesterone concentration of >6–9 mmol/l (>2–3 ng/ml) indicates the presence of luteal tissue and confirms the presence of ovarian tissue. If a simple ELISA progesterone assay is used, a result showing a higher concentration than the low standard (9 mmol/l; 3 ng/ml) is diagnostic of the presence of luteal tissue. It must be remembered that concentrations lower than this are found both at the end of metoestrus and during anoestrus; therefore a basal concentration does not eliminate the possibility of ovarian tissue being present if the time of sampling is outside the luteal phase.

An alternative diagnostic method is the use of GnRH (0.16 mg buserelin) to induce LH release which stimulates oestrogen production from the ovaries. The oestrogens can be measured in plasma or serum, but the laboratory which carries out the assay should be contacted for their recommended protocol.

Ovariectomy is the treatment of choice. If the bitch is attractive and therefore under the influence of oestrogens, the ovary will be enlarged and easier to find than in anoestrus. However, at this stage, the surgery may be more difficult because of the increased vascularity produced by the elevated oestrogen concentrations. After ovulation, the ovary is also large because of the presence of corpora lutea. There is a slight chance of a prolonged pseudopregnancy if the surgery is carried out later than 30 days after ovulation as prolactin concentrations are rising (see Pseudopregnancy). Possibly the best compromise is to perform the surgery about 2 weeks after attractiveness has ceased. The bitch should then be in metoestrus with basal oestrogens and prolactin, thus eliminating both the problem of haemorrhage and the risk of pseudopregnancy.

Spayed bitches are seen that apparently become suddenly attractive a number of years following ovariohysterectomy. It seems unlikely that these animals have ovarian tissue which suddenly becomes active after such a prolonged length of time, although this has been reported. The investigation discussed above should eliminate this possibility and will probably reveal the presence of a mild vaginitis which produces pheromones attracting male dogs.

**Ovarian tumour**

Ovarian tumours are not particularly common. Granulosa cell tumours have the highest frequency and are usually seen in older bitches.

The most commonly diagnosed tumours are inactive, with clinical signs related to the large abdominal mass, namely abdominal distension and/or ascites.

If the tumour is endocrinologically active, the presenting signs will depend on which hormone is produced. The most frequent are oestrogen-producing tumours presenting with a persistent pro-oestrus or oestrus. Vaginal cytological changes are similar to those normally found at these stages and comparable to those of a bitch with follicular cysts. The bitch with a tumour will not respond to therapy with hCG or progestogens, however, and will continue to present the same signs.

It is possible that bone marrow suppression may be induced by oestrogen-producing tumours. Early diagnosis, including haematology to assess the degree of anaemia, and tumour removal are therefore essential.

Some granulosa cell tumours can produce progesterone, and cystic endometrial hyperplasia and pyometra can be induced.

Other tumours which occur are ovarian cystadenomas, fibromas or adenocarcinomas. These do not normally present as reproductive problems, but either as large abdominal masses or occasionally with clinical signs of secondary spread.

Diagnosis is based on abdominal palpation, radiography and ultrasound investigation. Exploratory laparotomy allows both diagnosis and removal of the structure which is normally unilateral. The tumour is usually non-malignant and ovariohysterectomy will resolve the problem.

## Conditions of the luteal phase

**Pseudopregnancy (pseudocyesis or false pregnancy)**

Pseudopregnancy is a relatively common condition which presents between 6 and 14 weeks after oestrus. It is thought to be due either to excessively elevated plasma prolactin concentrations or to the bitch's increased response to this hormone. In all bitches, prolactin increases naturally after day 30 of metoestrus, acting both on the mammary glands in preparation for lactation and as the major luteotrophic hormone, maintaining the corpora lutea.

It has been suggested that all bitches have a pseudo-pregnant phase but that only in some is it overt. Because bitches in a pack or a kennel tend to come into heat synchronously, it has been suggested that pseudopregnancy allows several bitches to lactate so that the puppies may be suckled by bitches other than the natural mother. This would have advantages in allowing the dominant bitch who has produced the puppies to be able to hunt for food while her puppies were being fed and looked after by other members of the pack.

This common condition presents in a variety of ways:

- Mammary development with or without milk production
- Nesting and toy fixation
- Aggression or dullness
- A milky vulval discharge.

The history, clinical signs and the time after oestrus usually suggest the possibility of the condition. Pregnancy testing should be carried out and, if negative, the diagnosis of pseudopregnancy is obvious if there are lactational or nesting signs.

In an entire bitch with aggressive behaviour, pseudopregnancy can be suspected if she has shown similar behaviour following previous seasons. If a bitch that normally has a good temperament becomes aggressive about 2 months after her season, pseudopregnancy should be suspected.

Some bitches, supposedly spayed, show signs of aggression at 6-monthly intervals. These signs are the result of an ovary which has been left at spaying. They should be investigated for 'ovarian remnant' syndrome and the presence of ovarian tissue confirmed.

Finally, as will be discussed below, bitches that have been spayed during an overt pseudopregnancy may develop a prolonged, if not permanent condition. Indeed, bitches spayed at the time of rising or elevated prolactin (after day 30) which showed no clinical signs at the time of the surgery (covert pseudopregnancy) may develop the condition.

Occasionally, bitches are presented showing only a cloudy vulval discharge and in these cases the first investigation should be an examination of a smear of the discharge for the presence of polymorphonuclear leucocytes. In the absence of a heavy population, infection can be ruled out; however it should be borne in mind that a normal metoestral smear will show a certain population of polymorphonuclear leucocytes.

***Treatment:*** The treatment depends on the presenting signs (Table 4.1). The main approaches to the condition are conservative treatment, reproductive steroids or anti-prolactin drugs.

The physiology of the bitch must always be borne in mind, as about 3 months after oestrus, the bitch will be out of the luteal phase and prolactin concentrations should be basal or at least declining. Therefore by about 12 weeks, pseudopregnant signs should be regressing spontaneously. Treatment in many cases is therefore attempting to buy time until the bitch naturally passes out of the phase. However, it should be noted that in a small number of cases, pseudopregnancy may not be apparent until as late as 14 weeks after oestrus and the signs may then continue until as long as 5 months after the season started.

Conservative treatment includes the use of diuretics such as frusemide with restriction of food and water where lactation is present. Where nesting behaviour occurs, increasing exercise and trying to interrupt the behaviour pattern can be attempted.

Currently, the most widely used drugs in all types of pseudopregnancy are reproductive steroids. These function by producing a negative feedback on the pituitary release of prolactin. A variety of such hormones are used, the most common being progestogens (such as medroxyprogesterone acetate or proligestone) and oestrogens/testosterone combinations.

Anti-prolactin drugs have not been widely used in the UK. Cabergoline, which has been available in other European countries, is an extremely efficient drug with a low incidence of side effects and is now to be available in the UK.

A single course of 5 days treatment with cabergoline is effective in over 80% of cases of primary referrals. In post-spay cases and in those which have become persistent (see below) and in which treatment has been unsuccessful with reproductive steroids, cabergoline has a high success rate.

If reproductive steroids such as progestogens or oestrogen/androgen combinations are administered, only one type of drug should be used. If the one selected does not work or is only transiently effective, the same drug should be used again. It is assumed that a temporary suppression of the prolactin concentration occurs and when the treatment is terminated prolactin rises again and the condition reappears. It may help to reduce the dosage rate towards the end of the treatment period, as this may reduce the rebound release of prolactin. If this is not successful, repeating the same course of treatment often brings about a permanent cure.

It is likely that bitches that are pseudopregnant will have a recurrence after subsequent heats and for this reason, unless bitches are going to be used for breeding, they should be spayed during anoestrus. It is interesting to note that bitches which previously have had pseudopregnancies with aggressive signs tend not to have similar behavioural problems when they are pregnant.

It is vital to appreciate the potential side effects of the treatment selected. If handled incorrectly, a condition which in most cases is no more than relatively inconvenient can become extremely serious if it becomes permanent, and euthanasia may have to be contemplated, especially in the aggressive type.

| Type of presentation | Priority of treatments suggested |
|---|---|
| Cloudy vulval discharge | No treatment |
| Mammary development or milk production | No treatment; Diuretics; Anti-prolactin agent (if cabergoline); Reproductive steroids; Anti-prolactin agent (if bromocriptine) |
| Behavioural or nesting | No treatment; Anti-prolactin agent (if cabergoline); Reproductive steroids; Anti-prolactin agent (if bromocriptine) |
| Aggression | Anti-prolactin agent (if cabergoline); Reproductive steroids; Anti-prolactin agent (if bromocriptine) |

***Table 4.1:*** *Approach to therapy in false pregnancy*

The worst sequential combination of drugs appears to be one which contains oestrogens followed by a progestogen. This can exacerbate the condition, as it mimics the pattern of hormones at oestrus and almost appears to re-prime the release of prolactin. Under these circumstances, the condition may become much worse and the duration prolonged.

The side effects of the anti-prolactin drugs are less serious but more obvious; bromocriptine may cause emesis. As previously mentioned, a much better series of ergoline derivatives has been available in mainland Europe, including cabergoline; the frequency of side effects of these drugs is very low.

One treatment which must *not* be carried out is ovariohysterectomy when the signs of false pregnancy are present. If performed, the precipitous drop in progesterone allows prolactin to rise and bitches can develop permanent signs of pseudopregnancy. In addition, bitches are seen which develop long-term false pregnancies following spaying, even when there were no obvious signs of the condition at the time of the surgery. These were probably spayed during the luteal phase after prolactin concentrations started to rise and the effect was similar to spaying when signs were present. For these reasons, it would appear to be much safer to spay bitches about 4–5 months after their heat in the depth of anoestrus; this takes into account the fact that some false pregnancies do not show signs until almost 4 months after the end of oestrus.

The treatment for bitches that have been spayed and appear to have a 'permanent' false pregnancy is reproductive steroids or bromocriptine. Delmadinone acetate has also been found to be useful in such cases. This author has found that bitches with aggressive and behavioural signs after spaying have responded particularly well to cabergoline and this may be the method of choice when available.

## Conditions of the vagina and vulva

### Vaginitis

A bitch with vaginitis is normally clinically healthy but has a purulent vaginal discharge. The bacterial population in the discharge is similar to the normal vaginal flora. The bitch may or may not lick her vulval region. In some cases, the discharge may make her attractive to male dogs. From the aetiological point of view, it is convenient to sub-divide these cases using the age at first referral.

***Pre-pubertal or 'juvenile' vaginitis:*** This is a relatively common condition in pre-pubertal bitches, where as early as 2 or 3 months of age, they produce a copious purulent vaginal discharge. The bitch is fit and healthy. The condition appears to be the result of overactive vaginal glands, the products of which become contaminated with commensal bacteria.

Various congenital conditions, including intersex and vaginal strictures, may produce discharges, caused partly by blockage and contamination of vaginal secretions and partly by licking the vulva. Such conditions should be eliminated from consideration before assuming the bitch has a pre-pubertal vaginitis.

Vaginal swabs for bacteriology and cytology are best taken from the discharge at the vulval lips, in contrast to taking an anterior vaginal swab as for adults. There seems to be no advantage in causing the young bitch discomfort by swabbing an immature vagina when the pus at the vulva will give the same answer. Smears stained with Giemsa or Diff-Quik® show vast numbers of polymorphonuclear leucocytes (Figure 4.7). Bacteriology is normally uninformative, as are haematology and biochemistry. Vaginoscopy probably will require sedation in such young bitches and is not normally carried out, as the examination usually reveals only erythema and the discharge.

When the bitch reaches puberty and has her first oestrus, the condition spontaneously resolves due to elevated oestrogen and the flushing effect of the pro-oestral discharge (Figure 4.8). It is interesting that in spite of the fact that there is clearly bacterial contamination present, no specific organisms are associated with the condition and antibiotic treatment is commonly unsuccessful, in spite of using a series of different antibiotics. Temporary improvement tends to occur following antibacterial therapy but invariably the condition recurs. The probable reason is the failure to achieve an adequate concentration of antibiotics in the immature vagina.

There are circumstances where, in spite of assurances that the condition will resolve at the first heat, owners press for treatment. If it is felt that this is desirable, possibly a more effective treatment than systemic antibiosis is the flushing of the vagina with either water-soluble antibiotics or an extremely dilute disinfectant. It is likely that the flushing effect is as useful as the antimicrobials. Care should be taken with such treatment, as over-frequent therapy may alter the vaginal flora, resulting in overgrowth with fungi and mycoplasms. The use of low doses of systemic oestrogens or creams containing oestrogens has been advocated as being useful in some cases.

Cases that resolve at puberty do not tend to reappear nor do they appear to have any association with the later development of pyometra. The major difficulty in handling these cases is in convincing the owner that the condition is 'normal'. However, if the bitch has delayed puberty, the condition can last for a disturbingly long time and the temptation, under owner pressure, is to treat the bitch.

***'Adult' vaginitis:*** Adult vaginitis is invariably chronic and can result from a variety of causes, including congenital and acquired defects. Both types present with a mucopurulent or purulent discharge which is occasionally bloody or blood-tinged. These bitches have had at least one oestrus and there is no recognisable age distribution.

The condition is recognized by the presence of an overt discharge or a small amount of crusty material on the vulval lips; the bitch may lick her vulva excessively and male dogs may be attracted. Manipulation of the vulval lips usually produces some of the discharge from the vestibule. The bitch is healthy and not dull or lacking in appetite.

The possibility of pyometra or pregnancy must be excluded as differential diagnoses. A vaginal swab should be taken to determine whether the discharge is indeed purulent, this being confirmed by the presence of polymorphonuclear leucocytes. Haematology and biochemistry in such cases will tend to be within normal ranges. If no specific causes are found for the discharge (see below) and the condition is therefore primary, the use of local antibiotics or mild disinfectants is usually successful.

**Congenital defects of the vagina and vestibule**
Congenital defects are normally diagnosed:

- Because the owner has observed an enlarged clitoris
- At a pre-breeding examination
- After an attempted mating at which intromission could not be achieved
- Because the bitch has a vaginal discharge.

Almost certainly, many such defects go undiagnosed in unmated bitches if they cause no problems.

The vestibule should be examined for any structures protruding through the vulval lips and the clitoris should be palpated to investigate if it is enlarged. If nothing is seen, the lips should be parted to check for an enlarged clitoris. This is the cardinal sign of intersexuality. However, racing Greyhounds whose seasons have been suppressed with androgens will also commonly have an enlarged clitoris. This, on occasions, can allow vaginal secretions to accumulate and become purulent.

Congenital defects of the reproductive tract are normally found at the vestibulo-vaginal junction or the vulva. As such, they are relatively easily detected. Using a lubricated gloved finger, it is easy in all but the smallest bitch to insert a finger into the vestibule during pro-oestrus or oestrus. In the vast majority of bitches a finger is long enough to pass through the vulva, cranially and dorsally into the vestibule and then horizontally to the vestibulo-vaginal junction, the latter being about 10 cm from the vulva. The junction is easily detected in the normal bitch by a feeling of temporary restriction which usually dilates with gentle pressure. In the maiden bitch this constriction, nor-

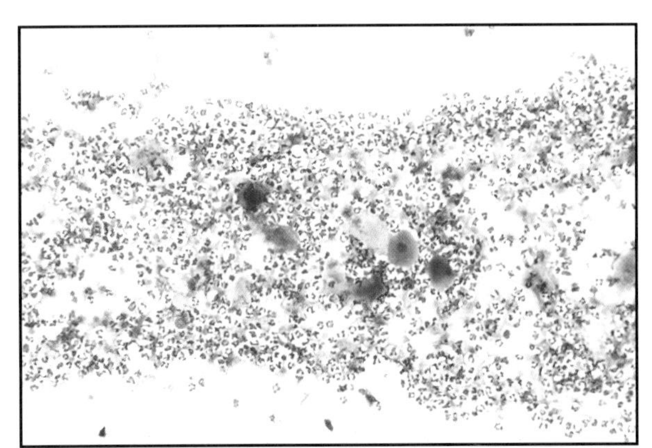

***Figure 4.7:*** *Vaginal cytology from a 7 month old bitch with 'juvenile vaginitis'. Note the large number of non-degenerate polymorphonuclear leucocytes.*

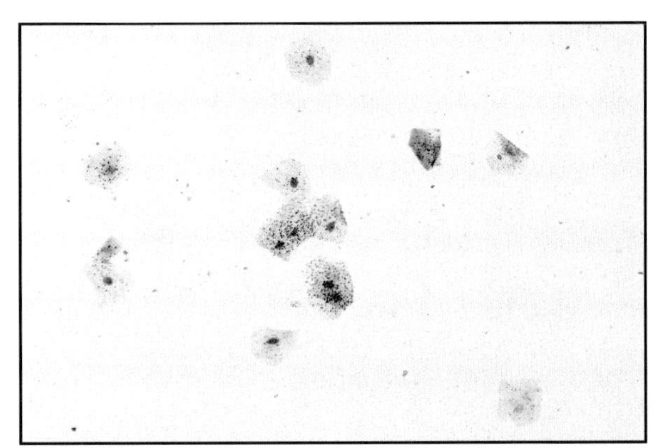

***Figure 4.8:*** *Vaginal cytology from same bitch as Figure 4.7 in pro-oestrus about 3 months later. Note the complete absence of polymorphonuclear leucocytes.*

mally referred to as the hymen, is more marked and breeders may try to distend this prior to the first mating.

Because of the additional advantage of feel, a digital examination is much more useful in determining the nature of these defects than instruments such as vaginoscopes. In spite of a relatively commonly held belief, it is not possible to palpate and determine the patency of the bitch's cervix, as the vagina may measure in excess of 20 cm.

It is interesting to note that the areas which tend to have congenital defects, namely the vulva, vestibule and vestibulo-vaginal junction, are formed from different embryological structures from those forming the vagina. This may help to explain why it is very uncommon to get a true congenital vaginal stricture and if the bitch's tract is normal as far cranially as the vestibulo-vaginal junction, problems are unlikely to be encountered.

**Vestibulo-vaginal strictures**

This is the most common site for strictures of the reproductive tract. Vestibulo-vaginal strictures are relatively easily diagnosed by digital palpation. The vestibulo-vaginal junction is felt in relatively large bitches about a finger's length into the tract, therefore in the majority of bitches if there is no stricture palpable it is unlikely that there is one present. Bitches examined prior to their first mating should have a routine digital examination to check that no such abnormality exists. Animals with vestibulo-vaginal strictures do not usually have a vaginal discharge.

Strictures can take an annular form which, on palpation, feels like a ring and not unlike the normal vestibulo-vaginal junction, but is rigid and impassable. The maiden bitch will probably have a hymen or remnant of hymen which can feel similar except that it is easily broken by gentle digital pressure. A single band of tissue is sometimes found which runs dorso-ventrally at the vestibulo-vaginal junction, and this is commonly a persistent mesonephric duct. This is easily diagnosed digitally or using a vaginoscope. The instrument can easily pass round the side of such a structure during introduction and thus the defect may be missed unless care is taken to continue observation until the vaginoscope has been completely withdrawn from the tract. In many cases these defects are found at autopsy, having caused no problems throughout the bitch's lifetime.

Annular rings of the vestibulo-vaginal junction are not easily corrected. Attempted distension is usually unsuccessful, as the fibrous nature of the ring normally reverts back to its original size following cessation of the distension. However in a proportion of cases distension is successful, in these cases the defect was probably not a true stricture. Proper strictures should not be surgically corrected because of the ethics of correcting a congenital defect. The single band type of defect is relatively easily cut following episiotomy, but again the question of altering congenital abnormalities must be considered.

**Vulval constrictions**

Occasionally bitches are seen in which the vulva is congenitally fibrosed and will not even allow the entry of a fingertip. Such animals may have already experienced at least one oestrus, indicating that even elevated oestrogen concentrations have not allowed distension of the vulva. These bitches should be re-examined during standing oestrus to confirm this fibrosis, as even a vulva which seems tiny may be patent under oestrogen domination. Greyhounds, for example, have very small vulval lips, but at mating time are patent to the dog's penis. Some bitches when seen early in pro-oestrus appear unlikely to have the potential to distend sufficiently to allow intromission, but by the time of mating are quite normal. The difference between the vulva which distends and the abnormal one is that the latter type feels fibrotic, whereas the other feels normal and pliable, but small.

Digital distension does not make any difference to a congenital fibrotic vulval stricture. Surgical correction is not acceptable on the grounds that it is changing the characteristics of the animal. Even if episiotomy were to be carried out, mating would be inhumane. It may also be that such conditions are inherited.

**Intersexes**

Intersexuality can exist in a variety of phenotypes. This is because the gonads can contain a variety of combinations of ovarian and testicular material, either functional or inactive. The most obvious indication of intersexuality is either an enlarged clitoris or an underdeveloped penis and prepuce. The genital tubercle is an embryological structure which is common to both sexes and is androgen sensitive, developing into either a clitoris or penis. In the absence of male hormones, the clitoris remains a vestigial structure. The presence of testicular material has the potential effect of causing enlargement of the clitoris. With more extensive stimulation, the result may be an improperly developed penis. By the time the animal develops physical features which allow the owners to realise that it is abnormal, it is part of the family and although the breeder offers a replacement, very few owners accept.

Most intersexes appear to be isolated cases with no obvious aetiology, although intersexuality has been reported as being inherited in some breeds such as the Cocker Spaniel. While chromosome analysis can be used in diagnosis, the presenting signs are usually diagnostic and indeed in many intersexes, the cytogenetic finding is a female karyotype (78XX) in an animal which has a female phenotype. In such cases, the test is uninformative. In other cases, karyotyping may be of help, in that it may show a difference from the phenotypic appearance, thus confirming the intersex state.

It has been reported that supplementation of pregnant bitches with some synthetic progestogens may give rise to tract abnormalities such as intersexuality,

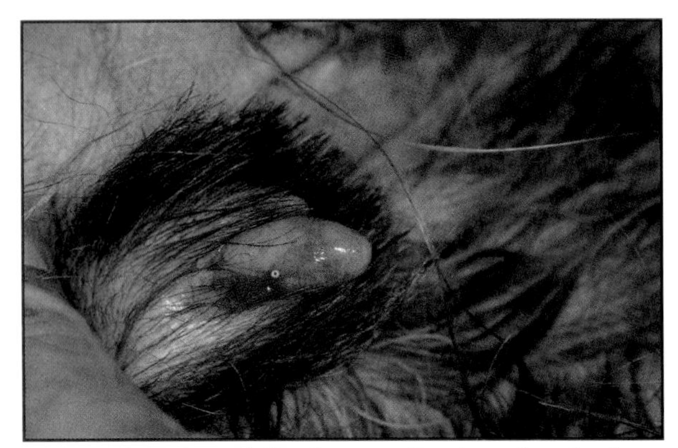

*Figure 4.9: Enlarged clitoris protruding from the vulva in an intersex animal presenting as a female.*

but no detailed studies have been carried out. For this reason, the use of progestogens to augment the bitch's own progesterone should only be carried out when there is clear evidence that the bitch has suffered from luteal insufficiency during a previous pregnancy. Even under these circumstances, full discussion should be carried out with the owner. It is probably safer to use progesterone rather than synthetic progestogens, although the former is only available in the UK as a short-acting injection. The injections should be stopped before the anticipated whelping date in order for parturition to commence. If the injections are not stopped, or if depot injectable progestogens are used, a prolonged gestation will result, with the subsequent death of the puppies and other sequelae.

***Intersex animals presenting as females:*** By far the most common diagnostic sign is an enlarged clitoris in an animal which appears to be female (Figure 4.9). It can either be seen protruding through the vulval lips or, if smaller, after parting the lips. The clitoris is palpable and commonly contains an os clitoris. The abnormality is normally first seen at about 3-4 months of age. The size of the clitoris can increase as the testicular tissue becomes endocrinologically functional. This commonly occurs at the time of expected puberty (4-6 months of age).

In some older animals, the primary presenting sign is a vulval discharge where the owner has not seen the enlarged clitoris. Virtually all of these bitches have not had a season, although occasionally if the gonadal tissue contains functional ovarian tissue, signs of oestrus may occur.

Gonadectomy should be carried out plus removal of whatever reproductive tract is present. The gonads invariably contain testicular material which may consist of ovo-testes or testes and in both situations there can be epididymal tissue present (Figure 4.10). In many cases removal of the testicular material causes some reduction in clitoral size. If the clitoris is then effectively covered by the vulval lips, reconstructive surgery is unnecessary. If the clitoris is still exposed, it remains vulnerable to damage, either self-induced or by external trauma. If this is the case, removal of the clitoris including the os is necessary.

***Intersex animals presenting as males:*** Masculinized cases can be seen in which the animal presents with an underdeveloped prepuce and penis (Figure 4.11). Occasionally there may be haematuria and attractiveness to other males, the result of pro-oestral bleeding resulting from ovarian and uterine tissue.

Gonadectomy is the treatment of choice. If the genital tubercle has developed to an extreme extent, the urethra runs through the underdeveloped penis and urine scalding can occur if the prepuce does not cover the penis. The owners will have to ensure that the skin of the abdominal floor and groin is washed and dried frequently to avoid skin problems. Petroleum jelly or emollient creams can be used to prevent irritation. Only if the scalding is an insurmountable problem does the extreme measure of penile amputation become a consideration.

It should be remembered that racing Greyhounds tend to be treated with testosterone compounds to suppress their oestrous cycles as progestogens tend to reduce racing performance. These drugs can cause clitoral enlargement together with a vulval discharge and this should not be confused with intersexuality. Once the medication is stopped, the

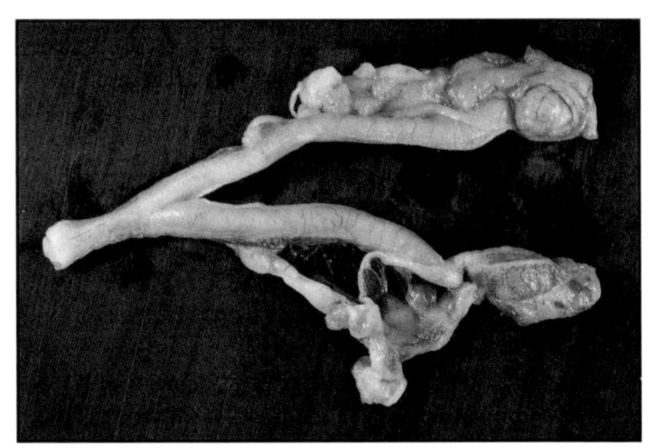

*Figure 4.10: Reproductive tract from an intersex animal, presenting as female. Both male and female gonadal tissue are present with an epididymis and uterine horns.*

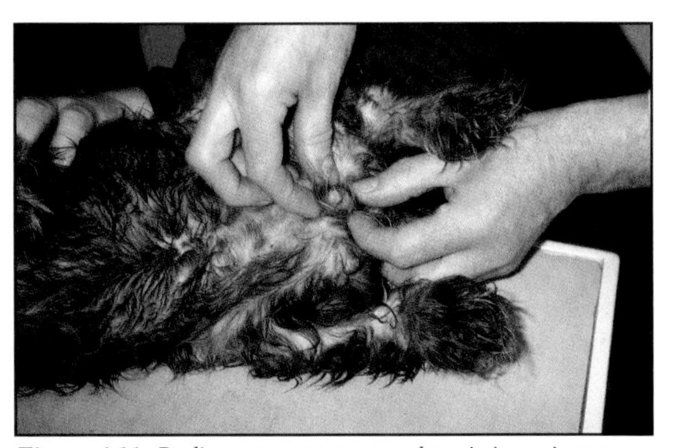

*Figure 4.11: Rudimentary prepuce and penis in an intersex presenting as more male than female.*

clitoris tends to regress in size. There appears to be no reduction in subsequent fertility following cessation of treatment.

**Acquired vaginal conditions**
A visual or vaginoscopic examination should reveal the presence of acquired vaginal conditions and cytology of any discharge will confirm if it is purulent in nature. If the vaginitis is a secondary condition, the primary problem must be corrected followed by the required treatment for the vaginitis.

***Vaginal ulceration:*** This condition, which is not common, presents as a haemorrhagic discharge that can become infected and purulent. It is most likely to follow a traumatic mating. Forcible separation of the tie can also cause vaginal damage including the possibility of gross damage or ulceration. Owners should be made aware that as the sperm-rich fraction is produced in the first 2 or 3 minutes of the mating, separation will not stop a potential pregnancy. Breeders carrying out excessively rough digital examination for vaginal distension may also be responsible for mucosal damage of the vagina.

Diagnosis is by vaginoscopy and at this examination, it should be clear whether there is damage present which is causing the bleeding or a pathological condition, such as a tumour. In most cases, ulceration will resolve spontaneously. If it persists, local therapy with antibiotics or emollient creams may help the healing process.

***Vaginal tumours:*** Vaginal tumours are normally benign, with fibroleiomyomas, fibromas and lipomas being the most common. Presenting signs are vulval bleeding or discharge. Diagnosis is normally by using vaginoscopy. A useful technique is contrast radiography, in which a Foley catheter is introduced into the vestibule and, after gently clamping the vulva with bowel clamps, contrast medium is infused and a lateral radiograph taken. The normal vagina has a very sharp margin, and the outline of the tumour, if present, is clearly visualized. Surgical removal is normally undertaken via an episiotomy. It has been advocated that ovariohysterectomy be carried out at the same time, as the tumours may be hormone-dependent. Following removal, recovery should be uneventful.

***Transmissible venereal tumour:*** In the UK, this tumour is rarely seen and then only in animals which have been mated overseas. It is an interesting tumour in that it is spread venereally by cell transmission. All cases appear to have a common origin, in that the chromosome picture of the tumour cells from most animals studied is similar. The TVT has a cauliflower-like appearance and is friable. Clinical signs include vulval bleeding or the appearance of the tumour at the vulva. The tumour can be removed following an episiotomy, although spontaneous regression may occur.

***Foreign body:*** Vaginoscopy should reveal the presence of any obvious focus of infection, such as grass awns. Even if missed, the usual local therapy of flushing with aqueous antibiotics or mild disinfectant should remove any foreign body.

**Vaginal hyperplasia**
Vaginal hyperplasia is often improperly referred to as vaginal prolapse. It can present in one of two ways:

- A circular doughnut-like structure (Figure 4.12) protruding from the vulva and representing a 360° hyperplastic area
- A single spherical swelling (Figure 4.13), which is a discrete portion of the vagina that has undergone hyperplasia.

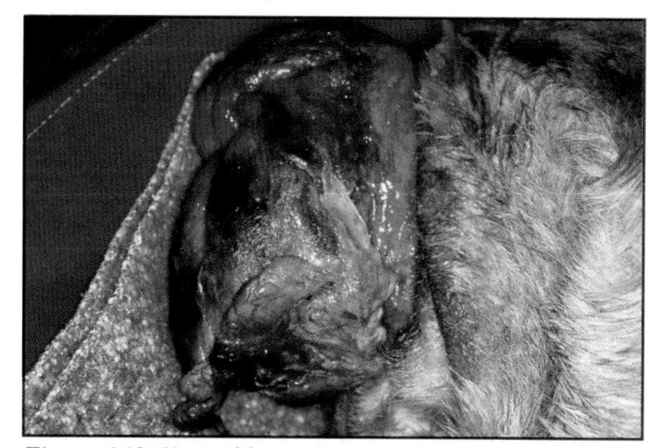

***Figure 4.12:*** *Vaginal hyperplasia in a bitch with 360° vaginal protrusion ('doughnut-type'). Photograph taken following episiotomy prior to surgery.*
*Courtesy of Dr Martin Sullivan.*

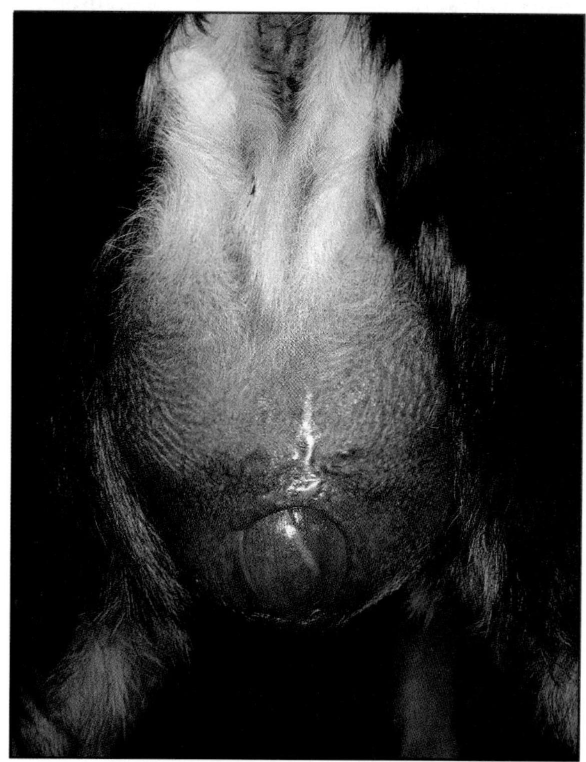

***Figure 4.13:*** *Vaginal hyperplasia with a single discrete swelling of part of the vagina.*

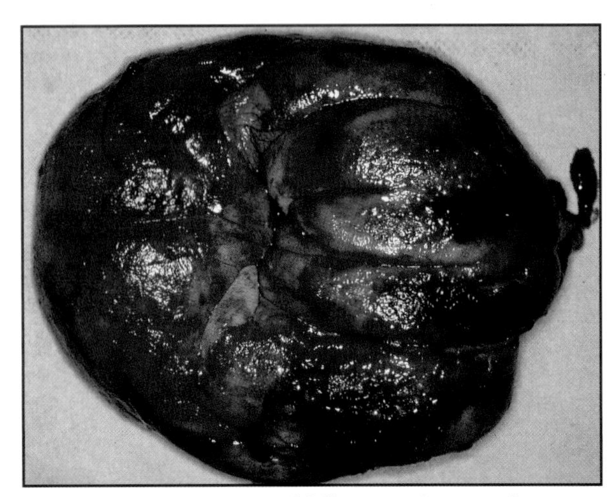

*Figure 4.14: Tissue removed following sub-mucosal resection of the bitch in Figure 4.12.*

Courtesy of Dr Martin Sullivan.

This condition is normally only seen in bitches in pro-oestrus and oestrus, where the high oestrogen concentration encourages the hyperplastic condition. As the vulva swells and then relaxes, the vagina distends and the vaginal wall protrudes through the vulva. The doughnut appearance is typical and diagnostic.

The condition resolves when the bitch goes out of oestrus and her oestrogen concentrations decline; however, the problem is likely to recur at the next oestrus.

Regression of the condition can be accelerated by the use of progestogens which cause follicular regression, but care should be taken to select one which is safe to use in pro-oestrus (see Chapter 16). Once the condition has regressed, the permanent solution is ovariohysterectomy when the bitch is in anoestrus.

If the animal is required for breeding, sub-mucosal resection (Figure 4.14) can be performed during oestrus but the problems of haemorrhage make the surgery difficult. It has been suggested that this condition may be inherited and therefore surgery may be considered unethical.

### Acquired fibrosis of the tract

Vaginal damage caused during dystocia may result in strictures forming, where one part of the repairing vagina adheres to another. Although congenital defects almost invariably occur caudal to the vagina proper, such acquired damage can occur anywhere within the birth canal. The history will be of a parous bitch which, at her next breeding, experiences a painful mating or where the male dog was unable to gain full intromission and no tie was achieved. There may also have been haemorrhage following the mating. Investigation is by digital and vaginoscopic examination and if there is marked distortion of the vagina, contrast radiography may is useful. If the damage interferes with mating, future breeding of the bitch is unlikely to be possible as surgical repair is unlikely to be successful.

## CONDITIONS OF THE UTERUS

### Cystic endometrial hyperplasia (CEH)–pyometra

This is the most serious condition of the bitch's reproductive system. It occurs in a uterus which has developed cystic hyperplastic changes after a series of oestrous cycles, commonly unaccompanied by pregnancy. During oestrus, the uterus becomes infected and, under the influence of progesterone from the corpora lutea, develops into a pus-filled organ which releases toxins into the circulation. These have marked deleterious effects both systemically and on specific organs, such as the kidney. The condition must always be considered as a possibility in any bitch which is unwell during the 2-month period after oestrus. It should be borne in mind that not every bitch that is unwell during this time is suffering from pyometra and the condition must be proven to exist rather than be assumed.

Pyometra is typically a condition of the middle-aged to older bitch which has not had puppies, although it may occur in both parous and young bitches. The bitch typically was in season about 6–8 weeks previously (Figure 4.15) and has normal elevated progesterone concentrations. The condition may, however, occur as early as one week after the end of oestrus. Pyometra may also occur in a bitch which has 'ovarian remnant syndrome' in which a portion of uterus has been left, the so-called 'stump pyometra'.

There may be a history of either oestrogen therapy for unwanted mating or, particularly in the pubertal bitch, progesterone therapy for oestrous control.

The bitch is typically dull, recumbent, inappetant, vomiting, polydypsic and polyuric. Most of the signs result from systemic toxaemia, with the polydipsia and polyuria being the result of glomerulonephritis. The rectal temperature may be elevated early in the course of the disease, but may become normal or subnormal as the condition progresses. Occasional

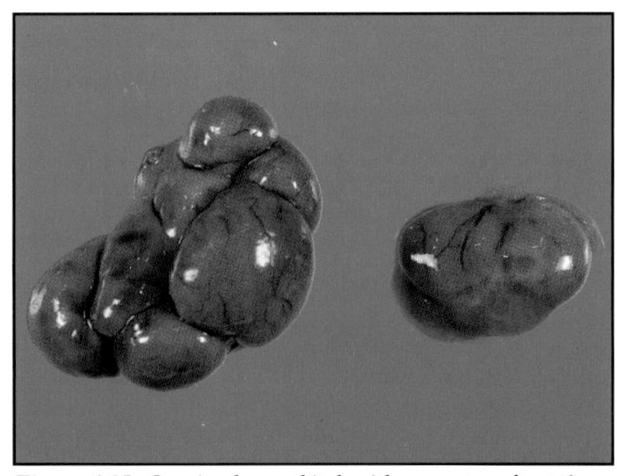

*Figure 4.15: Ovaries from a bitch with pyometra about 6 weeks after her season. Note the corpora lutea.*

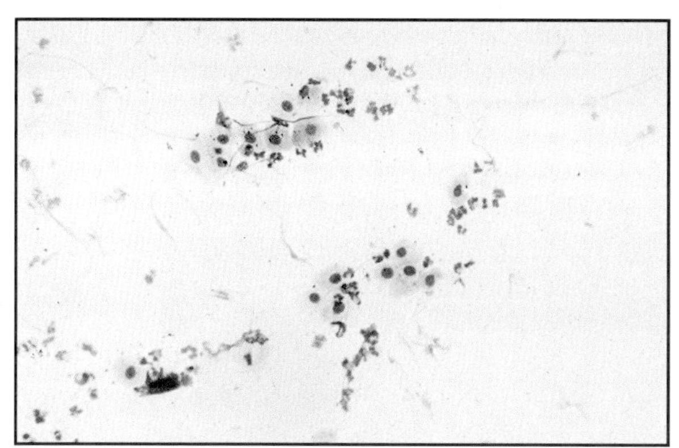

*Figure 4.16: Vaginal cytology from a bitch with open-cervix pyometra showing degenerate polymorphonuclear leucocytes.*

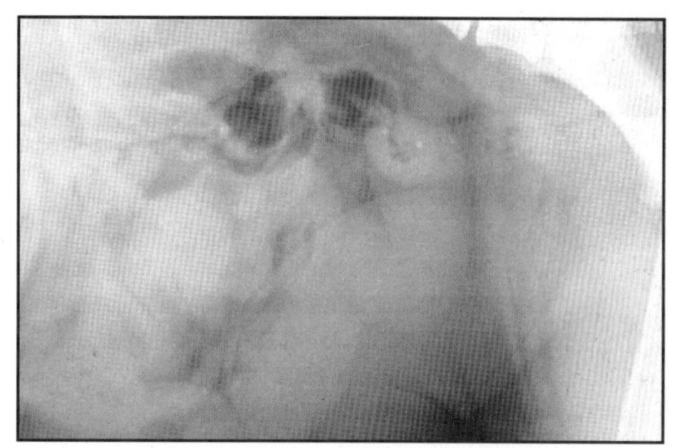

*Figure 4.17: Lateral radiograph of a bitch with pyometra showing uniform-contrast enlarged uterine horns.*

clinical signs include skeletal pain and acute uveitis. Not every bitch shows all of these signs; in fact none of these signs is invariable.

There may or may not be a purulent or blood-stained vulval discharge containing degenerating polymorphonuclear leucocytes (Figure 4.16). The presence of a discharge is the basis for classification as either 'open-cervix' or 'closed-cervix' pyometra. Abdominal distension is common as the uterus is markedly enlarged, weighing up to 10 kg in some large bitches. In some highly toxic bitches, however, the uterine diameter may be only modestly increased.

The work of Dow (1959) indicated that the repetitive injection of oestrogens and progesterone was most effective in inducing pyometra experimentally. It is normally considered that the condition starts as cystic endometrial hyperplasia, with ascending infection from the vagina through the open cervix at oestrus. The elevated progesterone concentrations encourage bac-terial growth. It is not venereally introduced, as most bitches with pyometra have never been mated. The most common bacterium is *Escherichia coli*, although a variety of other non-specific bacteria can be found in the discharge.

***Confirmatory diagnosis:*** Haematology is valuable, with most bitches having elevated polymorphonuclear leucocyte counts (greater than about 15 x $10^3$ per $mm^3$) with a marked left shift. Bitches with closed-cervix pyometra tend to have higher counts than open cases, with counts normally in excess of 20 x $10^3$ per $mm^3$, going as high as 100 x $10^3$ per $mm^3$. In contrast, occasional cases may be within normal range (7–10 x $10^3$ per $mm^3$).

Biochemistry may reveal an elevated blood urea nitrogen due to the reversible renal damage but frequently the biochemistry is unhelpful. Hepatocellular damage may be indicated by an elevation in serum alanine aminotransferase levels, but whether this is caused by the pyometra or is a coincident condition in an older bitch may be debatable.

On a lateral radiograph (Figure 4.17), the fluid-filled enlarged uterus has a soft tissue opacity and is imaged in the ventral abdomen. The intestines are displaced and there is commonly an increased distance between the bladder and rectum. Whenever a uterus is visualized on a radiograph, it is almost certain to be enlarged and it should be noted that a pregnant uterus between 4 and 6 weeks, the stage before the fetal skeletons become visible, can appear very similar to a pyometra. One marked disadvantage of radiography is the necessity to sedate a bitch that is already toxic.

The enlarged and fluid-filled uterus can be imaged by ultrasonography, often showing varying sized loops (Figure 4.18). The fluid is normally anechoic but may be floccular. Differentiation of pyometra from pregnancy can be made by the absence of fetuses, the floccular nature of the fluid and the fact that the loops are of varying sizes. The additional advantage of ultrasonography is that sedation is not required and frequent monitoring of the condition can therefore be carried out safely.

***Treatment:*** The treatment of choice is ovariohysterectomy, the only permanent method of curing the condition. Following analysis of blood electrolytes and acid–base balance, fluid replacement should be given. Immediate antibiotic therapy should be initiated and the surgery should be carried out as soon as is practicable using standard methods (see Chapter 15). The

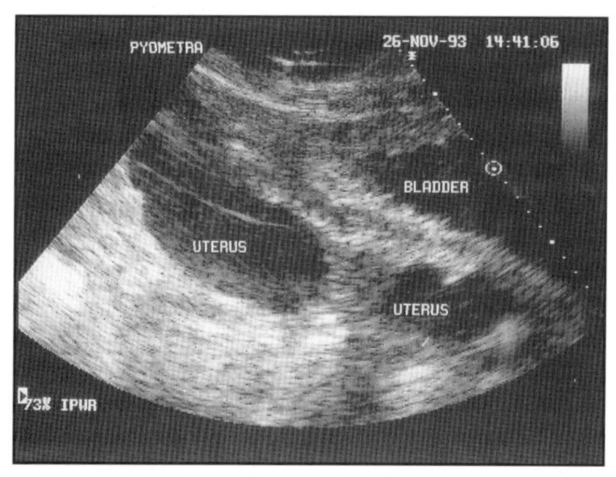


*Figure 4.18: Ultrasound scan of a bitch with pyometra.*

*Courtesy of Professor Jack Boyd.*

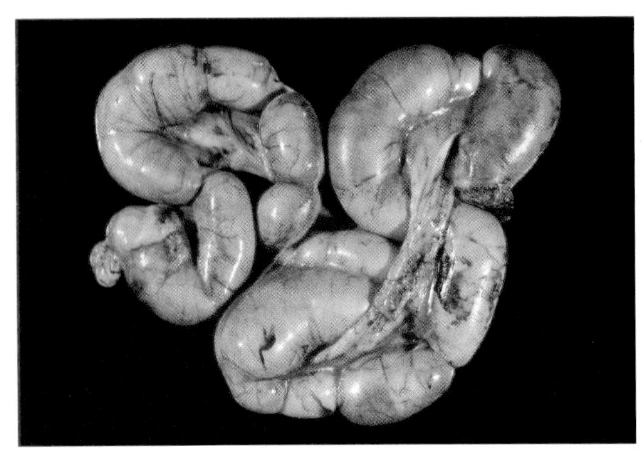

*Figure 4.19: Uterus of a bitch with pyometra.*

amount of time available to stabilize the bitch will depend on the severity of the condition.

Following surgery, in the absence of postoperative infection, the polymorphonuclear leucocyte count may continue to rise due to the continued production of the cells which cannot now migrate into the uterus, but the count will gradually decline. In bitches which do not appear to be making satisfactory postoperative recovery, the blood urea concentration can provide a useful prognostic guide. If this stays elevated for more than a few days, it is likely that the kidney damage may not be reversible and therefore recovery may not be made in spite of successful surgery.

Medical treatment with prostaglandins can be carried out under exceptional circumstances, for example in the high-risk surgical candidate, the extremely old bitch or the good quality young bitch from which the owner is very keen to breed.

In extremely mild open-cervix cases, pyometras may be treated with appropriate antibiotics. This may prevent disease progression while the animal passes out of dioestrus and into anoestrus. Once progesterone concentrations are basal, the condition will start to resolve. However, following the next season, the pyometra will inevitably recur, almost certainly in a more severe degree.

As with antibiotics, the use of prostaglandin $F_{2\alpha}$ can only be recommended under the following circumstances:

- The possibility of pregnancy is eliminated
- It is an open pyometra
- The condition is not life-threatening
- The well-being of the bitch is the primary consideration
- The welfare of the bitch is not jeopardized by the owner's wish to obtain a litter
- The candidate is an old bitch that is a very poor anaesthetic risk
- The candidate is a very young bitch that is intended for breeding
- The owner is aware that the drug is not licensed for use in the bitch
- The owner is aware of the potential side effects.

Prostaglandins have two effects, luteolysis and uterine contraction, hence their restriction to open-cervix pyometra. Natural $PGF_{2\alpha}$ should be used (see Chapter 16). It has been reported that splitting the dose and treating twice daily is more effective, but this is usually impractical under most circumstances in practice. The bitch should be hospitalized for at least one hour after treatment as the side effects can appear distressing and include panting, retching and vomiting plus signs of stress including scraping the kennel floor. In the majority of bitches, these signs resolve within 30 minutes and in many cases they appear to become less severe over the course of treatment. Appropriate antibiotic therapy against *E. coli* should be given in parallel with fluid replacement therapy. The response to the therapy can be rapid, with the bitch showing clinical improvement within a day or two. A copious discharge is not always seen, although this may have been present and removed by the bitch.

The main effect of prostaglandins is luteolysis, which is rapid; progesterone concentrations become basal within 2 or 3 days. Progesterone concentrations may therefore be measured to assess the efficacy of the treatment, and monitoring of the size and state of the uterus can be carried out using ultrasonography. The uterine diameter reduces over the first 2 or 3 days and if therapy has been successful and the progesterone concentration stays low, the condition will resolve. Treated bitches that are spayed 2 or 3 months later have uteri which appear grossly normal.

If the reason for medical treatment is the desire to breed from a young bitch, mating should be carried out at the next oestrus. Fertility tends to be surprisingly good and normal litter sizes are common. If breeding has been successful, ovariohysterectomy should be strongly recommended before the next expected oestrus, as there is a high risk of the condition recurring following a later, if not the next, season. This recurrence occurs in spite of the bitch having a normal litter which might have been thought to protect the uterus from further changes.

In theory, anti-prolactin drugs such as cabergoline should cause luteal regression and therefore result in resolution of the pyometra. Such studies do not yet appear to have been carried out in any numbers but this treatment would have the advantage of having relatively few side effects in comparison with prostaglandins. There has been some success using the progesterone receptor antagonist aglepristone to treat pyometra, but presently this drug is only available in France (see Chapter 16).

***Stump pyometra:*** Bitches that have had an ovariohysterectomy in which a small portion of uterus has been left may develop a 'stump pyometra'. Assuming that both ovaries have been removed, the bitch will present with a purulent vulval discharge, usually without the clinical signs of toxaemia. Radiography or ultrasonography will reveal the presence of a stump containing pus. The polymorphonuclear leucocyte count may be normal.

If ovarian tissue has been left together with the stump, the stump pyometra may produce mild clinical signs due to the presence and effect of circulating progesterone. Such a bitch would present as 'ovarian remnant syndrome' followed by the signs of the pyometra.

In both situations, surgical removal of the stump is necessary.

**Uterine neoplasia**

Tumours in the uterus are uncommon, though fibroleiomyomas are the most common. While most may not be diagnosed, some may cause haemorrhage with resultant vaginal discharge. In this situation, there will be no vulval swelling and a vaginal smear will reveal an absence of oestrogen stimulation, ruling out a normal or aberrant pro-oestrus. Ultrasonography, radiography or exploratory laparotomy will confirm the diagnosis and treatment is ovariohysterectomy.

## CONDITIONS OF THE NON-PREGNANT QUEEN

One of the most important facets of investigating reproductive problems is possessing a full knowledge of the reproductive cycle of the cat (see Chapter 2).

### Disorders presenting as cycling abnormalities

**Anoestrus**

***Puberty:*** Queens usually reach puberty during the year following their birth. However, the cat which is born late in the year may not cycle until she is about 15–18 months of age.

***Season:*** As the cat is a seasonal breeder, coming into season in the spring when there is an increasing light pattern, dark conditions and poor management may cause a marked delay in the occurrence of puberty or of the onset of the breeding season in the adult. The queen's environment should always be considered when investigating cases of delayed puberty.

***Intersexuality:*** Intersexes are found in the cat, but are uncommon. In most cases the diagnosis is made accidentally at spaying or in the course of an investigation of a cat with primary anoestrus. In the latter case, when other causes such as seasonal effects and management are eliminated, intersexuality or chromosomal defects are possible causes. Unlike the bitch, the primary sign of intersexuality is not normally an enlarged clitoris, but more often underdeveloped external genitalia in a cat showing either anoestrus or possibly aberrant reproductive behaviour.

***Chromosomal abnormalities:*** Cats with primary anoestrus have occasionally been found with sex chromosomal defects such as the XO syndrome. Laparotomy will confirm the permanent nature of the ovarian agenesis but the precise chromosomal aetiology can only be diagnosed using karyotyping which reveals the absence of one of the two X chromosomes. These cats are physically normal but are permanently infertile.

**Cycling disorders**

Unlike the bitch cycling disorders are rare in the queen, mainly because the cat is an induced ovulator. The queen which is mated normally ovulates and therefore terminates the oestrus. Those which are not mated normally do not go into a luteal phase; thus, conditions of the luteal phase are much less common in the cat.

**Cystic ovaries**

Follicular cysts can occur in the queen and the condition normally presents as prolonged oestrous behaviour. Surgical rupture of the cysts has been carried out, with mating recommended at the next heat as the condition will probably recur. In theory, the use of progestogens to cause follicular atrophy might be successful but in most cases ovariohysterectomy is the treatment of choice.

**Ovarian remnant syndrome**

The aetiology and approach to this condition are similar to those for the bitch. The spayed queen shows signs of calling with regular frequency similar to that occurring prior to surgery. The most likely cause is an ovary or part of an ovary not removed at surgery. It is therefore always worthwhile to inspect the tract following a routine ovariohysterectomy for the presence of two complete gonads.

The use of vaginal cytology to confirm oestrogenic stimulation is essential when a spayed queen is showing overt signs of oestrus or is attracting males. At that stage she will have elevated oestrogen concentrations and although oestrogen assays can be used, it is cheaper and easier to use vaginal cytology. The use of a paediatric bacteriology swab moistened with saline is an easy way to obtain the cells. The stage of oestrus is irrelevant and the diagnosis is made by the observation of cornified cells indicating elevated oestrogens.

If the queen is calling every 3 weeks, this suggests that she is not ovulating. In this case hCG or GnRH can be used to induce ovulation and a progesterone assay carried about 7–10 days later should reveal a progesterone concentration of >3 mmol/l (>1 ng/ml), confirming the presence of active ovarian tissue.

If the queen is calling approximately every 6 weeks, she is ovulating and therefore is either being induced by mating or less commonly ovulating spontaneously.

A blood sample taken about 7-10 days after the signs of oestrus will reveal an elevated progesterone concentration.

Following confirmation of the presence of ovarian tissue, surgical removal will resolve the problem. In the situation where the owner does not want surgery carried out for a second time, the use of oestrous suppressants on a permanent basis is an alternative.

## Ovarian tumours

As with the bitch, ovarian tumours are uncommon, the most common being the granulosa-theca cell tumour. Clinical signs are those of a large abdominal mass rather than of endocrine malfunction. These signs include abdominal distension, ascites, vomiting and weight loss. If the tumour produces steroid hormones, the result may be persistent oestrus ( if oestrogen-producing) or possibly endometrial hyperplasia (if progesterone-producing). A variety of other tumours have been reported, including dysgerminoma which may produce similar signs to the granulosa cell tumour. Most of these tumours are not malignant. The diagnosis can be made using either radiography or ultrasonography and the treatment is ovariohysterectomy.

### Vaginal tumours

Fibroma and leiomyoma of the vagina are found uncommonly. They may produce a haemorrhagic or purulent vulval discharge but affected cats are usually presented with constipation associated with pressure on the rectum. Ovariohysterectomy is normally recommended, combined with excision of the tumour, as it has been suggested that these tumours may be endocrine-related.

## Pseudopregnancy

The term pseudopregnancy in the cat applies to the normal prolonged luteal phase which follows a non-fertile ovulation. This may be due either to an infertile mating or because of spontaneous ovulation. During the 6-week period of this extended luteal phase, the queen normally does not show the behavioural signs associated with pseudopregnancy in the bitch (see above) although lactational signs are sometimes seen.

When lactation does occur in the queen, it can be extremely marked and is referred to as mammary hyperplasia. This can be observed following the physiological pseudopregnancy, ovariohysterectomy or following treatment with progestogens. For details and treatment see Chapter 5.

## Cystic endometrial hyperplasia (CEH)–pyometra

CEH-pyometra is much less common in the cat than in the bitch; as in the bitch it is more commonly found in the older animal. There is little difference in the frequency of the condition in cats that have had litters compared to nulliparous queens. The lower incidence of pyometras in cats is probably due to two factors. First, unmated cats tend not to ovulate, therefore the uterus is not under the influence of progesterone and thus cystic endometrial changes do not develop. Secondly, even if they do ovulate, the length of the progesterone phase is shorter in the queen than in the bitch (45 days compared with more than 60 days).

The systemic signs are similar to those in the bitch; dullness, inappetance and polydipsia/polyuria are seen together with vaginal discharges in cases of open-cervix pyometra. Open-cervix cases appear to be much more common than closed ones in the queen. In both types, abdominal distension is usually present, and this may be the most obvious clinical sign especially in closed-cervix pyometra. Elevated total white cell counts may be observed. Radiography and/or ultrasonography are essential diagnostic aids, with the latter being much more useful as it will differentiate pregnancy from pyometra and does not require sedation.

The treatment of choice is ovariohysterectomy with antibiotic cover and fluid replacement. As with the bitch, such cases are not ideal surgical candidates.

If the cat is young and the owners are keen to breed from her, or if she is old and a poor surgical risk, medical treatment may be attempted. Prostaglandin $F_{2\alpha}$ in a similar regimen to that used in bitches (see above) has been reported to be successful in resolving the condition. Such treatment should be restricted to cases of open-cervix pyometra. Evacuation of the contents occurs rapidly and the condition of the cat normally improves. The side effects are similar to those seen in the bitch. Young cats can become pregnant at subsequent heats, but they should be spayed at the earliest appropriate time as the condition is likely to recur, even if there has been a pregnancy.

The use of progestogens in a spayed queen in which a portion of uterus has been left can induce a stump pyometra. The approach to this situation is similar to that in the bitch.

## Uterine tumours

Uterine tumours are rare in the cat but a highly malignant adenocarcinoma is recognized as the most common, with fibromas and leiomyosarcomas also being found. The signs associated with a tumour are those deriving from its mass, obstructive effects or malignancy rather than from any direct reproductive effect. Occasionally vulval discharges may be found associated with uterine tumours. Ultrasonography will help with the diagnosis of a tumour but the precise diagnosis and organ involved relies on laparotomy and pathological findings. Ovariohysterectomy should be carried out.

## FURTHER READING


Arthur GH, Noakes DE, Pearson H and Parkinson TJ (1996) *Veterinary Reproduction and Obstetrics, 7th Edition.* WB Saunders, Philadelphia.

Dow C (1959) Experimental reproduction of the cystic hyperplasia-pyometra complex in the bitch. *Journal of Pathological Bacteriology* **78**, 267-278

England GCW (1998) *Allen's Fertility and Obstetrics in the Dog.* Blackwell Science, Oxford

Feldman EC and Nelson RW (1996) *Canine and Feline Endocrinology and Reproduction.* WB Saunders, Philadelphia

Harcourt-Brown N (1996) Down the endoscope; the reproductive tract. *Practice* **18**, 262-265

Harvey MJA, Cauvin A, Dale M, Lindley S and Ballabio R (1997) The effect and mechanism of the anti-prolactin drug cabergoline on pseudopregnancy in the bitch. *Journal of Small Animal Practice* **38**, 336-339

Herron MA (1983) Tumours of the canine genital system. *Journal of the American Animal Hospital Association* **19**, 981-994

Herron MA (1986) Infertility from noninfectious causes (in the cat). In *Current Therapy in Theriogenology, 2nd Edition.* WB Saunders, Philadelphia

Jeffcoate IA (1993) Gonadotrophin-releasing hormone challenge to test for the presence of ovaries in the bitch. *Journal of Reproduction and Fertility* Supplement **47**, 536-538

Johnson CA (1991) Diagnosis and treatment of chronic vaginitis in the bitch. *Veterinary Clinics of North America* **21**, 523-531

Wykes PM and Soderberg SF (1983) Congenital abnormalities of the canine vagina and vulva. *Journal of the American Animal Hospital Association* **19**, 995-1008

CHAPTER FIVE

# The Mammary Gland

*David J. Argyle*

## INTRODUCTION

This section provides a short synopsis of the basic anatomy and physiology of the canine and feline mammary gland, and will give an overview of the common disease processes that affect the lactating and the non-lactating mammary gland in these species.

## ANATOMY

The dog generally has five pairs of mammary glands spread along the ventral aspect of the thorax and abdomen. The two cranial pairs are termed thoracic, the next two abdominal and the most caudal are the inguinal glands. The cat has a similar arrangement, but there are generally four pairs of glands.

The blood supply (Figure 5.1) originates from the external pudendal arteries and the lateral and internal thoracic arteries. The inguinal glands and the caudal abdominal glands are supplied from the caudal superficial epigastric artery and branches of the cranial abdominal and deep circumflex iliac arteries. The thoracic and cranial abdominal glands are supplied from the lateral thoracic artery, the cranial superficial epigastric artery and branches of the intercostal arteries. It is generally understood that lymphatic drainage from the thoracic and cranial abdominal glands is to the axillary and sternal lymph nodes, while drainage from the inguinal and caudal abdominal glands is to the superficial inguinal nodes. However, the pathways of lymphatic drainage are erratic and some lymph may cross the midline.

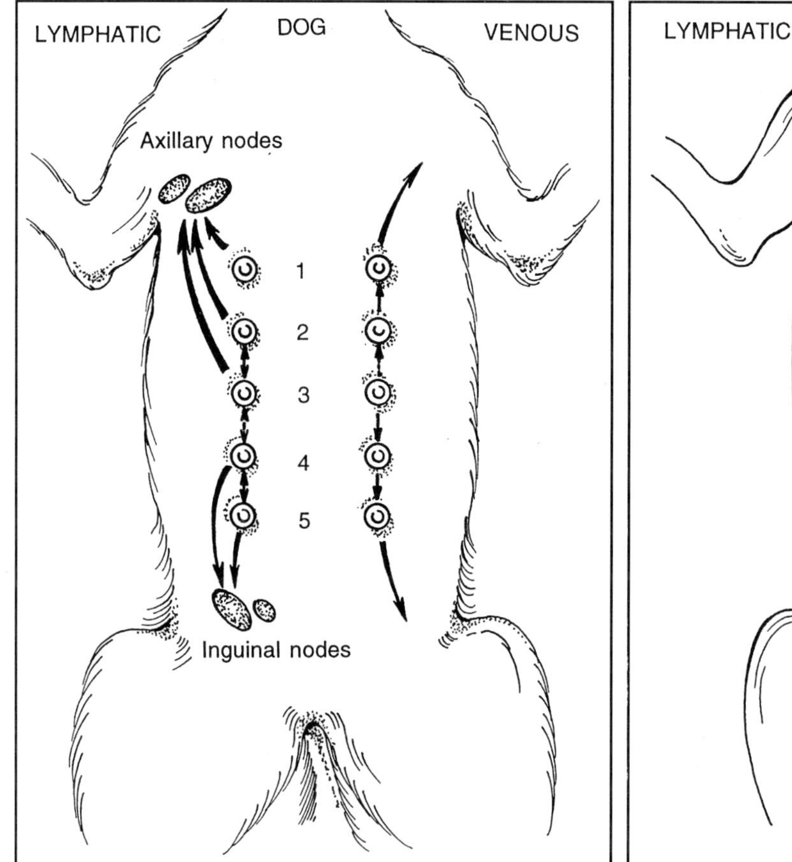


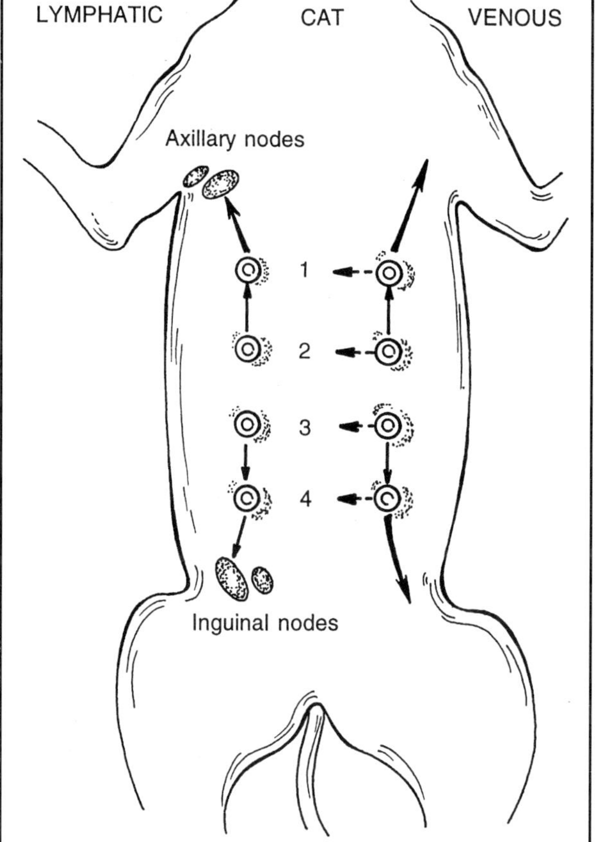


*Figure 5.1: Blood supply and lymphatic drainage of the mammary glands of the dog and cat.*

## CONDITIONS OF THE LACTATING MAMMARY GLAND

For the production of milk in adequate quantity and of adequate quality, the health and nutrition of the bitch or queen are of paramount importance. There are a number of conditions which affect the mammary gland in the periparturient period, the prompt diagnosis and treatment of which are essential for the health of the mother and her litter.

### Galactostasis (congestion)

This is a condition of uncertain aetiology which is seen just prior to parturition, or shortly afterwards. There is engorgement of the mammary glands, leading to pain and failure of milk let-down. It is more commonly seen in bitches on a high plane of nutrition. Animals show pain and discomfort and can be anorexic. Treatment involves fasting the animal for 24 hours, limited feeding for several days and the use of diuretics. Cold packs can be used to reduce engorgement, as can milking of the mammary glands.

### Agalactia

Agalactia may occur where there is a failure of mammary gland development, or there is failure of milk let-down. The former condition is uncommon, and the cause may involve complex interactions between environmental and hormonal factors. Failure of milk let-down is more amenable to therapeutic intervention, and can be suspected when the mammary glands are firm and swollen, but there is no milk present in the teat canal. This condition can occur in highly nervous females who fail to settle and allow the litter to suck.

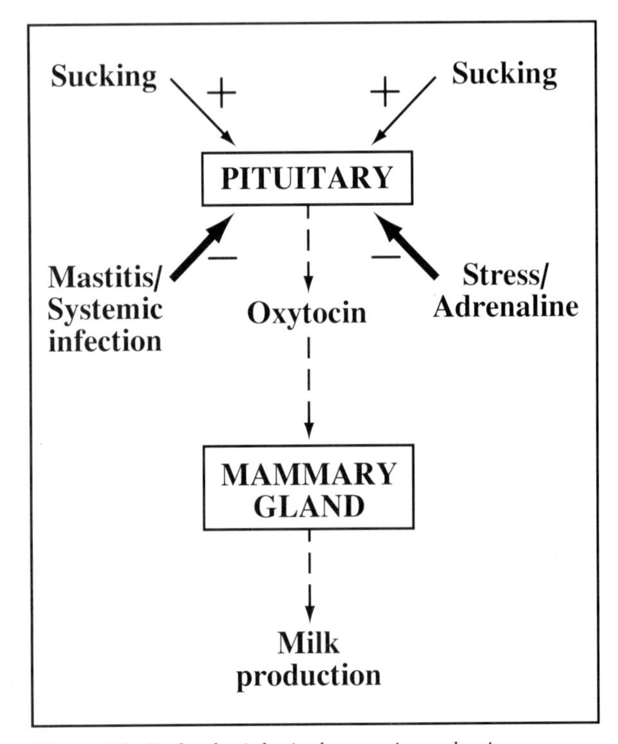


*Figure 5.2: Pathophysiological events in agalactia.*

In some highly stressed animals the production of adrenaline can block the release of oxytocin from the pituitary (Figure 5.2). In very nervous females, the judicial use of sedatives (acepromazine, 0.5–2 mg/kg sid or bid) can settle the dam into a suckling routine. Oxytocin (2–20 IU intramusculary or subcutaneously) can be used until milk production is established, and sucking by the litter should be encouraged. It must be remembered that conditions such as metritis, mastitis and systemic infections can also lead to agalactia and in these situations, treatment of the underlying cause is essential to re-establish lactation.

### Mastitis

Mastitis is not a common condition encountered in bitches and queens, but can be associated with prolonged galactostasis or where animals are kept in poor sanitary conditions. It can occur in both acute and chronic forms and is associated with pain, erythema and swelling of the mammary glands. In addition, the animal may be anorexic, pyrexic and show a marked neutrophilia, often with left shift, on haematology. In severe cases abcessation can occur and there may be a purulent exudate from the gland. Diagnosis is based upon clinical signs, haematology and cytological examination of the milk, which may reveal bacteria, white blood cells and erythrocytes. In an ideal situation, culture and sensitivity testing of the milk is performed before an antibiotic is chosen. However, it is often the case that antibiotic therapy needs to be instituted as soon as possible after diagnosis, and generally broad-spectrum antibiosis is administered immediately. Causal organisms include *Escherichia coli*, streptococci and staphylococci, and Gram staining may be helpful in choosing an antibiotic in the absence of culture and sensitivity. Where glands are frankly purulent, then sucking of that gland should be prevented by bandaging, but if the milk is grossly normal, the continued sucking may speed resolution. In severe cases of gland abcessation, surgical drainage may be required. If many glands are affected or the mother is severely systemically ill, then the litter may require to be hand-reared as orphans.

## CONDITIONS OF THE NON-LACTATING MAMMARY GLAND

### Tumours of the mammary gland (dog)

In the non-pregnant, non-lactating bitch the most commonly encountered abnormality of the mammary gland is neoplasia. Mammary neoplasia is the most common neoplasm in the bitch and accounts for over 50% of all tumours seen. Of these tumours, about half are malignant, and half of these have metastasized by the time of presentation. In contrast, this tumour is rare in the male dog and only accounts for around 1% of all tumours in males.

### Risk factors and aetiology

It is well recognized that the age at which ovariohysterectomy is performed is directly related to the risk of developing mammary neoplasia. Ovariohysterectomy prior to the first oestrus reduces the risk of developing mammary cancer to 0.05%, but this beneficial effect gradually reduces as the dog ages, and there is no apparent effect of ovariohysterectomy on mammary tumour development after 2.5 years of age. These data clearly demonstrate the prophylactic benefit of early ovariohysterectomy to the development of this disease.

The definitive role of sex hormones in the development of mammary neoplasia is unclear. However, about half of canine mammary tumours carry receptors for oestrogen or for oestrogen and progesterone, and these tumours tend to carry a better prognosis, as receptor-positive tumours tend to be benign. It has been suggested that the use of synthetic progestogens to control oestrus may contribute to the development of mammary tumours, but the exact role of exogenous hormones in mammary carcinogenesis has yet to be resolved. The age at which the bitch has her first litter, abnormal oestrous cycles and a history of pseudopregnancy are not associated with an increased risk of developing the disease.

Studies are currently underway to determine genetic alterations, particularly oncogene and antioncogene expression, in canine mammary tumours. Of particular interest is the loss of normal function of the tumour suppressor gene *p53*. These studies on the molecular mechanisms of carcinogenesis are particularly pertinent, as they not only provide information about prognosis, but may also lead to the development of improved and novel therapies.

### Tumour pathology

Mammary tumours can be divided simply into benign or malignant. Benign tumours include fibroadenomas, simple adenomas and benign mixed mammary tumours. Of the malignant tumours, carcinomas are most commonly encountered but complex carcinomas and sarcomas are also represented (Table 5.1).

### Clinical presentation

Many animals with mammary tumours are presented because the owner has noticed a solid mass, or the condition may be identified during a routine clinical examination. The clinical course of the disease may vary from being protracted in the case of slow growing tumours, to advancing quite rapidly, as is the case with many of the anaplastic carcinomas. The majority of canine tumours develop in the caudal mammary glands and often cause few overt clinical problems. However, in the more severe, poorly differentiated tumours there can be a marked inflammatory reaction leading to pain, ulceration and swelling. In some severe cases, bitches may present with marked limb oedema because of a severe inflammatory and neoplastic infiltrate into the dermis and lymphatic structures. These lesions carry a very poor prognosis and must be distinguished from severe mastitis. Other clinical signs associated with mammary tumours may be attributed to metastatic disease. In advanced cases, weight loss, tachypnoea/dyspnoea and lameness may indicate metastatic disease.

### Diagnosis

A diagnostic database should be constructed to determine the nature and extent of the disease. This database should include the following.

***History and clinical examination:*** A full history includes signalment, whether entire or spayed, onset and duration of clinical signs, evidence of weight loss, exercise intolerance, coughing, lameness and an investigation of the general health status of the animal.

Clinical examination includes a full general examination and, more specifically, examination of the mass in terms of size, mobility, ulceration and erythema, adherence to the overlying skin and evidence of pain. Often severely anaplastic carcinomas present as diffuse masses which have the appearance of a severe ventral dermatitis. Examination of the local lymph nodes should be performed, and include investigation of size, mobility and adherence to underlying or overlying structures.

***Haematology, biochemistry and urinalysis:*** These give an indication of the overall health status of the animal. Many cancer patients have a mild normocytic

| Tumour | Classification | Incidence (%) |
|---|---|---|
| Carcinoma | Papillary carcinoma<br>Tubular adenocarcinoma<br>Anaplastic carcinoma | 60 |
| Complex tumours | Mixed secretory and myoepithelial component | 30 |
| Sarcoma | | 10 |

***Table 5.1:*** *Incidence of malignant tumours in the bitch.*

non-regenerative anaemia associated with chronic disease. Infected mammary tumours may lead to increased neutrophil counts. Reduced platelet counts, bleeding tendencies and an increase in blood clotting times should alert the clinician to the possibility of disseminated intravascular coagulation. Serum biochemistry and urinalysis can indicate organ dysfunction which may or may not be related to the primary tumour.

***Diagnostic imaging (Figures 5.3 and 5.4):*** This is an invaluable tool in the overall assessment and staging of a tumour patient. Thoracic and abdominal radiography should be performed to indicate any evidence of metastatic disease. It is essential that inspiratory radiographs are taken of both the right and left lateral and ventrodorsal thorax to provide the optimum conditions to detect metastatic disease, as the most common site to find metastatic disease is the lungs. However, other sites recorded include liver, bone, pancreas, kidney, ovary and urethra. Abdominal ultrasonography can be very useful where transcoelomic spread is suspected from survey abdominal radiographs, or for evaluating abdominal viscera for metastatic deposits. Some animals present with a mild to moderate lameness because of metastatic spread to the long bones; these areas of bone are painful to manipulate and pathological fracture can occur. Radiography of affected limbs is essential in these cases.

***Cytology/biopsy:*** Where cytology of needle biopsy samples is performed, it is important to know that while false-positive cytology results are rare, false-negative results are more common and excisional or incisional biopsy should always be performed to confirm a diagnosis. In many cases cytology is not performed because surgical management is invariably indicated. In certain circumstances incisional biopsy may be utilised to rule out diffuse inflammatory disease.

**Staging**

The successful management of any neoplasm is based upon accurately defining the nature and extent of the disease. Anatomical staging of the disease allows for an accurate prognosis and correct selection of a treatment protocol. The TNM (Tumour, Node, Metastasis) classification system is used to stage canine mammary tumours (Figure 5.5).

**Prognosis**

Table 5.3 details prognostic factors that help predict survival and disease-free interval.

**Treatment**

The treatment of choice for mammary neoplasia is surgical excision. It is important to recognize that survival or disease-free interval is not influenced by the extent of the surgical procedure. However, several studies indicate that partial or regional mastectomy, with resection of local nodes for staging, is the procedure of choice. Because of its position, the inguinal lymph node is always removed when gland five is being resected. There is little evidence to suggest that ovariohysterectomy at the time of mastectomy alters survival times. Table 5.4 presents a summary of the surgical procedures applied to mammary neoplasia but the reader is referred to other texts for a detailed account of the surgical techniques involved.

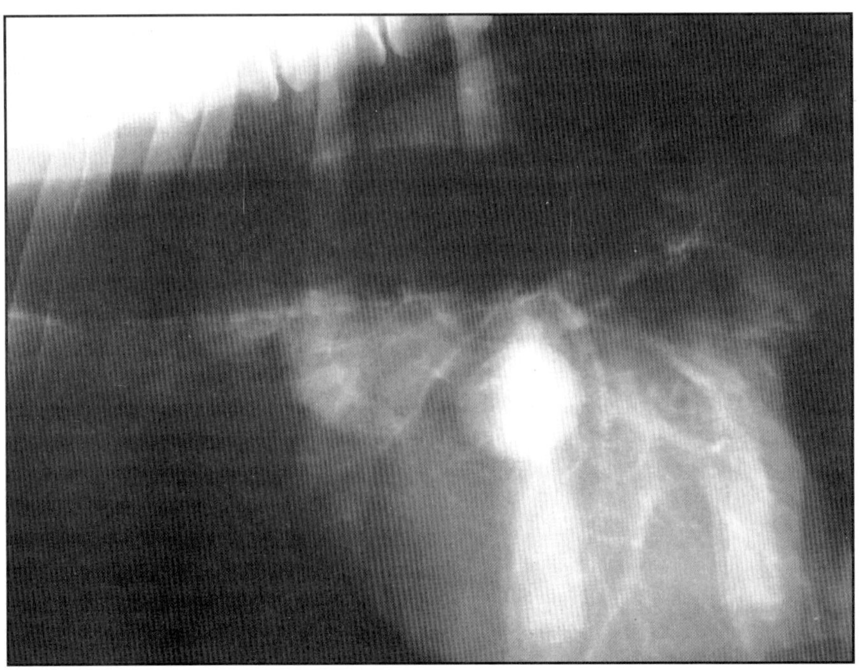

***Figure 5.3:*** *Lateral thoracic radiograph of a canine patient with evidence of pulmonary metastatic disease.*

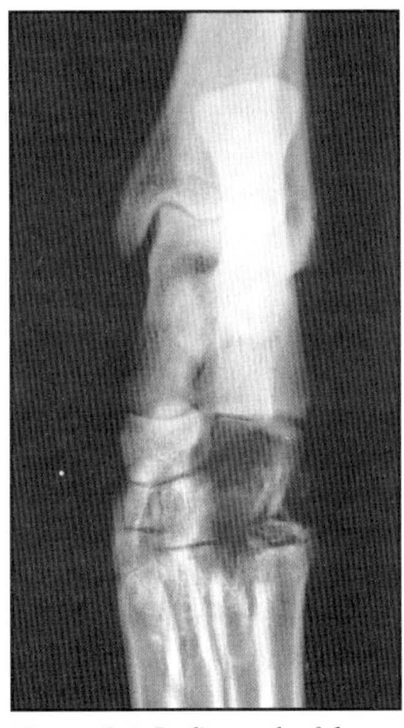

***Figure 5.4:*** *Radiograph of the right tarsus of a dog with evidence of metastatic disease affecting the hock joint and leading to pathological fracture.*

| Factor | Effect on prognosis |
|---|---|
| Tumour size | Tumours <3 cm carry a better prognosis than larger tumours |
| Degree of invasion and ulceration | Invading, ulcerating tumours carry a poorer prognosis |
| Lymph node involvement | Metastasis to local lymph nodes increases the risk of tumour recurrence. There is no evidence that removal of affected lymph nodes increases survival or disease-free interval |
| Histopathological grading | Poorly differentiated tumours carry a poorer prognosis. Presence of an immunological infiltrate correlates with increased survival times |
| Hormone receptor status | Progesterone/oestrogen receptor-positive tumours are often benign |

***Table 5.3:*** *Prognostic factors for survival in mammary tumours.*

| Surgical procedure | Comments |
|---|---|
| Nodulectomy | Used for masses <0.5 cm, but generally considered to be a biopsy procedure only |
| Mammectomy | Removal of a single gland. This is an ideal procedure for confirmed benign masses |
| Regional mastectomy | Removal of sets of glands according to lymphatic and venous drainage |
| Unilateral mastectomy | For multiple lesions affecting one mammary chain |
| Bilateral mastectomy | Applied when there are multiple tumours affecting both mammary chains. This is carried out as two unilateral mastectomies, 2–3 weeks apart |

***Table 5.4:*** *Surgical procedures for mammary neoplasia.*

| T categories for canine mammary tumours (Owen, 1980) | | | | |
|---|---|---|---|---|
| **Characteristic** | **T1** | **T2** | **T3** | **T4** |
| Size | <3 cm | 3–5 cm | >5 cm | >5 cm |
| Skin | Minimum involvement | Minimum involvement | Minimum involvement | Major involvement |
| Fascia/muscle | As above | As above | As above | As above |

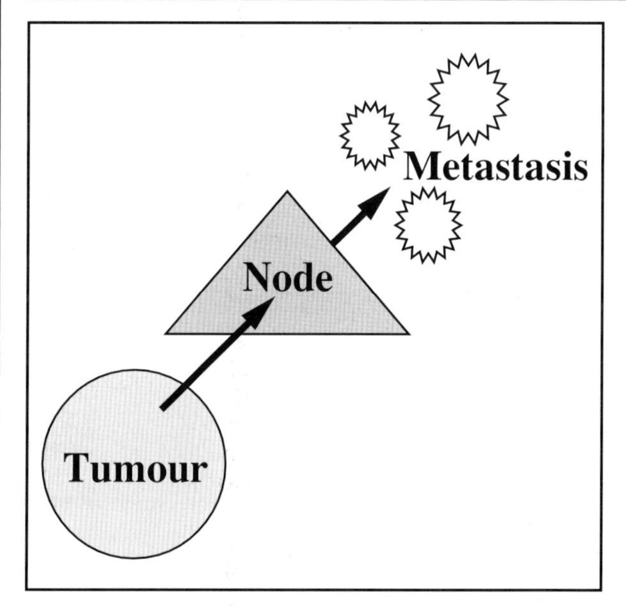


| N categories for canine mammary tumours | |
|---|---|
| N0 | No evidence of regional lymph node (RLN) involvement |
| N1a | Ipsilateral involvement (not fixed) |
| N1b | Ipsilateral involvement (fixed) |
| N2a | Contralateral involvement (not fixed) |
| N2b | Contralateral involvement (fixed) |

A + or — designation indicates whether the node is histologically positive or negative, respectively.

| M categories for canine mammary neoplasia | |
|---|---|
| **Stage** | **Degree of spread** |
| M0 | No evidence of metastatic disease |
| M1 | Metastatic disease present |

***Figure 5.5:*** *The TNM classification system allows an objective assessment of the extent of disease.*

### Adjunctive treatment

***Chemotherapy/hormonal manipulation:*** Chemotherapy and/or hormonal manipulation as an adjunct to surgery is now common in the treatment of mammary tumours in women. While attempts have been made to control mammary neoplasia in the bitch with chemotherapy, there is no ideal drug or protocol which has proved particularly effective. It has been suggested that combinations of adriamycin and cyclophosphamide provide the most effective method of adjunctive treatment but, as yet, there are no data from large clinical trials using these drugs to substantiate these claims.

The use of the anti-oestrogen drug tamoxifen has proved particularly effective as an adjunct in the management of mammary tumours in post-menopausal women. Tamoxifen is a synthetic triphenylethylene, which acts by binding to oestrogen receptors. However, the pharmacology is complicated by conflicting agonist and antagonist actions between species and target organs within species. Several clinical trials have been carried out in bitches with varying results. In one trial the majority of bitches given tamoxifen developed oestrogenic side effects, notably vulval bleeding and swelling and attractiveness to males, which were attributable to the tamoxifen. Consequently, these patients were removed from the trial. The oestrogenic effects seen with tamoxifen suggest that the drug in this situation may be acting as an agonist rather than an antagonist, or that the drug may be acting as an antagonist on mammary tissue but an agonist on endometrial tissue. In addition, the fact that tamoxifen has a beneficial effect in women with receptor-negative tumours suggests an alternative mode of action of the drug is possible. It is evident that further work is required in this area to assess fully the possible beneficial effects of this class of drug.

***Radiotherapy:*** There is little evidence in veterinary medicine to prove or disprove the effectiveness of radiotherapy in the management of mammary neoplasia. However, it may be useful in the pre-operative situation to reduce the size of the primary mass prior to surgery.

***Immunotherapy/gene therapy:*** There is a large body of data to suggest the existence of an immune response to certain tumour types. Attempts have been made to manipulate or enhance the immune response, particularly the cell-mediated response, against mammary tumours. Original trials using non-specific biological response modifiers met with little success, but methods using substances such as liposome encapsulated muramyltripeptide phosphatidyletholamine (L-MTP-PE) appear to be more promising. A particularly exciting field is that of gene therapy. With an increasing understanding of the molecular mechanisms in carcinogenesis and significant advances in DNA manipulation and gene delivery, it is feasible that gene therapy will eventually become part of clinical veterinary practice.

In summary, the primary treatment modality for mammary neoplasia remains as surgical excision. Combination chemotherapy, as an adjunct to surgery, may have beneficial effects if it is felt that surgical excision has not been complete or the histopathology suggests that the tumour is very aggressive. The use of oestrogen-blocking drugs in the dog is still experimental and the beneficial effects are controversial. More work is required to study the role of reproductive hormones in the development of canine mammary tumours to allow them to be exploited in treatment.

## Tumours of the mammary gland (cat)

### Incidence and aetiology

While the incidence of mammary tumours in cats is half that in bitches, the majority of tumours presented are malignant adenocarcinomas (80–90%). As in dogs, the aetiology of feline mammary tumours is uncertain. It has been shown that intact cats are more likely to develop mammary neoplasia but the exact role that the reproductive hormones play in the development of the disease is uncertain. It has been suggested that there is an increased risk of the development of mammary masses with prior use of synthetic progestogens.

### Clinical presentation and diagnosis

Cats are usually presented when the disease is very advanced and metastasis may have already occurred. In many cases multiple glands are affected and the masses are often ulcerated. The most common differential diagnosis for mammary neoplasia is mammary hypertrophy (fibroepithelial hyperplasia) but this usually occurs in young intact females, whilst the mean age for occurrence of mammary tumours is 9–12 years. A full diagnostic work-up and staging procedure should include haematology/biochemical profile, urinalysis, and thoracic radiography. Pre-surgical biopsy should only be performed if it will alter the surgical procedure or the owner's willingness to treat the disease.

### Treatment

Because of the high incidence of malignancy in these lesions, aggressive therapy should be adopted and surgery is the treatment of choice. Surgical management of this condition is often hampered by the degree of invasion and ulceration, so the procedure of choice is radical mastectomy with removal of the draining lymph nodes. If a bilateral mastectomy is indicated then two procedures are performed 2–3 weeks apart. The use of chemotherapy as an adjunct to surgery has not been extensively studied in the cat. Some workers have advocated the use of combinations of adriamycin and cyclophosphamide for feline mammary cancer but further studies with larger clinical trials are required to establish the exact role of adjunctive chemotherapy in this disease.

## Benign mammary gland hyperplasia (fibroepithelial hyperplasia)

This is not an uncommon condition and is seen in young oestrous cycling cats, pregnant cats, or neutered cats which are on prolonged treatment with the progestogen megoestrol acetate. The lesions are composed of non-encapsulated growths of ductal structures within actively proliferating connective tissue. The similarities between the lesions seen in young cycling females and those in neutered cats on long-term progestogen therapy suggest that the lesion may result from excessive endogenous or exogenous progesterone stimulation. The condition often affects multiple glands and these can be so large that they affect walking. The affected glands may be oedematous and painful but ulceration is not a feature. In young cycling females the treatment of choice is ovariohysterectomy, bearing in mind that the lesions can take up to 5–6 months to regress completely after surgery. This condition can, however, regress spontaneously without spaying. The lesions seen in this condition are often misdiagnosed as malignancies and are treated as such. In cases where the lesion has been caused by prolonged administration of megoestrol acetate, treatment is based upon cessation of medication. In refractory cases there are reports of success using the anti-prolactin drug, cabergoline (M. J. Harvey, personal communication).

# FURTHER READING

Ettinger S (1995) *Textbook of Veterinary Internal Medicine, 4th edn.* WB Saunders, Philadelphia

Ogilvie G and Moore AS (1995) *Managing the Veterinary Cancer Patient.* Veterinary Learning Systems, Trenton, New Jersey

Owen LN (1980) *TNM Classification of Tumours in Domestic Animals.* WHO, Geneva

Slater DH (1985) *Textbook of Small Animal Surgery.* WB Saunders, Philadelphia.

Theilen GH and Madewell BR (1987) *Veterinary Cancer Medicine.* Lea & Febiger, Philadelphia

White RAS (1991) *Manual of Small Animal Oncology.* BSAVA, Cheltenham

CHAPTER SIX

# Physiology and Endocrinology of the Male

*Denise Hewitt*

## INTRODUCTION

A familiarity with the endocrinology and physiology of male reproduction is essential for the clinician performing an andrological examination of the male animal. Basic knowledge of normal male reproductive function and the concentrations of relevant hormones in the peripheral circulation is required, as is an understanding of how these parameters can be affected by the age and size of the animal, environmental factors and any medication given. Consequently, abnormal reproductive function can be recognized as deviation from these characteristics.

Male physiology and endocrinology are well described in text books, but it is essential to be familiar with interspecies variations that occur. Specific characteristics of the structure and function of the male reproductive system have been well documented in the dog, but less information has been published for the tom.

The male reproductive system performs three physiological functions. These are: the production and maturation of spermatozoa in the testes; the maturation, storage and transport of spermatozoa in the duct system; and the deposition of the spermatozoa into the female reproductive system via the penis. The endocrinological role of the system is also three-fold, controlling the production of spermatozoa and the development of the masculine body form, and enhancing masculine behaviour (libido and aggression).

## ANATOMY OF THE MALE REPRODUCTIVE SYSTEM

The testes are the primary organs of male reproduction, producing the male gametes (spermatozoa) and steroid hormones (androgens, testosterone and oestradiol). The testes differ from the female primary reproductive organs (the ovaries) in that all the potential gametes are not present at birth. Instead, germ cells undergo continual cell divisions, forming new spermatozoa throughout the reproductive life of the male.

Development of the testes is similar in all species. In the fetus, the primordial germ cells move to a location near to the kidneys. At this stage, it is impossible to differentiate between an ovary and a testis. Once it can be identified as a testis, the development of the epididymis and the ductus deferens occurs. A gubernaculum develops at the caudal end of the testis. This enlarges and, in doing so, relocates the testis to the inguinal canal. Once this repositioning has occurred, the gubernaculum degenerates and the testis moves into the scrotal cavity where the gubernaculum was located. In most of the domestic species, testicular descent is complete by birth. Testes descend within an outpouching of the peritoneum, the vaginal tunic which forms within the scrotum. In the dog, the scrotum is pendulous and the testes are positioned almost horizontally, whereas in the tom, they are held closer to the body. Dog testes vary in size in relation to the size of the animal, but on average are 3 x 2 x 1.5 cm. A correlation between body mass and testicular mass has been shown in the dog. Cat testes measure 13 x 8 x 6 mm. In both the cat and the dog, the scrotal skin is thin, with sweat glands, no subcutaneous fat and an efficient system to allow cooling of the arterial blood. Thermoregulatory mechanisms prevent increases in testicular temperature which would be detrimental to the seminiferous epithelium. An additional temperature control system in the dog is the cremaster muscle, which can alter the distance of the testes from the body. In the dog, the descent of the testes often occurs in the fetus so that the puppy is born with descended testicles, although there are reports of descent as late as 6–8 months after birth. Spermatogonia have been found in testicular biopsies at 8 weeks of age. Kittens' testicles are usually descended by birth, with maturation of the Leydig cells occurring at approximately 5 months, and spermatozoa in the seminiferous tubules by 6–7 months of age.

The testis itself is composed of two tissue types, the seminiferous tubules and the interstitial tissue. The proportion of the two tissue types varies considerably between different species. The seminiferous tubules open into collecting ducts called the vasa efferentia which in turn open into the epididymis. This is a highly convoluted structure. The epididymis is divided into the head, body and tail, which arise medially and are located on the dorsolateral surface of the testis in the dog. Movement of the spermatozoa along the epididymis is the result of peristalsis. The tail of the epididymis acts as a store for spermatozoa awaiting ejaculation and leads into the vasa deferentia which also acts as a reservoir for spermatozoa. The epididymal tail is pea-sized and can be palpated caudally within the scrotum. The vas deferens runs within the vaginal sac and conveys the spermatozoa from the testes to the penis.

Accessory glands add secretions to the spermatozoa and in most species make up a large volume of the ejaculate. They may bestow important properties, such as the correct pH, on the seminal plasma. There is considerable variation in the accessory glands in different species. The four types which can be present in domestic species are ampulla, prostate, vesicular gland and bulbourethral gland. The ampullae act as reservoirs of spermatozoa in the vasa deferentia prior to opening into the urethra. They are absent in the tom and present in the dog, but not anatomically significant.

The prostate is present in the dog and the tom and is the only anatomically significant accessory gland in the dog, where it is large and contributes a large volume of watery secretion to the ejaculate. This secretion contains lactate, cholesterol and enzymes, but only very low concentrations of reducing sugars compared to those found in other species. The source of metabolizable energy which provides an environment suitable for sperm motility is therefore not known in this species. In the normal dog, the prostate is positioned near to the cranial rim of the pelvis and surrounds the terminal portion of the ductus deferens, the proximal part of the urethra, and the neck of the bladder. The prostatic fluid is constantly secreted into the prostatic excretory ducts which open into the prostatic urethra. The prostate is symmetrical and divided into two lobes by the median septum. It is usually about 2 cm in diameter, although this varies with the size and weight of the animal. The cat prostate is 2 mm in diameter and is again divided into two symmetrical halves, the right and left lobes. In contrast to the dog, the cat prostate does not cover the ventral aspect of the urethra.

The major accessory gland in the tom is not the prostate but the paired bulbourethral glands. These glands are positioned craniolateral to the base of the penis and are approximately 3 mm in diameter. Bulbourethral glands are absent in the dog. Vesicular glands are absent from both the dog and the tom.

A bone (os penis) is present within the penis of the dog and tom. This allows the male to achieve intromission before the attainment of a full erection. The os penis of the dog has a deep groove within it, housing the urethra (Figure 6.1).

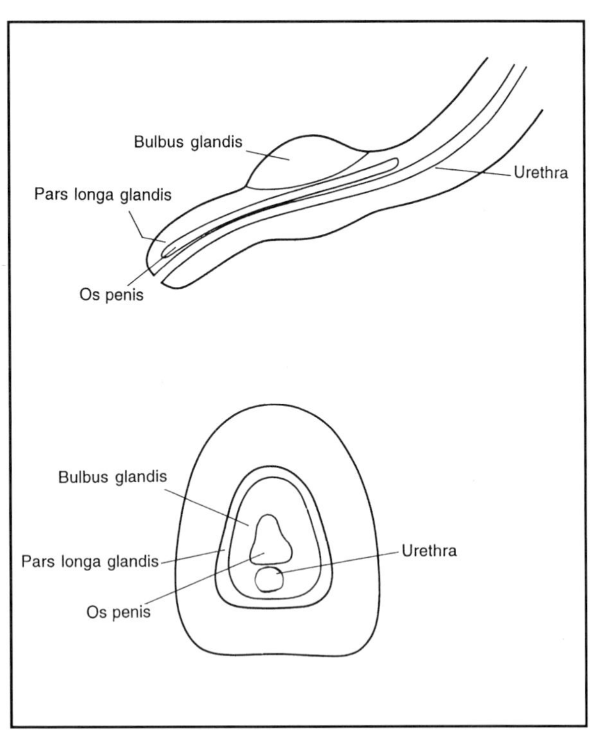


*Figure 6.1: Parasagittal and cross section of dog penis, showing the pars longa glandis, bulbus glandis, os penis and urethra. (Redrawn after Christiansen, 1954).*

The glans penis is composed of two parts, differentiated by the location of erectile tissue. The bulbus glandis is composed of erectile tissue surrounding the os penis and urethra, whereas the pars longa glandis has erectile tissue dorsally and longitudinally to the os penis and urethra only (Figure 6.1). The penis of the cat is unusual in that it is directed caudally, and small spines or papillae are present along the glans penis. These become fully developed at puberty. The spines are thought to stimulate a pituitary-mediated luteinizing hormone (LH) release in the female upon intromission.

## PHYSIOLOGY OF THE MALE REPRODUCTIVE SYSTEM

The processes of spermatogenesis (the production of spermatozoa) and steroidogenesis (the secretion of hormones) are closely related and yet performed by spatially distinct areas of the testis. This has been termed 'functional compartmentalism'.

Spermatogenesis occurs within the seminiferous tubule compartment. This compartment, which is composed of a basal and adluminal section, contains two cell types, the Sertoli cells (the somatic cells) and the germ cells. The basal section contains spermatogonia dividing by mitosis, whilst the adluminal section contains primary spermatocytes undergoing meiosis, resulting in secondary spermatocytes and spermatids. The basal and adluminal compartments are separated by a boundary of Sertoli cells which form junctional complexes, to create a blood–testis barrier. This barrier excludes blood macromolecules and interstitial cell fluid from the adluminal section of the seminiferous tubules, thus creating the environment suitable for meiosis to occur.

Steroidogenesis occurs within the interstitial tissue compartment of the testis, which is composed of Leydig cells closely associated with blood and lymphatic vessels. Leydig cells are the only testicular cells with receptors for LH. LH binds to the receptors on the Leydig cells and, in response, synthesis of several steroids, including testosterone, occurs. Testosterone is essential for the development of the secondary sex characteristics, normal behaviour, function of the accessory glands, production of spermatozoa and maintenance of the male duct system. The interstitial cell compartment surrounds the seminiferous tubule compartment which is consequently bathed in a fluid rich in testosterone.

In contrast to the dog, in which semen collection and analysis is well established, this procedure is rare in the tom except within research establishments. As a result, there is a comparative lack of information on spermatogenesis in the tom. There is no information on the changing concentrations of gonadotrophin releasing hormones, gonadotrophins or reproductive steroids in the tom. It is therefore assumed that the pattern of hypothalamic, pituitary and testicular function occurs in a similar way to that described for the dog and other domestic animals.

## Hormonal control of the male reproductive system

Male reproductive physiology is controlled endocrinologically by two gonadotrophins, LH and follicle stimulating hormone (FSH) secreted by the anterior pituitary gland. Gonadotrophin secretion is under the positive control of gonadotrophin releasing hormone (GnRH), which is released in an episodic fashion from the hypothalamus. GnRH binds to specific receptors on the plasma membrane of pituitary gonadotrophs and stimulates the release of LH and FSH. GnRH is under the negative control of testosterone and its active metabolites oestradiol and dihydrotestosterone. The negative feedback is thought to occur at both the hypothalamic and the pituitary level. This intricate system therefore allows communication between the hypothalamus of the central nervous system, the anterior pituitary and the testis (Figure 6.2). There may also be some negative control by oestradiol produced via local aromatization of testosterone. The negative feedback system acting upon LH and FSH release is common to both gonadotrophins. This is an important consideration when agents are used clinically in an attempt to modify reproduction (Figure 6.3). Despite this, LH and FSH concentrations do not always rise proportionately. As a consequence, the existence of an additional inhibiting factor called inhibin, solely responsible for the control of FSH secretion from the pituitary has been proposed. There may also be other products of the Sertoli cell which have the opposite effect and stimulate FSH secretion (activins).

The negative feedback effect of the testis upon the pituitary gland has been demonstrated by the fact that castration results in a rise in both LH and FSH concentration. The stimulatory effect of GnRH upon the Leydig cell can also be demonstrated by administra-

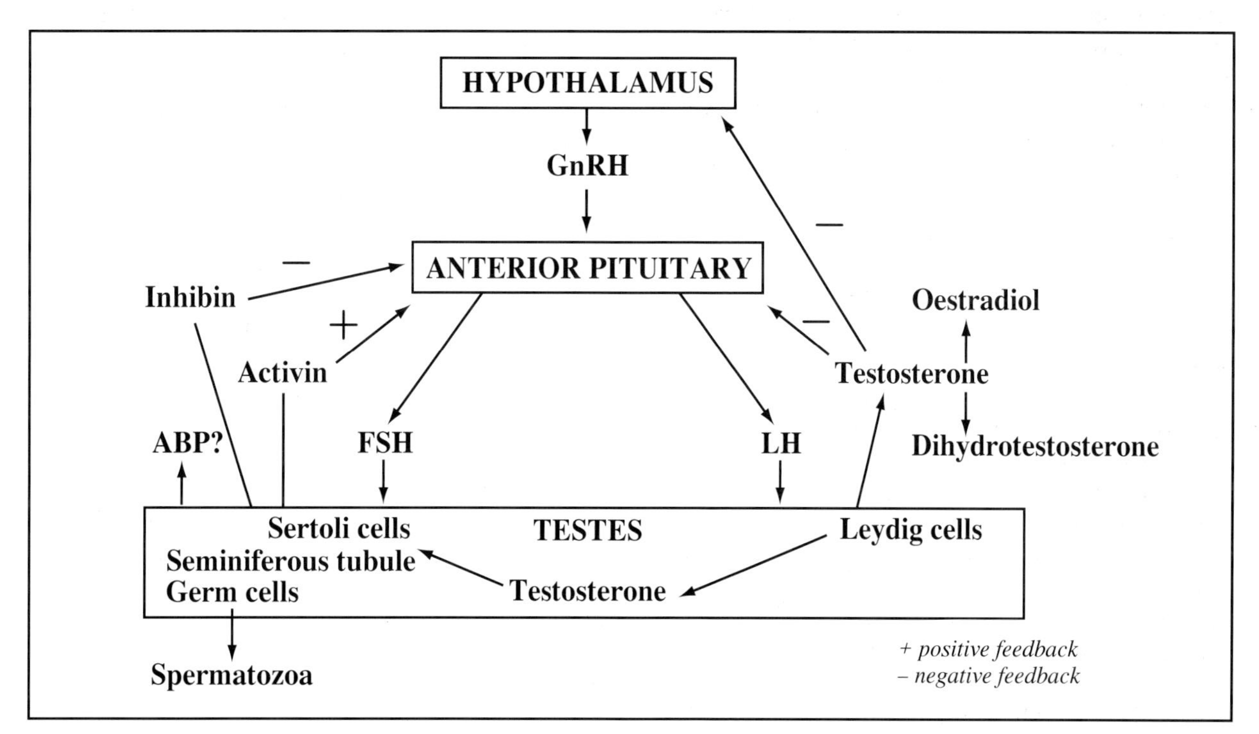


***Figure 6.2:*** *The hypothalamic–pituitary–testicular axis, showing the influences of steroid and gonadotrophic hormones on testicular function.*

*ABP = Androgen binding protein*

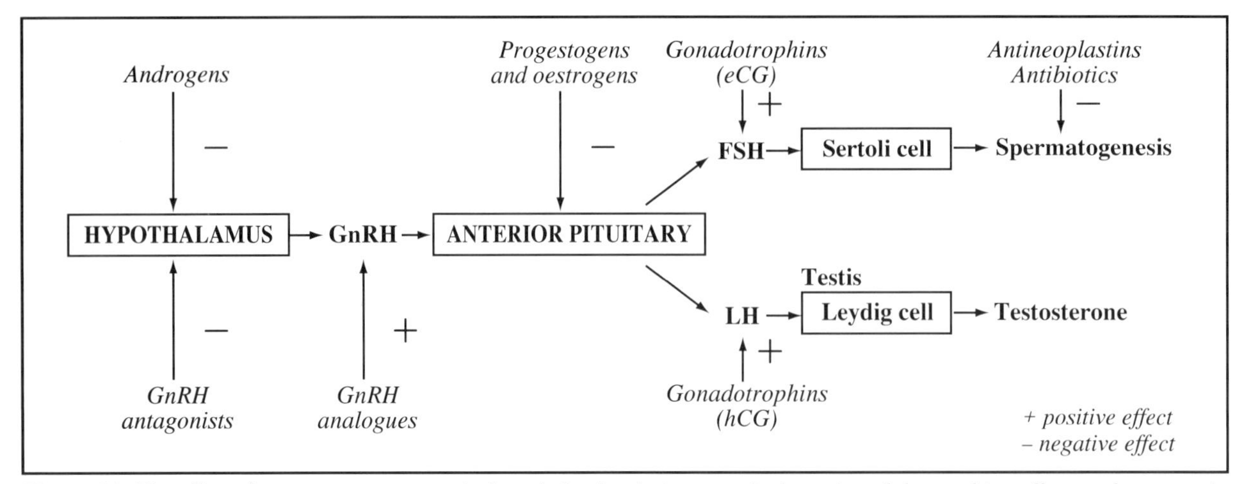


***Figure 6.3:*** *The effect of exogenous agents on the hypothalamic–pituitary–testicular axis and the resulting effect on the two main functions of the testis.*

*hCG = human chorionic gonadotrophin; eCG = equine chorionic gonadotrophin*

tion of exogenous GnRH, which causes a rise in the plasma concentration of testosterone. This can be used as a test for functionality of the pituitary and Leydig cells.

Following stimulation with GnRH, LH release is immediate and transient, with concentrations rising steeply, but quickly falling to baseline and rapidly clearing from the peripheral circulation. Between 5 and 20 LH pulses occurring randomly over a 24-hour period have been measured in the dog. Testosterone production by the Leydig cells occurs approximately 50 min after the LH surge (Figure 6.4). Testosterone secretion occurs both locally and into the general circulation. Peripheral circulating testosterone is important for the maintenance of secondary sexual characteristics, sexual behaviour and negative feedback control of gonadotrophin secretion, whilst its local function within the testis is important for spermatogenesis. High concentrations of testosterone within the testis are maintained partly by testosterone binding to androgen binding protein (ABP) which is produced by the Sertoli cells.

FSH acts on the seminiferous tubules together with endogenous testosterone to stimulate Sertoli cell support of the germ cells. This in turn aids spermatogenesis, in particular the development of the spermatid. FSH receptors are present on the Sertoli cells and possibly also on the spermatogonia within the seminiferous tubules. Binding of FSH results in the stimulation of adenylate cyclase activity. This increases the production of proteins which are possibly important for the regulation of spermatogenesis via ABP and transferrin, feedback control of FSH via inhibin, and possibly Leydig cell function. It is not known whether FSH is required for the maintenance of spermatogenesis once it has been initiated. This mechanism requires investigation and it is possible that it may be different between species. It is known,

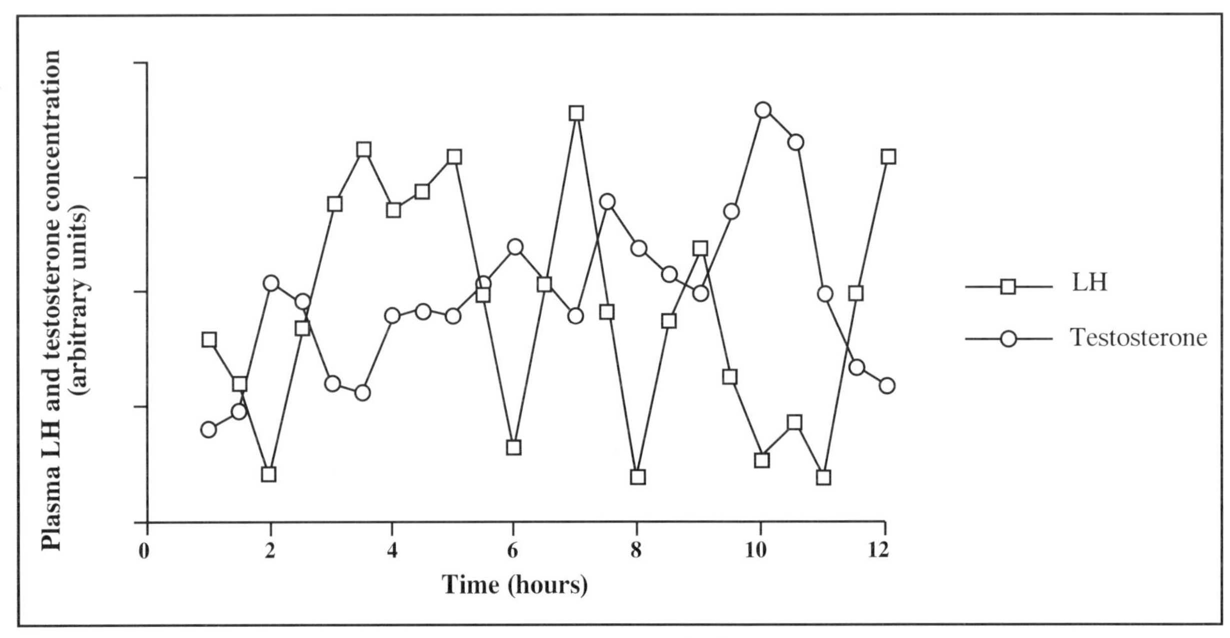


***Figure 6.4:*** *The relationship between LH and testosterone secretion in the dog.*

however, that within a species, the spermatogenic cycle is constant, and unable to be accelerated, so it seems unlikely that FSH further stimulates spermatogenesis once initiated. It may be that FSH takes its effect by decreasing the number of germ cells that normally degenerate. Like LH, FSH secretion is pulsatile, but its secretion occurs less rapidly after GnRH stimulation, and more gradually. This is reflected in the smaller amplitude of FSH variation seen in the peripheral circulation.

Prolactin is thought to act synergistically with LH in regulating testosterone production by the Leydig cells; prolactin receptors are present on the Leydig cells. Once spermatogenesis reaches the stage of spermatid production, inhibin production occurs to create the negative feedback for FSH secretion.

Sertoli cells are the only testicular cells that bind FSH. FSH stimulates spermatogenesis indirectly by acting on the Sertoli cells, which in turn provide the environment necessary for germ cell development. The normal functioning of the Sertoli cells is dependent on the concentration of testosterone being considerably higher in the testis than in the peripheral circulation. Sertoli cell function differs depending on the developmental stage of the germ cells with which they are associated. Sertoli cells also secrete substances responsible for the control of steroidogenesis and spermatogenesis. ABP is thought to act by either increasing the concentration of testosterone within the seminiferous tubules or aiding the transport of testosterone to the epididymis. Inhibin may act on the anterior pituitary to suppress FSH production and subsequently spermatogenesis.

The action of these hormones, where they are released from and the site at which they act are summarized in Table 6.1.

**Hormone concentrations in the dog and cat**

The pulsatile nature of the release of LH makes its measurement difficult, and a single sample is therefore virtually meaningless. Only serial sampling provides a profile which is a true reflection of the secretion of this gonadotrophin. Basal serum concentrations of LH in the dog are approximately 1.0–1.2 ng/ml, with surges reaching 3.8–10 ng/ml. Basal testosterone is usually between 1.7 and 5.2 mmol/l (0.5–1.5 ng/ml), rising to a peak of 12.1–20.8 mmol/l (3.5–6.0 ng/ml). Seasonal shifts have been reported in the concentrations of both LH and testosterone, yet although there is a relationship between the secretion of the two hormones, seasonal variations are independent. The secretion of LH in the tom has been found to range between 3 and 29 ng/ml. The pulsatile release of testosterone has been shown to vary from 0.3 to 11.4 mmol/l (0.1 to 3.3 ng/ml) within a 6-hour period.

Administration of exogenous GnRH allows the clinician to test the pituitary release of LH and subsequent testicular release of testosterone. In normal males, GnRH administration causes an increase in LH secretion in 30 minutes and an increase in testosterone in 60 minutes.

## SPERMATOGENESIS

Spermatogenesis is the sum of the transformations that result in the formation of spermatozoa from spermatogonia whilst spermatogonial numbers are maintained.

In the male fetus, primordial germ cells differentiate into gonocytes which undergo mitosis during fetal and prepubertal life and differentiate into spermatogonia. Germ cell development is then arrested in the seminiferous tubules until the onset of puberty.

| Hormone | Released from | Action |
|---|---|---|
| GnRH | Hypothalamus | On anterior pituitary for release of LH and FSH |
| LH | Anterior pituitary | On Leydig cells to stimulate steroidogenesis |
| FSH | Anterior pituitary | On Sertoli cells to stimulate spermatogenesis |
| Testosterone | Leydig cells | On Sertoli cells to stimulate spermatogenesis<br>Negative feedback on hypothalamus and anterior pituitary to control release of GnRH and gonadotrophins, respectively |
| Inhibin? | Sertoli cells | Negative feedback on anterior pituitary to control release of FSH |
| Activin? | Sertoli cells | Positive feedback on anterior pituitary to control release of FSH |
| Prolactin | Leydig cells | Regulates testosterone production by the Leydig cells |
| Androgen binding protein | Sertoli cells | Increases testosterone in seminiferous tubules or epididymis? |

***Table 6.1:** Summary of reproductive hormones and their actions.*

## Spermatocytogenesis

Initially, relatively undifferentiated stem cell spermatogonia located along the basement membrane of the seminiferous tubules multiply by mitosis. This occurs within the basal compartment of the seminiferous tubule. Spermatocytogenesis allows both the cyclical production of primary spermatocytes and the maintenance of stem cell number. In addition to this proliferating pool, there is a reserve of spermatogonia which do not proliferate and are extremely resistant to damage by radiation and toxins. These cells survive even after severe trauma to the testes. The remainder of the process occurs in the adluminal compartment where primary spermatocytes undergo meiosis to produce secondary spermatocytes. These undergo a further meiotic division to produce spermatids.

## Spermiogenesis

The final morphological transformation is termed spermiogenesis and involves the differentiation of spherical spermatids into mature spermatids which will be released into the lumen of the seminiferous tubules as spermatozoa.

As spermatogenesis proceeds through these different stages, the developing gametes migrate from the basement membrane of the seminiferous tubules towards the lumen.

## Spermiation

The process which involves the release of the germ cells into the lumen of the tubule after spermatocytogenesis and spermiogenesis is known as spermiation. The released germ cells are now considered to be spermatozoa.

Sertoli cells are associated with germ cells throughout their development and their functional morphology has been shown, in some domestic species, to differ according to the developmental stage of the germ cell with which they are associated. This study has, however, yet to be performed in the dog or tom. The role of the Sertoli cell in spermatogenesis includes providing support and nutrition, phagocytosing waste products and secreting essential luminal fluids.

Leydig cells also have a role in supporting spermatogenesis. The interstitial cell compartment surrounds the seminiferous tubule compartment which is consequently bathed in a fluid rich in testosterone. In addition, myoid cells which form the boundary tissue of the seminiferous epithelium help in the propulsive movements of spermatozoa and fluid in the seminiferous tubules. Like Leydig and Sertoli cells, they are affected by testicular growth factors.

By viewing histological transverse sections of seminiferous tubules, the different organizational patterns that occur may be seen (Figure 6.5). These patterns depend on the germ cell type, morphology and stage of development and their arrangement in layers within the tubule. Each layer contains the cells from a different generation of germ cells which become increasingly differentiated as they near the lumen of the tubule.

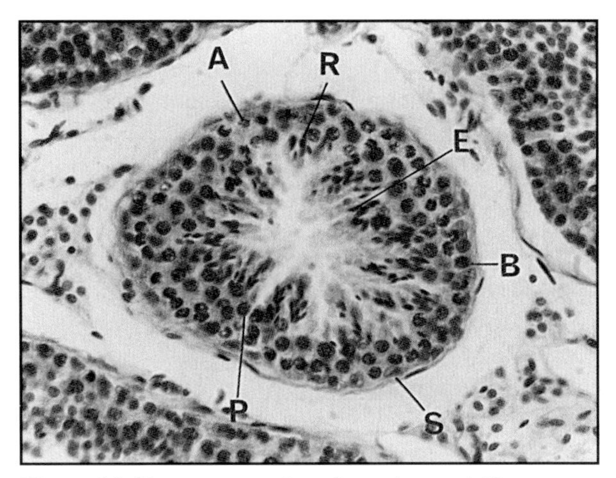


***Figure 6.5:*** *Transverse section of a canine seminiferous tubule: A spermatogonia (A), B spermatogonia (B), rounded spermatids (R), elongated spermatids (E), primary spermatocytes (P), Sertoli cells (S). H&E.*

At any one point in the seminiferous tubule, germ cell associations of each type appear in sequence. The whole series of these associations is called the spermatogenic cycle. The interval for one cycle is termed the duration of the spermatogenic cycle. It is the length of time between two consecutive releases of spermatozoa. This is 13.8 days in dogs. An average of 62 days is required for spermatogenesis in the dog. This information is currently unavailable in the literature for the tom, but it is assumed to be similar.

The spatial sequential order of the germ cell development along the length of the seminiferous tubule at any given time is known as the spermatogenic wave. This spatial arrangement may serve to allow a constant release of spermatozoa, reduce the competition for the use of hormones and metabolites for a given stage of development, reduce the congestion that would otherwise occur if spermiation was simultaneous along the length of the tubule, and generally facilitate sperm maturation and transport within the tubule.

## Epididymal transport, maturation and storage

From the seminiferous tubules, spermatozoa are passed through the rete testis and the vasa efferentia into the epididymis. Here they undergo the final stages of maturation: the ability to become motile; membrane changes; and the loss of the cytoplasmic droplet. Mature spermatozoa are stored in the tail of the epididymis. At ejaculation, they pass along the vas deferens and various secretions are added by the accessory glands. The important accessory glands in the dog and tom are the prostate and the paired bulbourethral glands, respectively. Further changes of the spermatozoa occur within the female tract, with capacitation and acrosome reaction enabling fertilization of the oocyte.

The events of spermatogenesis are summarized in Figure 6.6.

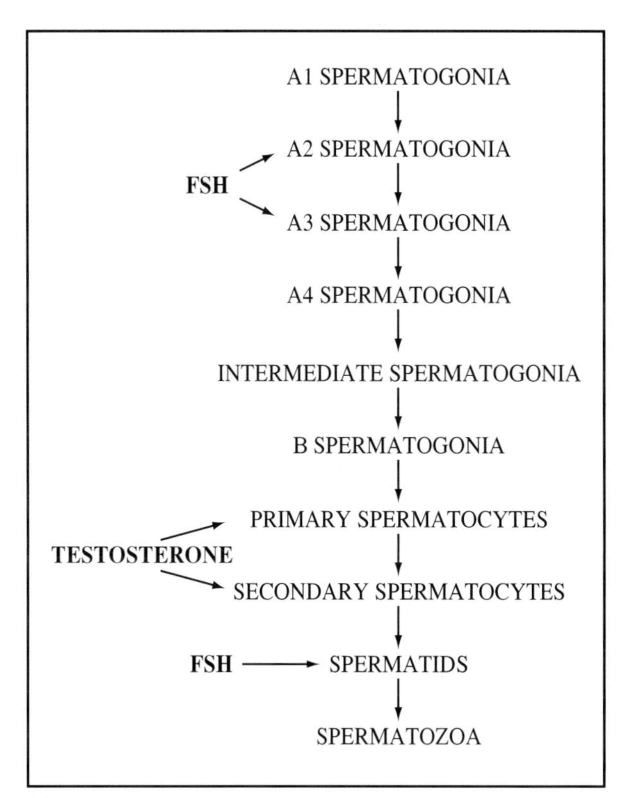


***Figure 6.6:*** *Summary of spermatogenesis.*

## SPERMATOZOA

Structurally, the spermatozoon is divided into: a head, containing the nucleus and the acrosome with the acrosomal enzymes; the mid-piece, containing the mitochondria for the metabolism of the spermatozoon; and the tail, whose flagellar movement allows sperm movement to occur (Figure 6.7).

Dog and cat spermatozoa both have flattened spade-shaped heads. This is similar to the human, bull and rabbit spermatozoa. In the dog, the spermatozoon has the following dimensions (±SD): total length 68 ± 0.3 μm; head length 7 ± 0 μm; head width 5 ± 0.1 μm; mid-piece length 11 ± 0.2 μm; and tail length 50 ± 0.3 μm. In the tom, less detailed measurements have been made: the length of the spermatozoon is 55-65 μm, the length of the head is 6.5 μm, and the width of the head is 3 μm.

## SPERM PRODUCTION

Examination of the ejaculate provides essential information about sperm production. The number of spermatozoa in the ejaculate is related to the frequency of ejaculation and in a sexually rested male, sufficient numbers of ejaculates are required before epididymal spermatozoa are stabilized. Only then will the daily sperm output in the ejaculate reflect the daily sperm production. A good correlation exists between total scrotal width, paired testes weight, weight of testicular parenchyma, daily sperm production and sperm output in dogs. The daily sperm production in decapsulated testis in the dog is 16 x $10^6$ per gram of testis. This appears to be an inefficient rate compared with other species, but is probably related to the long duration of spermatogenesis in the dog. Large dogs in general ejaculate greater numbers of spermatozoa than small dogs. Values for sperm production in the tom cat are not available.

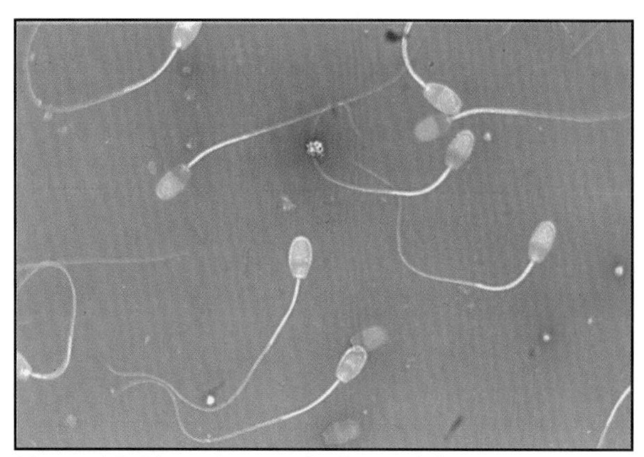

***Figure 6.7:*** *Photomicrograph of normal dog spermatozoa. Nigrosin & Eosin.*

The age of the male also affects the number of spermatozoa produced. The male dog reaches puberty at an average age of 9 months, several weeks later than the female. Similarly, the tom cat reaches puberty at 8-10 months, again later than the female. High or low temperatures can be detrimental to sperm production, as can radiation, some drugs, vitamin A deficiency, limited diet and toxic chemicals. The length of the period of abstinence influences the number of spermatozoa in the ejaculate. If it is long, the male duct system becomes congested with older spermatozoa, and these are eventually expelled from the system, either in the semen or in the urine.

## EFFECT OF EXOGENOUS AGENTS UPON STEROIDOGENESIS AND SPERMATOGENESIS

Spermatogenesis may be affected by the action of some therapeutic agents on the Sertoli cells. Drugs may also influence spermatozoal maturation and the storage of spermatozoa in the epididymis. Some agents may affect androgen production (on which meiosis and epididymal spermatozoal development are dependent), either by directly affecting the Leydig cells, or indirectly by altering the production of LH. These effects are summarized in Figure 6.3.

### Androgens

Androgens affect the development of the secondary sexual characteristics and play an important role in spermatogenesis and the maintenance of libido. Androgens, including testosterone, have a negative feedback effect on the release of GnRH by the hypothalamus which in turn suppresses the release of gonadotrophins by the pituitary gland. Exogenous androgens therefore produce a significant reduction in semen quality.

It has been proposed that low endogenous testosterone concentration in the dog is the cause of poor libido or impotence. Androgens (testosterone) should not be used to treat this condition, however, because of their effect upon semen quality. In many cases, impotence occurs as a result of musculoskeletal pain or psychological problems and therefore androgen supplementation is not the treatment of choice.

### Progestogens

Progestogens are used for their anti-androgenic effects. They work by inhibiting the release of gonadotrophins. Progestogens administered to dogs at the same dose as used routinely in bitches to control cyclical activity have no effect on sperm production or libido; at an increased dosage, they do have an effect, but one which is not sufficient to render the animal infertile. Progestogens alone are therefore not suitable for contraception in the male dog. In combination with testosterone, however, progestogens can be detrimental to semen quality for about 50 days after treatment, although libido and spermatogenesis are maintained.

Progestogens appear to reduce peripheral circulating androgen concentrations, which may be useful in certain clinical conditions. Both entire and castrated dogs, for example, may exhibit antisocial behaviour, including aggression, territory marking, copulatory activity, excitability and destructiveness. Progestogens have been used successfully in conjunction with behaviour modification training to control some of these problems. Raised concentrations of testosterone in the peripheral plasma are not thought to be related to antisocial and aggressive behaviour.

Reducing plasma testosterone concentrations (using progestagens or via castration) may be useful in other androgen-dependent conditions such as benign prostatic hyperplasia, circumanal adenomas and some epileptic seizures. Some androgens are used in ageing dogs for their anabolic effects.

### Oestrogens

High doses of oestrogen affect the physiology of male reproduction by suppressing the release of gonadotrophins.

Oestrogens have been used successfully in controlling some behavioural problems in dogs. The enlargement of the prostate seen in benign prostatic hyperplasia may also be reduced by oestrogen therapy. Benign tumours in the perineal region, circumanal adenomas, can also be treated with oestrogens.

### Gonadotrophins

Human chorionic gonadotrophin (hCG) has been shown to produce an effect similar to LH. It should therefore cause an increase in testosterone secretion by the Leydig cells of the testes. Also, equine chorionic gonadotrophin (eCG) acts in a similar way to FSH and should therefore act to stimulate spermatogenesis in the seminiferous tubules. In general, however, these two agents cannot be given in a physiological manner, and therefore tend to produce short-term physiological effects. They are not useful for treating testicular hypoplasia or cryptorchidism. An hCG stimulation test is however useful for demonstrating the presence or the absence of testicular tissue.

### GnRH analogues

GnRH analogues stimulate the release of gonadotrophins from the pituitary gland. These effects, however, are not physiological, and prolonged stimulation is followed by receptor down-regulation. GnRH analogues may therefore be used as contraceptive agents, causing cessation of ejaculation and a decrease in the proportion of normal motile spermatozoa after 3 weeks of treatment, with a decrease in the proportion of normal motile spermatozoa. Normal sperm production returns rapidly after the termination of treatment.

### GnRH antagonists

GnRH antagonists oppose the effects of GnRH and may be useful as contraceptive agents in the future.

### Other agents

Careful attention should be paid when therapeutic agents are administered to male animals, since these may adversely affect reproductive function. Agents with a known adverse effect include:

- Antineoplastic drugs, which cause aplasia of the germinal epithelium of the testis
- Nitrofurantoin and amphotericin antibiotics, which arrest spermatogenesis
- Griseofulvin, which in large doses causes oligospermia.

## CONCLUSION

Recent studies have improved the understanding of male reproductive function. A thorough understanding of the normal physiology and endocrinology will enable the clinician to diagnose reproductive pathology and institute appropriate treatment.

## REFERENCES AND FURTHER READING


Allen WE and England GCW (1990) Reproductive endocrinology of the dog. In: *Manual of Small Animal Endocrinology*, ed. M Hutchinson, pp. 121-126. BSAVA, Cheltenham

Amann RP (1986) Reproductive endocrinology and physiology of the dog. In: *Current Therapy in Theriogenology 2. Diagnosis, Treatment and Prevention of Reproductive Diseases in Small and Large Animals*, ed. DA Morrow, pp. 532-538. WB Saunders, Philadelphia

Bedford JM and Hoskins DD (1990) The mammalian spermatozoon: morphology, biochemistry and physiology. In: *Marshall's Physiology of Reproduction. Volume 2. Reproduction in the Male, 4th edn*, ed. GE Lamming, pp. 379-568. Churchill Livingstone, Edinburgh

Burke TJ (1986) *Small Animal Reproduction and Infertility, A Clinical Diagnosis and Treatment*. Lea and Febiger, Philadelphia

Christiansen GC (1954) Angioarchitecture of the canine penis and the process of erection. *American Journal of Anatomy* **95**, 227-262

Christiansen IBJ (1984) *Reproduction in the Dog and Cat*. Baillière Tindall, London

Depalatais L, Moore J and Falvo RE (1978) Plasma concentrations of testosterone and LH in the male dog. *Journal of Reproduction and Fertility* **52**, 201-207

Feldman EC and Nelson RW (1996) Canine male reproduction. Clinical and diagnostic evaluation of the male reproductive tract. In: *Canine and Feline Endocrinology and Reproduction, 2nd edn*, pp. 673-690. WB Saunders, Philadelphia

Foote RH, Swierstra EE and Hunt WL (1972) Spermatogenesis in the dog. *Anatomical Record* **173**, 341-351

Schmidt PM (1986) Feline breeding management. *Veterinary Clinics of North America: Small Animal Practice* **16,** 435-451

Soderberg SF (1986) Canine breeding management. *Veterinary Clinics of North America: Small Animal Practice***16,** 419-433

Woodall PF and Johnstone IP (1988) Dimensions and allometry of testes, epididymes and spermatozoa in the domestic dog (*Canis familiaris*). *Journal of Reproduction and Fertility* **82**, 603-609

CHAPTER SEVEN

# Conditions of the Male

*John P. Verstegen*

## THE MALE DOG

Male dogs are much more frequently presented for investigation of infertility or reproductive tract disease than male cats. Most conditions of the male dog are related to prostatic or testicular disease. However, congenital or acquired abnormalities of the prepuce, penis, scrotum, testes and prostate are also relatively frequent in this species.

### Congenital and genetic conditions

#### Immaturity of the genital tract – hypoplasia/aplasia

Aplasia of the epididymis and the vas deferens is rarely reported in the dog but its frequency may be underestimated. If bilateral, this condition will render the animal sterile.

Penile hypoplasia may be seen in female pseudohermaphrodite animals (see Chapter 4).

#### Hypospadias

Hypospadias is a congenital anomaly of the male external genitalia in which the urethral orifice opens somewhere ventrally on the prepuce or in the perineal area. This condition is often associated with other congenital defects and is particularly observed in intersex animals. The degree of expression of hypospadias is variable, ranging from the opening occurring a few centimetres from the tip of the penis (glandular hypospadias), at the pars longa glandis (penile hypospadias), at the scrotum (scrotal hypospadias) or subischially (perineal hypospadias). Hypospadias can be induced in fetuses by the administration of exogenous progesterone or oestrogen during pregnancy. The treatment of hypospadias is essentially surgical (see Chapter 15). Reconstruction is often difficult and urethrostomy and penile amputation may be required.

#### Persistent penile frenulum

This condition is rare. It is characterized by the ventral persistence of a thin sheet of fibrous connective tissue between the prepuce and the tip of the penis. It is related to a failure of separation between the surface of the penis and the preputial mucosa which should normally occur under the influence of testosterone just before or after whelping. Persistent penile frenulum, which is usually present on the ventral mid-line of the penis, prevents protrusion of the penis or deviates it when erect. The treatment is surgical with a good prognosis.

#### Intersexuality

Intersexuality refers to congenital abnormalities characterized by different degrees of impaired sexual differentiation. An abnormal sexual phenotype then results. The condition is less commonly encountered in dogs than in some other domestic species, e.g. goats or pigs. When observed, it prevents reproduction. Classification of the condition is by the type of gonadal tissue present.

A true hermaphrodite animal is characterized by the presence of both male and female gonads in a combined structure (ovotestis) or as separate organs (one testis and one ovary). The phenotypic sex of true hermaphrodite animals is generally female when they are young, but their sex becomes more ambiguous as they mature. True hermaphrodism may be associated with XX/XY or XX/XXY chimaeras.

Pseudohermaphroditism is characterized by the presence of gonads of one sex and a phenotype ambiguous in appearance or of the other sex. The phenotypic expression may vary greatly, so that the diagnosis may sometimes be difficult prior to puberty or surgery. In some males, the only signs are underdevelopment of the penis and prepuce, or total absence

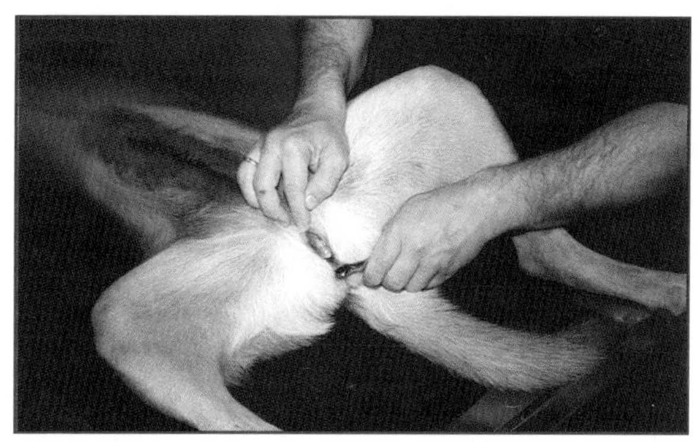

***Figure 7.1**: Presence of a hypertrophied clitoris in a hermaphrodite bitch.*

of the external genitalia. In some females, clitoral hypertrophy is observed which reveals the presence of a small penis including an os penis (Figure 7.1). In other females, abnormalities of oestrous cycles are observed, usually prolonged anoestrus. Often, the exact anatomical details can only be observed at surgery.

Many aetiologies are suggested for intersexuality, with a genetic basis being primarily indicated; however, treatment with hormones during pregnancy (particularly between days 25 and 35) when fetal gonadal differentiation is occurring has been implicated. Male pseudohermaphroditism can be linked to an inadequate synthesis of fetal androgens, a defect in the production or abnormality of the androgen receptors in the target cells or an ineffective response to Mullerian inhibiting factor leading to the persistence of the Mullerian duct. This is well documented in the Schnauzer breeds. Female pseudohermaphroditism is mainly the consequence of hormonal treatment during early pregnancy, for example administration of progestogens or interference with cortisol production. Surgical removal of the reproductive tract is advisable to prevent reproductive tract disease. Intersex disorders are more often seen in certain breeds such as Cocker Spaniels, Miniature Schnauzers, Kerry Blue Terriers, Beagles, German Shepherd Dogs, Australian Blue Terriers or German Short-haired Pointers.

**Cryptorchidism**

In dogs testicular descent is observed soon after birth, normally between days 3 and 10. However, accurate palpation of the testis is only possible at 2–4 weeks of age, and is usually carried out around weaning at 5–8 weeks. If the testes are not palpable within the scrotum by 8–10 weeks of age, a diagnosis of cryptorchidism may be suggested; however, to make a definitive diagnosis of cryptorchidism, it may be necessary in some animals to wait until puberty at around 5–10 months. In large breeds, however, where testicular descent may be apparently impaired by the rapid development of the animal, and where it is not uncommon therefore to see testes moving into and out of the scrotum, it is generally considered that a testis which is not present at 4 months of age is unlikely to reach its normal position. For this reason it is highly advisable that at the first visit to the veterinary surgeon at 6–8 weeks of age, all male dogs are carefully examined for integrity. If testes are not palpated, the owners should be informed of the probability of cryptorchidism. If the animal is to be used for breeding, it may be wise to change the animal or at least the owner should know that fertility might be compromised if the diagnosis is confirmed.

A cryptorchid animal may have one (unilateral cryptorchidism rather than monorchidism) or both testes (bilateral cryptorchidism and not anorchidism) retained somewhere along the normal path of testicular descent. Retention is usually classified as: subcutaneous, when the testis is palpated between the scrotum and the inguinal ring; inguinal or abdominal or according to whether the testis is in the inguinal ring or in the abdomen. It is important to point out that monorchidism (only one testis) and anorchidism (no testes at all) are extremely rare. For this reason, in all cases of absence of one or both testes, animals should be considered as cryptorchid. A hormonal stimulation test with human chorionic gonadotrophin (hCG)/luteinizing hormone (LH) or gonadotrophin releasing hormone (GnRH) could be used to demonstrate the presence and production of testosterone (see Chapter 11).

In dogs, the reported incidence of cryptorchidism varies between 0.8 and 9.8%. The reasons for failure of descent are not yet clear but it is likely that endocrine (insufficient gonadotrophin stimulation) and mechanical abnormalities are involved. The origin of these abnormalities are to some extent inherited and related to multiple gene influences. The degree of inbreeding appears to be higher in bilateral than unilateral cases of cryptorchidism. Heredity is almost certainly involved, as cryptorchidism is more common in certain breeds (for example Toy and Miniature Poodle, Yorkshire Terrier, Chihuahua, Boxer, Miniature Schnauzer) and in certain families. However, this does not explain all cases, as some unilateral cryptorchid dogs have been used for breeding for years and produced many litters with no evidence of defects in their offspring. The exact mode of inheritance is not known. Whereas the possibility exists of a pattern of multiple genetic inheritance, the simplest model is a sex-limited autosomal recessive mode, where both male and female carry the gene and can pass it on to their offspring.

All dogs of more than 6–8 months of age with absence of one or both scrotal testes should be castrated as cryptorchidism is responsible for at least four main clinical conditions:

- Testicular neoplasia, usually after 4–7 years
- Behavioural modifications characterized by hypersexuality, excitability, irritability and in some animals aggressiveness
- Decreased or modified fertility
- An increased risk of spermatic cord torsion.

In a normal animal, testes are located in the scrotum where the temperature is estimated to be 2–3 °C below body temperature. When testes are retained, they are submitted continuously to an increased temperature. This is responsible for the impairment of spermatogenesis, and thus the retained testis is generally sterile, while additionally the increased temperature in some animals stimulates steroidogenesis. Those animals become hypertestosteronaemic and the increased production of testosterone may be responsible for both a reduction of the spermatogenesis in the contralateral testis (by central feed-back mechanisms) and behavioural modifications characterized by satyriasis, excitation and sometimes increased aggression. During the first year of life of a cryptorchid animal with one retained testis, fertility may be maintained. However, bilateral cryptorchid animals in general become rapidly aspermic and sterile. After some years, the retained testes degenerate and may often be the site of neoplastic development. It is important to note that behavioural signs associated with cryptorchism are often difficult to control and castration in many cases is no longer efficacious. For this and the previously described reasons, knowing cryptorchidism has a high probability of being of genetic origin, it is wise not to use these animals for breeding. The author considers that all cryptorchid animals should be castrated before other clinical signs are expressed (thus before puberty). If there is a possibility that the dog will mate, any descended testis should also be removed or a vasectomy performed.

No treatments aimed at producing testicular descent have been demonstrated to be efficacious. Testosterone is not recommended because the dose used is supraphysiological and may lead to developmental abnormalities, particularly in joint development, and growth arrest. GnRH and hCG have been used, but the apparent success is low and probably occurs in those dogs which had inguinal or subcutaneous testes and in which descent would have occurred irrespective of treatment. However, since there is a high probability that cryptorchidism is inherited, treatment to induce testicular descent or orchidopexy should be discouraged and these treatments are considered unethical.

## Acquired conditions of the genitalia

### The external genitalia

***Male reproductive system trauma:*** Trauma due to fighting is a common feature in the dog, and licking, secondary to generalized skin problems, may also occur. Trauma to the reproductive tract is often very painful and is associated with swelling and haemorrhage. Lesions may affect the prepuce or both the prepuce and the scrotum with or without penile damage. These lesions are in general difficult to treat and require topical therapy combined with systemic antibiotics and anti-inflammatory drugs. Cleaning of the wounds and debridement may be necessary, with open wounds requiring surgical closure under general anaesthesia. Castration with the total removal of the scrotum is sometimes the only solution for severe lesions.

***Testicular abnormalities:*** The size and general appearance of the scrotum may differ dramatically from one breed to another, with some animals carrying their scrotum very close to the body wall, whereas in others the scrotum is more pendulous. The testes are roughly oval in shape, firm but not hard on palpation. The testes should be of similar size, with sometimes one slightly larger than the other. A marked disparity between the size of the testis should be regarded as abnormal.

The testes may be the primary or secondary site of disease. Primary abnormalities include cryptorchidism, hypoplasia, orchitis and neoplasia. Secondary conditions result from systemic or endocrine diseases. Dysfunction of one testis may impair the function of the other one.

***Orchitis and epididymitis:*** Orchitis and epididymitis are not commonly reported in European countries but are more frequent in Africa and in South and North America due to the presence of *Brucella canis* which specifically affects the reproductive system. *Brucella abortus*, *B. suis* or *B. mellitensis* can also affect the dog's reproductive system. Other causes of orchitis include viral (canine herpes virus, canine distemper virus), ascending or haematogenous bacterial infections, autoimmune disease or most commonly trauma. Aerobic bacteria (*Escherichia coli*, *Proteus* spp., *Staphylococcus* spp. and *Streptococcus* spp.) are commonly implicated. The raised temperature may temporarily impair fertility, but if the condition lasts for more than one day, resulting in local or general hyperthermia, testicular degeneration with fibrosis and atrophy can occur. Autoimmune orchitis is rare but the invasion of the testis by lymphocytes will result in infertility. Autoimmune orchitis has been shown to be a hereditary problem.

Clinical signs of orchitis/epididymitis are stiff gait, scrotal and testicular enlargement with oedema and redness, severe pain which can be associated with vomiting, anorexia, prostration and local or general hyperthermia. The scrotal skin may be inflamed and excessive licking may occur. However, the acute signs may disappear as the condition becomes chronic. At this stage, the scrotum may be normal but the testes may become soft or firm, irregular and small and the epididymis may appear enlarged and firmer than normal. Adhesions between the scrotum and the testis are common.

The treatment is initially medical with appropriate antibiotics (starting with broad-spectrum agents which may be changed following bacterial culture and sensitivity assessment); anti-inflammatory drugs to reduce swelling and pain are also administered. Antibiotic treatment should last a minimum of 2–4 weeks. As fertility is often lost after orchitis, surgical removal of the testis alone or with the scrotum should be advised.

***Hydrocoele:*** Clinical hydrocoele is rare in dogs; however, ultrasonographic examination may demonstrate the presence of cysts inside the testicular parenchyma. The significance of these is not known.

***Testicular neoplasia:*** Testicular tumours are more common in the dog than in other domestic animals. This is the second most common type of tumour in male dogs after skin tumours. Three types of testicular tumours have been described and have been classified following the cellular type involved: Sertoli cell, Leydig cells, and germ cell tumours (seminoma, teratoma and embryonal carcinoma). The incidence of testicular neoplasia is increased in cryptorchid dogs, Sertoli cell tumours and seminomas being 23 and 16 times, respectively, more frequent in cryptorchid than in normal animals, with Sertoli cell tumours having the highest incidence. Leydig cell tumours are most common in normal animals. More than one-third of dogs presenting with testicular tumours have two or three different types of histological lesion present simultaneously. Tumours may be unilateral or bilateral, with the mean age at first diagnosis being about 10 years. There are breed differences, with Boxers being the most likely breed to develop tumours of any type. Weimaraners and Shetland Sheep Dogs appear to have the highest risk of developing Sertoli cell tumours and German Shepherd Dogs a high risk of developing seminomas.

***Interstitial or Leydig cell tumours:*** Interstitial cell tumours are relatively frequent in old animals and are not related to retained testes. This tumour is rarely diagnosed except at post mortem following macroscopic and pathological analyses of the testes. Indeed many interstitial cell tumours are small and well circumscribed. They are often multiple and bilateral, coexisting with other lesions and some induce degeneration of non-neoplastic germ cell parenchyma. Fertility may thus be affected. No clinical signs are associated with this condition but occasionally signs related to hyperproduction of testosterone (behaviour modifications, perineal tumours or prostatic hyperplasia) may be detected. The testes are normal in size but may be harder with small nodules.

***Germ cell (seminoma) and mixed origin germ cell tumours (teratoma and embryonal carcinoma):*** Seminomas are often found in middle-aged and old dogs. The testes are homogeneously enlarged, being up to 5–10 cm in diameter and often softer than usual. Tumours are frequently unilateral and in general not associated with clinical signs. Some cases of hyperoestrogenism due to transformation of testosterone into oestradiol, or hyperandrogenism due to the absence of testosterone breakdown have been described. Seminomas are not generally malignant but a small number of seminomas have been described which have locally invaded or occasionally metastasized, particularly in retained abdominal testes (6–11%). Fertility may be reduced.

In some canine testes, a dual population of mixed germ cell and gonadal-stromal tumours (gonadoblastoma) has been identified. Teratoma and embryonal carcinoma are exceptional in dogs.

***Sustentacular or Sertoli cell tumour:*** These tumours are the most clinically significant testicular neoplasm of the dog. They are often observed in intact animals of more than 7–12 years of age, though earlier in cryptorchid animals, where Sertoli cell tumours may be detected as early as 4–5 years of age; the mean age for development of the tumour is about 10 years. The testis is generally enlarged, irregular and hard, particularly at the tumour site. Some of these tumours develop multilocular cysts and may undergo thrombosis and haemorrhage. Sertoli cell tumours are themselves in general 1–5 cm in diameter, although the size of the testis with this tumour can range from normal to more than 30 cm in diameter (Figure 7.2a). We recently removed an abdominal testis in a 6-year-old Bichon Maltese, the size of which was greater than a child's head. These tumours are generally not malignant, but metastases have been described particularly in cases with abdominal retention. Between 10 and 20% metastasize to the lumbar or iliac lymph nodes. Sertoli cell tumours are typified by the production of oestrogen. Clinical signs associated with oestrogen production are: feminization, preputial and penile atrophy, pendulous sheath, gynaecomastia and enlargement of the nipples, atrophy of the contralateral testis, bilateral non-pruritic symmetrical alopecia and hyperpigmentation of the skin (Figure 7.2b,c), prostatic squamous metaplasia and attractiveness to other males. An increase in the plasma oestrogen concentration or an attenuated oestrogen/testosterone ratio is generally observed. Fertility is usually affected, particularly in the advanced stages. Due to the increase of oestrogen production in chronic cases, bone marrow suppression can be observed in 10–15 % of dogs with Sertoli cell tumours. The main signs of this are leucopenia, pancytopenia, thrombocytopenia characterized by haematoma, petechia, haemorrhage and anaemia and occasionally infection, and septicaemia.

Adenomyosis (epithelial muscular invasion) of the epididymis is a common complication of Sertoli cell tumours in dogs and seems to be a consequence of increased oestrogen production. However, adenomyosis of the epididymis also occurs in dogs without Sertoli cell tumours.

Because of the high incidence of testicular neoplasia, physical examination of all middle-aged and older dogs should always include an examination of the testes. In all middle-aged or old animals with no scrotal testes and no history of castration, testicular tumours should be considered where there are signs of dermatological, urinary or digestive diseases. It is particularly important to check for retained bilateral testes and to consider the possibility of their neoplastic transformation, the incidence of which is certainly underestimated. We have seen many cases of presumed splenic tumours which turned out to be retained testicular tumours at surgery.

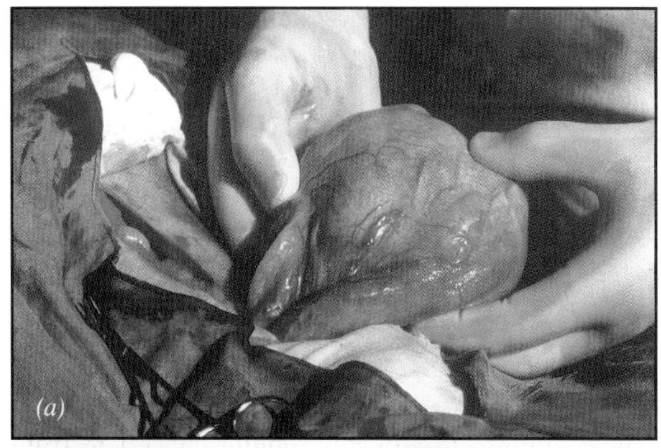


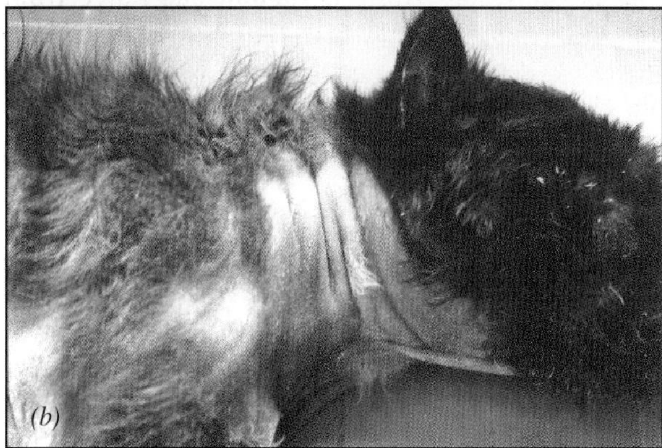


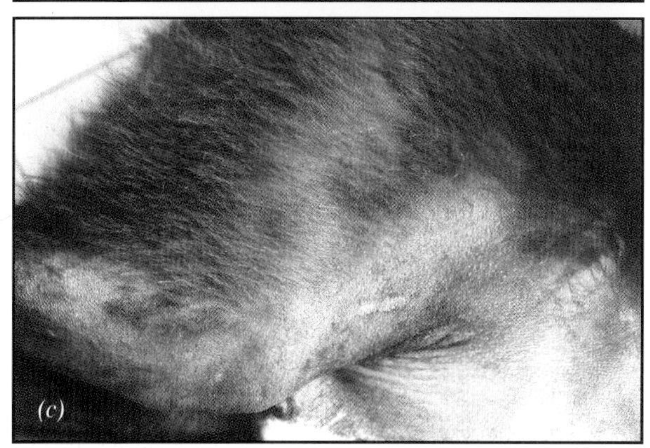


***Figure 7.2:*** *(a) Testicular tumour in a cryptorchid dog. (b, c) Dermatological lesions associated with the tumour.*

The only treatment for all these tumours is surgical removal. When a testis becomes neoplastic, the contralateral testis undergoes atrophy and the dog becomes sterile. If surgery is carried out early enough, fertility can recover and the dog may continue to be used for breeding.

***Testicular torsion:*** Torsion of the spermatic cord, which is also referred to as testicular torsion, is a rare condition that may occur in dogs of any age. It is seen more commonly in retained testes, probably due to instability of the suspensory tissue which allows the gonad to rotate. This leads to vascular engorgement and testicular infarction, especially when a large tumour is present in a retained abdominal testis. The degree of torsion may range from 360° to more than 720° and clinical signs will depend on the degree of torsion. There is often acute pain of sudden onset with a reluctance to move. The dog rapidly becomes shocked, anorexic, and vomits. In the acute form, this is a surgical emergency. In rare cases where the condition occurs in a normal non-cryptorchid dog, the scrotum shows swelling, oedema and redness. The condition is likely to recur and may be potentially fatal. In the cryptorchid animal, this is a clear indication for the early surgical removal of the retained testis.

### Penile and preputial conditions

***Balanoposthitis:*** Inflammation of the mucosa of the penis (posthitis) and of the prepuce mucosa (balanitis) is called balanoposthitis. This is a common condition in male dogs. Two types can be described: balanoposthitis in the young dog and balanoposthitis in the adult. In young prepubertal dogs or those close to puberty, yellow to green preputial discharges varying in quantity are common. This condition, except for the problem it causes for the owner, has no clinical significance. Topical therapy may be given but the condition often recurs a few days or sometimes weeks after treatment. Examination of the prepuce and the glans penis usually shows no local inflammation. The discharge contains many neutrophils but specific bacteria are rarely seen. The condition seems to be related to hormonal changes related to puberty.

In the adult entire male, balanoposthitis may also occur. The clinical signs are usually a purulent or haemorrhagic preputial discharge with frequent licking and possibly biting and self-mutilation and pain on palpation. The mucosa is usually inflamed, ulcerated with vesicles, or there may be other local lesions related to the cause (e.g. trauma, neoplasia, burns). Identification of the possible cause will determine the treatment required. Bacterial culture is usually of little value as it will reveal the normal extensive and varied commensal bacterial flora of the prepuce. *Staphylococcus aureus* is the most commonly observed bacterium; however, many others have been reported. The presence of *Proteus* or *Pseudomonas* could be considered significant when in pure cultures. *Mycoplasma canis* and *Ureoplasma* have been incriminated; however, there is no difference in the prevalence of mycoplasms between normal animals and those with genital infections. *Ureoplasma* is more frequent in the prepuce of infertile dogs. The presence of small papules or vesicles can be related to herpesvirus infection (also sometime called dog pox). As the herpesvirus can be responsible for genital disease, abortion or fetal resorptions, care must be taken with these animals. Many vesicle-like lesions are not due to herpesvirus but are simple lymphoid hyperplasia, a finding that is not significant; however, mating should be prevented as a precaution.

In the adult, treatment for balanoposthitis is always necessary and the therapy will depend on the cause. Before treatment, the penis should be carefully examined to exclude the presence of foreign bodies, neoplasia, ulceration or inflammatory nodules. Local or broad-spectrum antibiotics, local flushing with anti-inflammatory drugs or their systemic administration are generally used. Castration may be useful in reducing the amount of preputial secretion in the non-breeding animal.

***Phimosis and paraphimosis:*** In phimosis the dog is unable to extrude its penis because of a small preputial opening. The condition is generally congenital and is resolved by surgery (see Chapter 15). In severe cases, urine cannot escape and therefore remains within the prepuce. This condition is then often associated with a balanoposthitis.

In paraphimosis, the penis is unable to retract into the prepuce. This is most often due to a small preputial opening (congenital, traumatic or post-surgical) which is large enough to allow the normal and non-erect penis to protrude but when erection is complete acts as a tourniquet preventing detumescence. It may also be observed in animals with spinal lesions, foreign bodies in the prepuce, chronic balanoposthitis, swelling of the penile soft tissue from trauma or fracture of the os penis, or due to long hair around the opening of the prepuce causing obstruction. Indeed, inversion of the preputial skin and hair may prevent the penis from returning into the preputial cavity. This condition, if untreated, may lead to severe penile trauma as the penis rapidly becomes dry, cyanotic, oedematous and painful. Stranguria, haematuria and anuria are often observed and finally gangrene or necrosis of the penis may occur. Tranquillizers and particularly alpha-2 adrenergic agents are useful in inducing rest and hypotension which allow detumescence and manual replacement. After cleaning of the penile surface, gentle manipulation using cold water is often sufficient if the condition is recent. It is also advisable to give progestogens with anti-androgenic potency (delmadinone acetate, megoestrol acetate) to reduce sexual activity and excitability. These have to be given at high doses to be clinically efficacious (delmadinone acetate minimum of 2 mg/kg every 2 weeks). Simultaneously, acepromazine can be given for a few days post-treatment to maintain a degree of hypotension and to reduce the risk of erection.

Recurrence is not uncommon and if the animal is to be kept for breeding, surgical enlargement of the prepuce has to be considered. However, this procedure must be carefully performed, particularly to prevent too large an opening which may result in a chronic paraphimosis with permanent penile extrusion. In some extreme cases, the surgical ablation of the penis together with castration is the only solution.

A similar chronic exposure of the tip of the penis is also sometimes observed in dogs with a short prepuce or too long a penis. The protrusion may be permanent or only observed when the dog is sitting. This may lead to a persistent protrusion of the penis. Often the condition causes no problem, but drying of the mucosa and trauma may necessitate local treatment or sometimes amputation. A suture may also be placed at the preputial orifice to reduce its size and the risk of protrusion. However, as with the various techniques described to pull the prepuce forward, this technique is often unsatisfactory and unsuccessful. Castration, inducing organ involution may be indicated.

***Penile bleeding:*** Penile bleeding may occur spontaneously or following trauma or fighting. It is important to clearly localize the origin of the bleeding which may be the penis itself (e.g. trauma, herpes virus, tumours, TVT) but could also originate from a urethral prolapse (rare but sometimes observed) or from the urethra, bladder or prostate.

Penile bleeding is often observed in excited animals and may be due to lesions on the penile integument. Suturing or cauterizing the wound is generally effective when the lesion is fresh and not too extensive. Alpha-2 adrenergic agents, inducing hypotensives and tranquillizers, are useful to reduce the risk of excitement and erection, both of which will increase the haemorrhage. In some cases amputation is required. Castration or large doses of progestogens may be of use to prevent postoperative excitement and erection which could result in haemorrhage.

***Fracture of the os penis:*** The os penis, which is within the bulbus cavernosus, may occasionally be fractured due to a traumatic mating or attempts to separate animals which are tied at mating. The main signs are pain, bleeding, urethral haemorrhage and sometimes dysuria. Symptomatic treatment should be undertaken, but if urethral constriction occurs, penile amputation or urethrostomy may be necessary.

***Transmissible venereal tumour (TVT):*** This mucosal tumour of the dog (Figure 7.3) affects the external genitalia and other mucous membranes. It is mainly seen in dogs and bitches in tropical and subtropical countries. However, due to the increase in exchanges between countries, it is becoming less rare to have animals presented with this condition. TVT is now more common in Mediterranean countries and in the southern part of France, and may result from stud dogs that have been taken abroad for mating. In the male, the lesion typically occurs on the penis and prepuce but may also be seen on other sites such as the mucosa of the head following licking. After a period of incubation of a few weeks (occasionally months) the animals develop reddish-grey friable tumours. At first, the nodules are small (a few millimetres) but they rapidly grow and have a cauliflower-like appearance, being papillary and pedunculated. After a few months spontaneous regression may occur.

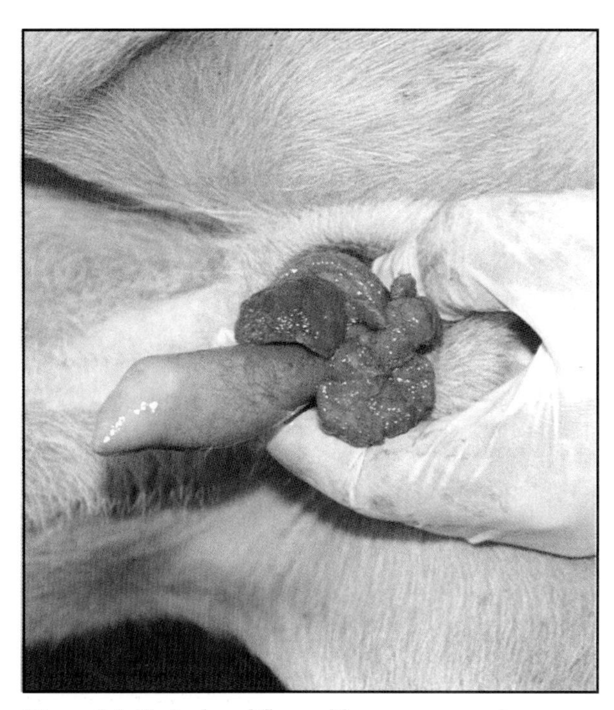

*Figure 7.3: Typical cauliflower-like appearance of a transmissible venereal tumour in a male dog.*

The first clinical signs detected by the owner is a preputial discharge which is often haemorrhagic, as the friable nodules bleed easily. The condition is not initially painful but the discharge may rapidly become infected and induce a balanoposthitis with associated pain and licking. The tumours may ultimately protrude from the preputial orifice. The lesions are usually multiple and a careful investigation of the preputial cavity will reveal nodules of different size in the same animal. The tumour has no effect on fertility but could prevent mating for mechanical reasons or due to secondary infections. Infected animals should not be allowed to mate, as the tumour is transmitted at coitus. However, the exact aetiology of this condition is still not established. Viruses have been implicated, especially as in many animals spontaneous regression is observed. However, specific viruses have not yet been identified. Neoplastic cells are clearly of an origin other than the dog, as they have a chromosome number of 59 ± 5, which is distinctly different from the normal canine chromosome number of 78. The mode of transmission seems to be via infected cells which are easily transplanted during coitus, or following sniffing, nuzzling or other social interactions. In the primary infection, there is invasion of the mucosa and the tumour behaves as an infection with involvement of the regional lymph nodes in some cases. General spread has not been described, but tumours have been detected in the brain, eye, thoracic and abdominal viscera. The treatment is multiple involving: hygienic local treatment using antiseptics and disinfection of the inflamed penis and prepuce; surgical removal of the largest tumours (cryotherapy, cauterization by electrosurgery). A complete surgical cure is unlikely as in general the tumours are multiple and may recur if a single cell remains. Recurrence occurs in 50% of cases. Chemotherapy or radiotherapy may achieve a cure as the tumour is responsive to these two treatments. As the condition is transmitted mainly by coitus and is highly contagious, all breeding animals should be treated by chemotherapy or radiotherapy. Vincristine administration once weekly for around 4-6 weeks induces complete remission in more than 90% of the animals without any recurrence.

**Prostatic disease**

Disease processes in the prostate are common and may be infectious, hormonal, anatomical or embryological in origin. An increase in the incidence is observed with age and could be related to the physiological hyperdevelopment of the organ observed under the influence of androgens.

The prostatic secretions, which are acidic, play a bactericidal role during rest, preventing ascending bladder infections, and have a major role during ejaculation in the production of the seminal fluid. Indeed, in dogs the prostate is responsible for most of the volume of ejaculate produced (more than 90-95%).

During the dog's life, the development of the prostate can be divided into three periods. The first one corresponds to the period of embryogenesis and immediate post-natal development. This phase ends when the animal is around 2-3 years. The second phase is one of exponential hypertrophic development. This phase is clearly androgen dependent and terminates when the animal is approximately 12-15 years old. The last phase is senile involution and begins when, in geriatric animals, the production of androgens slowly begins to decrease. The clear distinction between these different periods is difficult to make and is indeed highly subjective, with variation between animals. However, it is generally accepted that after 5 years of age, nearly all dogs have a certain degree of prostatic hypertrophy which may continue to develop, leading in some animals to the pathological features known as benign prostatic hyperplasia (BPH). This clinical condition is observed in 60% of dogs older than 5 years and in nearly 100% of dogs over 10-12 years old. The position of the prostate changes slowly as it increases in size. Indeed, from being pelvic in young animals the prostate slowly becomes more abdominal and finally after 8-12 years becomes fully abdominal.

It is beyond the scope of this text to review all prostatic diseases, clinical signs, investigational methods and treatments. The reader is asked to consult the reference list below.

The prostate gland may be affected by infection (acute and chronic prostatitis and abscess), endocrine dependent disease (prostatic hyperplasia and metaplasia) or tumorous processes (adenocarcinoma). Cysts are also described, and are congenital (paraprostatic cysts), primary (prostatic cysts) or secondary (to

other diseases particularly benign prostatic hyperplasia and squamous metaplasia).
The main problems with prostatic diseases are:

- They are often complex, with different types of lesion observed simultaneously
- Different aetiological agents can occur independently or simultaneously
- A 'haemato-prostatic barrier' has been described which reduces treatment penetration and efficacy
- Their incidence is often underestimated.

Indeed, many clinical signs associated with prostate disease are non-specific (haematuria, anuria, pain, constipation or locomotor dysfunction) and are often first attributed to other organs (bladder or digestive diseases, or orthopaedic problems).

In all entire male animals of more than 4–5 years where one or several of the following clinical signs are observed, the prostate should be carefully investigated.

***Urinary symptoms:*** These include haematuria which is not necessarily associated with urination (blood is often observed passively in resting animals but also during urination), urinary incontinence with a full bladder, dysuria and anuria.

***Digestive symptoms:*** These are not constant, but are sometimes the only clinical signs related to prostatic disease. They are caused by the hypertrophy of the prostate in the pelvis, inducing constipation with sometimes characteristic flat faeces. These signs are highly variable in intensity and depend on the degree of prostatic enlargement and the position of the prostate. In general, the enlarged prostate moves cranially into the abdominal cavity and no digestive signs are observed. When inflammation is present (due to infectious processes or abscess), the digestive signs may be overexpressed and pain may be one of the dominant features of the disease.

***Locomotor symptoms:*** Stiffness, posterior lameness, paresis and oedema of the legs can sometimes be observed in dogs with prostatic disease. These symptoms may appear suddenly (abscesses or acute prostatitis) or progressively (hyperplasia, tumours) and are related to local compression related to the enlargement of the prostate gland or to the presence of local metastasis.

***General systemic signs of no specific origin:*** General signs such as fever, pain, prostration, anorexia, emesis and septicaemia may be associated with the condition. They sometimes may be the only symptoms detected by the owner, the more specific signs being more discrete. In all entire adult male dogs presenting with systemic clinical signs of no specific origin, a careful complete investigation of the prostate should be carried out.

***Infectious prostatic diseases:*** Bacterial prostatitis is probably the second most common prostatic disorder in dogs after benign prostatic hyperplasia. Infectious causes of canine prostatitis include *E. coli*, *Proteus* spp., *Pseudomonas* spp., *Brucella canis*, *Mycobacterium* spp., staphylccocci, streptococci and occasionally distemper virus and *Blastomyces dermatitidis*. Microscopic examination of centrifuged prostatic fluid allows the observation of normal prostatic cells with numerous polymophonuclear cells, red blood cells and sometimes bacteria (Figure 7.4).

Prostatitis can be acute and associated with many clinical signs, or chronic and more discrete. During acute prostatitis, the haemato-prostatic barrier is no longer functional and antibiotic therapy has a good chance of being effective. A broad-spectrum antibiotic should be used until antibiotic sensitivity has been assessed. Chronic prostatitis is always more difficult to cure due to the haemato-prostatic barrier, reduced blood supply and organ fibrosis. The antibiotic choice has to be made taking into account the results of antibiotic sensitivity and the pH of the prostatic secretion. High dosages and prolonged (minimum 4–6 weeks) treatments are the keys to success. Since other conditions such as BPH or metaplasia are often observed simultaneously, castration, antiandrogen drugs, progestogens or 5-alpha reductase inhibitors can be used concurrently. When abscesses

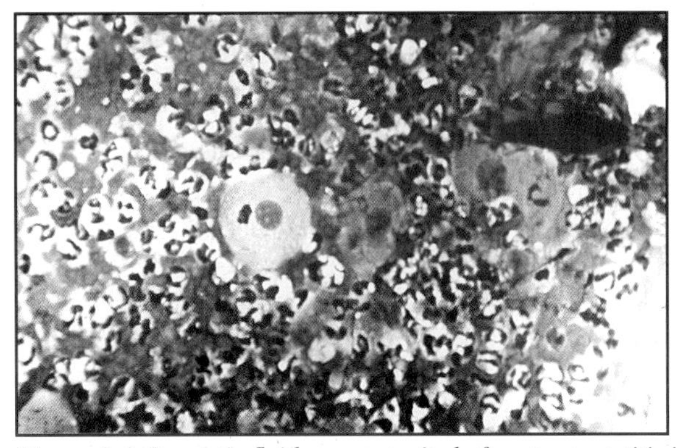

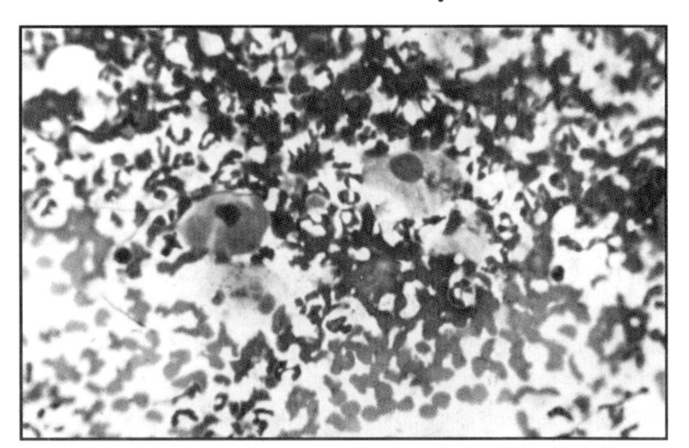

***Figure 7.4:*** *Prostatic fluid smears typical of acute prostatitis in dogs. Many inflammatory neutrophils, bacteria and cell debris can be identified.*

are present, antibiotics should be used and omentalization with prostatic curettage may be necessary if cavities more than 1 cm in diameter are detected by ultrasonography.

#### Endocrine prostatic conditions

***Benign prostatic hyperplasia:*** Benign prostatic hyperplasia is observed in entire adult males or animals that have been treated with androgenic hormones. The prostate is usually homogeneously enlarged. The exact aetiology of this condition in the dog is not yet established but the condition is closely related to the effects of testosterone after transformation into dihydrotestosterone by the 5-alpha reductase enzyme. Some effects of oestrogens have also been proposed. Prostatic hyperplasia is considered to be normal in animals older than 5–7 years and is only considered a disease when clinical signs related to the hyperplasia are present. Prostatic hyperplasia may be present in some animals without any clinical signs, but may be associated with mild to acute symptoms in others. The symptoms are as previously described. Castration is the most efficient treatment if the animal is not required for breeding. Progestogens, anti-androgens, 5-alpha reductase inhibitors or GnRH agonists or antagonists may be useful since they block the action of testosterone and of dihydrotestosterone. These have all been shown to be clinically efficacious.

***Prostatic metaplasia:*** Exogenous oestrogens administered as therapeutic agents (this is no longer accepted) or endogenous oestrogens from Sertoli cell tumours may cause prostatic metaplasia. The prostate is no longer homogeneous in shape and often cysts or abscesses develop. Numerous squamous cells are observed in the prostatic fluid. Castration (in case of Sertoli cell tumours) or withdrawal of exogenous oestrogens is curative.

***Prostatic tumours:*** Prostatic tumours (mainly adenocarcinomas) are fortunately rare in dogs. They may be observed in both young adults and old animals and are not directly androgen related in this species. Their appearance and subsequent growth is rapid and in general metastases are already present when a diagnosis is reached. Multinucleate cells can sometimes be observed in the prostatic fluid. Adenocarcinoma tends to metastasize through the external and internal iliac lymph nodes to the vertebral bodies as well as to the lungs. The colonic and pelvic musculatures, the pelvis and urethra may also be invaded by direct extension of the tumours. The prognosis is poor and no satisfactory treatment has yet been found. Anti-androgens, castration, partial prostatectomy or chemotherapy have been proposed but without major evidence of success.

#### Other prostatic conditions

***Prostatic cysts:*** These are often the consequence of prostatic gland obstruction related to calculi, cellular desquamation in metaplasia, or hyperplasia. They are of no clinical significance as long as their size is small and infection does not occur. As with abscesses, omentalization of the cavity is the best treatment and has replaced drainage or marsupialization.

***Paraprostatic cysts:*** Paraprostatic cysts are found laterally, outside the prostate and urinary bladder. Some may become confluent and contain large amounts of fluid. The origin of the cysts remains unclear, but suggests vestigial (Mullerian ducts of the female) abnormalities. Surgical removal and omentalization are recommended.

Prostatic calculi and calcification are sometimes observed but are of no clinical significance if no complications such as cysts, abscesses or infection are observed.

## THE MALE CAT

Male cats are not commonly presented for infertility investigations or disease of the internal or external genitalia. Abnormalities of the genitalia other than the consequences of fighting or bite wounds are not often reported. Indeed, most male cats are neutered at a young age, thereby dramatically reducing the incidence of reproductive system abnormalities.

The male genitalia are composed of: the testes, epididymides, prostate, bulbourethral glands and the penis. The cat does not possess seminal vesicles.

### Congenital and genetic conditions

Cats have 19 pairs of chromosomes. Not all congenital defects are related to gene mutation and therefore it is important to separate the two concepts of genetic and congenital disease. In genetic disease abnormality of the genome results in a congenital disease being present at birth. Congenital disorders of sexual development may be related to abnormal chromosome numbers or abnormalities of gonadal or phenotypic sex.

A wide variety of congenital defects have been described in cats. These may affect the nervous, cardiovascular, gastrointestinal, ocular, urinary, integumentary, musculoskeletal or reproductive systems. A complete discussion of each defect is beyond the scope of this chapter and the reader is encouraged to consult the reference list and other textbooks. Cats with heritable defects should not be used for breeding, and breeding of cats with defects whose hereditability is not proven should also be discouraged. The main congenital or genetic defects are related to abnormal chromosome numbers (XXY, XO, chimaeras and mosaics), abnormalities of gonadal sex (XX sex reversal, XX true hermaphrodite) or abnormalities of phenotypic sex.

**Hermaphroditism and pseudohermaphroditism**
True hermaphroditism and abnormality of gonadal differentiation are very rare in cats. Abnormal testes at the tip of the uterine horns of a 1-year-old phenotypic female blue tabby (male pseudohermaphroditism) have been reported. The testes contained seminiferous tubules lined with Sertoli cells but no spermatogonia.

**Cryptorchidism**
The absence of one or both testes in the scrotum is called cryptorchidism. This condition must be distinguished from monorchidism or anorchidism which represent the agenesis of one or both testes and are extremely rare in this species. The normal cat testis descends into the scrotum prenatally. However, testes may move freely up and down in the inguinal canal prior to puberty so the definitive diagnosis of cryptorchidism should not be made in cats less than 7–8 months of age. Indeed, most cats presented for cryptorchidism at 4 months have normal testes at 5 or 6 months. The incidence of cryptorchidism in adult cats ranges from 0.07 to 1.7% and is lower than in dogs. The hereditary nature of the condition is uncertain although it seems to occur more commonly in certain breeds such as Persians or in certain families. The mode of the possible inheritance is not yet clear although a simple autosomal recessive sex-linked model has been proposed.

Unilateral cryptorchidism of either testis is more than three times as prevalent as bilateral cryptorchidism.

Unilateral cryptorchid cats are usually normal and fertile; only rarely are there behavioural or pathological complications.

The presence or absence of testes in a male without scrotal testes can be assessed by the observation of the six to eight rows of keratinized penile spines, which will be absent if the male has been castrated. A hormonal evaluation can also be undertaken by measuring testosterone concentrations before and after hCG or GnRH injection. This stimulation test is required because the testosterone concentration is often basal or close to minimum detectable values in many normal male cats. Low basal values are therefore not indicative of the presence or absence of testes.

It appears that medical therapy (using GnRH or exogenous gonadotrophins) does not induce testicular descent, and since the condition may be inherited, treatment is unethical and the animal should be prevented from breeding.

When the condition is considered from a therapeutic or clinical point of view, since neoplasia in the cat is uncommon (only described once in a cryptorchid cat), and testicular cord torsion has not been documented, treatment is not essential and it is the veterinarian's choice either not to do anything or to propose castration. Castration will, however, prevent breeding and will modify urination and reproductive behaviour and may reduce male odour. When performing surgery, because monorchidism is rare it should always be assumed that the animal is cryptorchid and has two testes. Often, when not detected in the abdomen, the missing testis can be found in the inguinal ring or under the skin close to the inguinal ring.

The median laparotomy is preferred by the author since the testes might be anywhere in the abdomen from caudal to the kidney to the inguinal ring. Location of the vas deferens at the bladder neck is probably the easiest way to find the testis. When the vas deferens is observed gentle traction allows the testis to be withdrawn from its position and to be surgically removed.

## Acquired conditions of the genitalia

**Trauma**
Injury to the external genitalia of male cats is often observed and usually results from bite wounds sustained during fighting or mating. Clinical signs include pain, swelling and haemorrhage. Possible urethral damage may have to be investigated along with any penile wounds. In these cases, catheterization and contrast radiography may frequently be useful. Superficial penile lesions will resolve with only the topical application of disinfectant or emollient preparations. More serious lesions may require surgical resection of the penis and perineal urethrostomy. For scrotal lesions such as abcesses, lacerations or inflammation, the general principles of wound management have to be followed. Local disinfection, repair and drainage, with suitable antibiotics, are indicated. However, if the animals are not required for breeding, castration with scrotal ablation is the treatment to be recommended.

**Infections**
Bacterial cultures of the prepuce of normal entire males reveal the presence of bacteria in more than 90% of animals. These bacteria are often: *Escherichia coli*, *Pseudomonas aeruginosa*, *Proteus mirabilis*, *Klebsiella oxytoca*, *Streptococcus* spp., Enteroccoci, *Bacillus* spp. or *Staphylococcus* spp. Similar organisms can be detected in semen, but are probably of urethral origin. These bacteria are commensals and do not cause disease. Bacterial infections in cats are usually secondary.

**Orchitis**
Orchitis is rare in the cat but has been described following testicular infection (tuberculosis, feline infectious peritonitis virus or aerobic bacteria). Bacterial infections may follow ascent from urinary tract disease, but are most often due to injuries or wounds, especially bite wounds. One or both testes may be involved and testicu-

lar, spermatic cord and epididymal enlargement may occur. Brucellosis is rare in cats which seem to be resistant to natural infection with *Brucella* spp.

Feline infectious peritonitis (FIP) virus infection may cause scrotal distension due to extension of the peritoneal infectious process through the inguinal ring into the scrotum. There may be bilateral scrotal swelling, fever and lethargy.

Orchitis due to bacterial inflammation is associated with heat, pain, swelling, redness and anorexia. Furthermore, licking of the lesion is usually observed and may lead to an aggravation of the condition, sometimes with self-mutilation.

Broad-spectrum antibiotics administered for at least 2 weeks may be efficacious; however, if not, curative castration may be required.

Herpes virus infection is most commonly observed as a mild respiratory disease in cats older than 12 weeks of age and may rarely be transmitted by coitus. In tomcats, preputial vesicles have been observed in experimentally infected males but this has not been reported with naturally occurring FHV infection.

**Testicular hypoplasia**

In a normal animal, the testes are approximately the same size and are firm but not hard in consistency. Testicular hypoplasia is rarely reported in cats since the vast majority are not kept entire. It may be observed following infection with panleucopenia virus in the pre-pubertal cat, in cases of cryptorchidism, and in some conditions associated with chromosomal abnormalities. It may occasionally be observed in some generalized acute diseases associated with persistent hyperthermia or in chronic metabolic disease, both of which could cause the destruction of the germinal layer. In these cases, the condition is often bilateral, with the testes being small and soft. Similarly, poor libido and variable testicular degeneration leading to azoospermia have been associated with malnutrition, obesity, hypothyroidism and hypervitaminosis A.

**Neoplasia**

Transmissible venereal tumours do not occur in cats. Feline leukaemia virus infection has been suggested to be related to some tumours including testicular tumours and tumours of the skin overlying the genital region.

Sertoli cell tumours have been described but are rare and when observed are often associated with an intra-abdominal retained testis. No significant clinical signs are observed in this species.

**Testicular torsion**

Torsion of the spermatic cord has occasionally been observed. This condition is most often unilateral and is observed in both descended and undescended testes in animals of all ages. The cause is unknown. When observed, acute pain is the main clinical sign.

**Balanoposthitis**

This condition, which is relatively common in dogs, is rare in cats, but may follow trauma at mating or occur associated with penile-preputial adhesions (see later).

**Penile hypoplasia**

Penile hypoplasia has been described in cats and is often associated with early castration or certain genetic conditions. The relation of penile hypoplasia and early neutering to urethral obstruction will certainly be a major concern for the future. No specific treatment exists but urethrostomy can be carried out if the condition is associated with urethral obstruction.

**Hypospadias**

Hypospadias (characterized by an opening of the urethra on the ventral surface of the penis or in the perineal area) has not been described in cats.

**Phimosis and paraphimosis**

Inability to extrude or to return the penis into the prepuce are two rarely observed conditions in this species. If present, constriction of the preputial orifice may be congenital or acquired following infection or trauma. The preputial hairs of long-haired cats may entangle the preputial orifice, causing clinical signs similar to phimosis or paraphimosis. Clipping the hair should resolve this condition.

**Persistent penile frenulum**

Under the influence of testosterone, the surfaces of the penis and the preputial mucosa separate before or within a few weeks of birth. Failure of separation results in persistence of connective tissue between the penis and the prepuce. Adhesions of the prepuce and penis have been reported in cats castrated at, or earlier than, 5 months of age. This condition is often not clinically detectable unless the complication of balanoposthitis occurs. It might be advisable to check that penile-preputial separation has occurred before castration.

**Prostatic diseases**

With the exception of a few reports of prostatic neoplasia, prostatic disease is not described in cats. This may be related to the high incidence of castrated animals but also due to the difficulty in examining the organ in this species. Clinical signs of prostatic neoplasia (often adenocarcinoma) include haematuria, dysuria, pollakiuria and outflow obstruction. Clinical signs of prostatic adenocarcinoma in cats are sudden in onset and most often rapidly fatal. Generally, because of the rapid evolution of the disease, this condition is only detected at necropsy.

## Endocrinological and behavioural conditions

**Endocrinological conditions**

Plasma testosterone concentrations in adult intact

male cats has been reported to range from 0 to 23.5 ng/ml (0–81.6 nmol/l). Castration causes an immediate decline in testosterone to basal (0–0.05 ng/ml; 0–0.2 nmol/l), suggesting that in this species testosterone is of testicular origin only. In intact toms hCG stimulation induces a 3–10 times increase after 4 hours and GnRH stimulation a 3–15 times increase after 1 hour. Plasma oestradiol concentrations in the normal adult male range from 12 to 16 pg/ml (44.0–58.7 nmol/l) and plasma androstenedione from 0 to 35 ng/ml. Plasma androstenedione is not totally suppressed by castration, suggesting another source for this hormone.

Endocrine disorders may directly or indirectly involve the sex hormones. Hypothyroidism and adrenocorticoid disorders may affect both libido and fertility.

### Behavioural conditions

As much of the normal breeding behaviour of the cat is acquired learning, management of the first mating is important to prevent acquired behavioural problems which could reduce the ability of the male to mate. Acquired behavioural abnormalities of reproduction in the cat include the following.

***Lack of libido:*** This may be related to gonadal abnormalities, immaturity and insufficient expertise, senility, management conditions, health problems, timing of mating, or a painful or stressful previous experience. Excessive socialization with humans may lead to reduced libido, or libido directed towards humans. In certain breeds, particularly Persians and their relatives (Exotic, British Blue), sexual maturity may occur later (3 years or more) so that apparently adult males presented for poor sexual development are in fact only prepubertal.

***Inability and/or refusal to mate:*** Inability and/or refusal to mate may be a consequence of the previously described lack of libido. Refusal to serve is often observed in males which have previously had a stressful or painful mating. Some males with poor libido will refuse to mate an unknown queen or a queen with excessive or aggressive reproductive behaviour. However, this could also be related to: inability to extrude the penis (rare), nervous damage in the lumbar and sacral regions of the spinal cord, painful conditions, infections of the genitalia, orchitis and/or epididymitis, or other painful or degenerative conditions affecting reproduction.

## FURTHER READING


Allen WE (1992) *Fertility and Obstetrics in the Dog.* Blackwell Scientific Publications, Oxford

Aronson LR and Cooper MI (1967) Penile spines of the domestic cat: their endocrine-behavior relations. *Anatomical Record* **157**, 71–78

Barsanti JA and Finco DR (1993) Canine prostatic diseases. In: *Textbook of Veterinary Internal Medicine*, ed. SF Ettinger and EC Feldman, pp. 1662–1685. WB Saunders, Philadelphia

Basinger RR, Robinette CL, Hardie EM and Spaulding KA (1993) The prostate. In: *Textbook of Small Animal Surgery*, ed. D Slatter, pp. 1349–1367. WB Saunders, Philadelphia

Bloom F (1962) Retained testes in cats and dogs. *Modern Veterinary Practice* **43**, 160

Brearley MJ (1991) The urogenital system. In: *Manual of Small Animal Oncology*, ed. RAS White, pp. 297–314. BSAVA, Cheltenham

Herron MA (1988) Diseases of the external genitalia. In: *Handbook of Small Animal Practice*, ed. RV Morgan, pp. 673–678. Churchill Livingstone, New York

Hornbuckle WE and Kleine LJ (1980) Medical managment of prostatic disease. In: *Current Veterinary Therapy VII*, ed. RW Kirk, pp. 1146–1150. WB Saunders, Philadelphia

Johnston SD (1986) Disorders of the canine penis and prepuce. In *Current Therapy in Theriogenology*, ed. DA Morrow, pp. 549–550. WB Saunders, Philadelphia

McEntee K (1990) *Reproductive Pathology of Domestic Mammals*, pp. 224–374. Academic Press, Boston

Mickelsen WD and Memon MA (1995) Inherited and congenital disorders of the male and female reproductive systems. In: *Textbook of Veterinary Internal Medicine*, ed. SJ Ettinger and EC Feldman, pp. 1686–1690. WB Saunders, Philadelphia

Millis DL, Hauptman JG and Johnson CA (1992) Cryptorchidism and monorchidism in cats. *Journal of the American Veterinary Medical Association* **200**, 1128

Onclin K, Silva LDM and Verstegen J (1994) Physiology, investigational methods and pathology of the prostate in the domestic carnivores parts 1 & 2. *Annales de Médicine Vétérinaire* **138**, 529–549

Sojka NJ (1980) The male reproductive system. In: *Current Therapy in Theriogenology*, ed. DA Morrow, pp. 821–844. WB Saunders, Philadelphia

Stein BS (1975) The genital system. In: *Feline Medicine and Surgery 2nd edn*, ed. EJ Catcott, pp. 303–354. American Veterinary Publications, Santa Barbara, California

Wallen VN and Patterson DF (1986) Disorders of the sexual development in the dog. In: *Current Therapy in Theriogenology*, ed. DA Morrow, pp. 567–574. WB Saunders, Philadelphia

CHAPTER EIGHT

# The Infertile Male

*Laurence R.J. Keenan*

## INTRODUCTION

An 'infertile' dog is one that has failed to impregnate one or more fertile receptive bitches presented to him at the most favourable time during oestrus. Types of infertility may vary in severity and duration. The infertility may be complete and permanent, and the dog is then considered to be sterile.

However, infertility may also be incomplete, partial, or temporary and the dog may show a reduced capacity to breed characterized by low conception rates and/or small litter size. While the effects of the condition may be obvious as has been indicated, identification of causal factors, specific diagnoses and treatment of infertility may often be problematical.

Infertile dogs may essentially be divided into two types:

- Those unable to achieve normal mating
- Those unable to achieve normal fertilization.

## INABILITY TO ACHIEVE NORMAL MATING IN THE DOG

Dogs unable or unwilling to mate may suffer from lack of libido and/or physical defect(s). These may lead to lack of interest in mating or mating being disagreeable to the dog, or mating may be impossible to perform.

### Lack of libido

This may be defined as an individual's unwillingness to perform coitus, a process involving co-ordination of several behavioural, psychological and physiological steps, i.e. sexual interest and arousal, erection, mounting, intromission and ejaculation. Clinical signs vary from complete lack of sexual interest and inability to mate, to slowness or delay in exhibiting libido or attaining an erection, and an increasing inability or disinterest in mounting the bitch and achieving intromission. The condition may be *inherent*, manifesting itself at the first time of attempted breeding in an unproved dog or *acquired* in a proven dog, when it usually develops slowly. In the latter instance, loss of libido may be related to destruction of the Leydig cells by an extraneous agent.

Although sexual desire may be determined largely genetically, environmental influences play an important role in modifying its expression. For this reason it is difficult to know the frequency of truly genetic causes of lack of libido. Rare congenital conditions such as pituitary dwarfism caused by a defect in the hypothalamic-pituitary-gonadal axis may result in low testosterone levels and poor libido. However, many cases of dogs with poor libido may not have a genetic basis. Dogs with an innate strong sex drive, however, require more severe physical insults, or more protracted adverse environmental influences, to affect their performance, than those with a weaker sex drive.

Some factors affecting stud performance include:

- Management practices such as attempted breeding with a bitch at the incorrect time of oestrus, overuse at stud, isolation from their female peers in rearing leading to timidity, breeding initially with inexperienced possibly aggressive females, 'correction' of sexual interest during racing or show career, and mating in inappropriate or alien surroundings
- Injuries sustained or a 'bad experience' at a previous breeding or attempted breeding, even though no physical injuries incurred at that mating remain
- Age: younger dogs may show poor libido due to delay in onset of puberty and therefore low levels of testosterone. A similar condition may occur in older dogs due to senility, overuse or disease
- Systemic disease including those which result in fever, anorexia and debility but also conditions such as hypoadrenocorticism or hypothyroidism
- Drugs: many are responsible for decreased testosterone production and consequently poor libido. They include the use and over-reliance on testosterone administered for correction of poor libido. This causes a negative feedback on the pituitary axis, causing a reduction in secretion of LH and consequently in the production of endogenous testosterone. Other drugs may decrease testosterone by either direct or indirect effect; these include glucocorticoids, oestrogens, progestogens, anabolic steroids, cimetidine and ketoconazole

- Poor nutrition may delay puberty while over-fat males may become lazy or their obesity may exacerbate other conditions which affect performance, e.g. arthritis.

## Physical defects

These conditions may arise from problems related to the dog or the bitch. Depending on the cause, they have a variable effect on the dog's subsequent reproductive performance. If correction is possible, many dogs will recover to have a normal reproductive life. In more chronic conditions, the affected dog will suffer or associate pain with attempted coitus. The prognosis is poor in these cases.

Problems encountered include the following:

### Inability to mount the bitch

This may result from painful conditions of parts of the body unrelated to the reproductive tract such as orthopaedic disease of the back, hips or stifle joints or painful prostatic disease.

### Failure to achieve intromission

Congenital defects of the penis or prepuce, such as persistent penile frenulum, penile hypoplasia or stenosis of the prepuce, may interfere with normal coitus. Unsound general conformation may make mating impossible in certain breeds e.g. English Bulldog.

Acquired defects, such as trauma, lacerations and neoplasms of the penis, and/or prepuce, may make mating difficult or impossible due to pain or inability to protrude the penis or achieve penetration.

Premature full attainment of erection may occur in overexcited young inexperienced dogs. Dogs normally only attain engorgement of the bulbus glandis after penetration of the vagina and if this occurs too early, intromission is impossible.

Persistent, or a thicker than normal hymen, or where an abnormal slope of the female pelvis ('up-and-over vagina'), may make penetration difficult. Failure to achieve the normal tie may also arise from constriction of the female tract due to failure of the bitch to relax during the mating process. This may result from frigidity or conformational defects but is more likely to occur when the time of mating is inappropriate.

# INABILITY OF THE DOG TO ACHIEVE NORMAL FERTILIZATION

These dogs have normal libido and do not appear to have any difficulty in mating. Infertility or reduced fertility arises because there is:

- Failure of or incomplete ejaculation
- Absence or reduction in the number and quality of spermatozoa
- Abnormal seminal plasma.

## Failure of or incomplete ejaculation

This may occur where there is an inadequate tie, as aforementioned. It may also arise where there is fright or discomfort during mating or semen collection. Retrograde ejaculation may occur where there is a disorder of the sympathetic nervous system or an incompetence in the internal urethral sphincter muscle.

## Absence or reduction in the number or quality of spermatozoa

One or more of the following abnormalities may be encountered in semen specimens from affected dogs:

- *Azoospermia:* An absence of spermatozoa in the ejaculate in dogs with an otherwise normal ejaculate

- *Oligozoospermia:* A lower number of spermatozoa in the ejaculate than in the ejaculate of a normal dog of similar size

- *Asthenozoospermia:* An ejaculate that contains a high percentage of spermatozoa that do not possess normal progressive motility

- *Teratozoospermia:* An ejaculate that contains an abnormally high percentage of spermatozoa with morphological defects.

These conditions may result from *congenital* or *acquired* factors.

*Congenital defects:* These may be related to the gonads themselves, such as testicular hypoplasia, or result from abnormal development of the reproductive system.

Testicular hypoplasia is a hereditary or congenital disorder, resulting from a lack of, or marked reduction in, the number of spermatogonia in the gonads. To avoid confusion with poor sperm function associated with immaturity, diagnosis of this condition should not be made until the dog has attained full maturity. The condition is usually bilateral but may be unilateral. Because of reduction in the number of seminiferous tubules, the testicles of such dogs are considerably smaller than is average for the breed. Leydig cell function and libido are normal. Semen specimens are often watery and clear in appearance because of absent or severely reduced sperm numbers. Less severely affected dogs may be partially fertile but because of the hereditary nature of the condition they should not be used for breeding.

Abnormalities in spermatogenesis may have a genetic origin. Motility of sperm may be affected, as in the immotile cilia syndrome (Kartegener's syndrome), or specific sperm abnormalities may occur, such as double tails, coiled tails, double heads and retained proximal droplets. Chromosomal abnormalities include the XXY syndrome which is characterized by a phenotypic male with small soft hypoplastic testicles and complete aspermatogenesis. The XX syndrome, resulting in complete sterility, has been described in several breeds including Cocker Spaniel, Kerry Blue, Weimaraner and Short-haired German Pointer. Testicles may be undescended in affected dogs, and some have penile malformations. Congenital bilateral anomalies of epididymides and ducti deferentes will result in azoospermia and possibly the development of sperm granuloma and spermatocoeles. Oligozoospermia and reduced fertility may be present in dogs with unilateral cryptorchidism or unilateral segmental aplasia of the duct system.

*Acquired defects:* These result in some degree of testicular degeneration or atrophy. Depending on the cause, duration and severity, this disorder will result in either incomplete or complete shutdown in spermatogenesis or production of abnormal spermatozoa. Dogs with infertility due to acquired defects of the reproductive tract have, in general, been previously fertile. They may then have suffered some insult to the reproductive system which changed the characteristics of the semen to such an extent that they are now either incapable of producing offspring, or have a reduced capacity to do so. Conditions giving rise to such a clinical picture may have an acute onset or develop slowly. Severely affected animals will show azoospermia, and an increased percentage of poorly motile and abnormal sperm in their ejaculates. The type of abnormalities noted include head, mid-piece and tail defects. Some animals will recover from the original gonadal dysfunction and, in time, recommence the formation of sperm. Unfortunately, and more frequently, the condition will be progressive and permanent, and such animals are, or in time will become, completely sterile. Affected animals have variable clinical signs. In acute painful conditions, such as orchitis, the animal is systemically ill. More often affected animals are symptomless, apart from the local changes in the testicles and the ensuing failure of reproductive function. Testicles of affected dogs vary from being soft and flabby, in protracted chronic cases, to hard and shrunken in males where evidence of a deterioration due to age is evident. In some cases of infection or acute trauma, the testicles are swollen and painful. Neoplastic changes may result in enlargement of the affected testicle, this condition is usually unilateral. In such animals, the opposite testicle may be normal, or may show some degree of degeneration.

Factors associated with the acquired infertility include:

- Infectious agents producing azoospermia by directly affecting the testicle, causing orchitis and/or epididymitis. Severe inflammation may result in blockages to the ductal system. Local damage due to infections may interfere with the blood–testicle barrier and result in the immune system producing antibodies against sperm antigens. This immune-mediated disease may cause fibrosis, sperm granulomas and obstruction in the duct system. Possible aetiological agents include non-specific bacteria, *Brucella canis* and genital mycoplasms. Any increase in testicular temperature will have adverse effects on sperm production, the duration depending on the severity of the insult. Examples include high systemic fever, scrotal dermatitis, orchitis in the other testicle and possibly excessively high environmental temperatures
- Local trauma, dog bites, lacerations, kicks or blows may affect local testicular temperature, or be responsible for disruption of the blood–testicle barrier resulting in immune-mediated disease
- Drugs and environmental toxins. These may act on the germinal epithelium directly affecting spermatogenesis. Examples include anti-neoplastic agents (e.g cyclophosphamide, chlorambucil, cisplastin), excessive environmental, diagnostic or therapeutic radiation. Other agents may have an indirect effect, resulting in failure of spermatogenesis by interfering with the hypothalamic–gonadal axis (e.g. androgens, anabolic steroids, glucocorticoids)
- Neoplasms of the testicle, which may reduce sperm production by invasion and destruction of normal tissue and also by producing steroids with a negative feedback effect on the hypothalamic–pituitary axis, and subsequent aspermatogenesis.

## Abnormal seminal plasma

Abnormal seminal plasma constituents may originate from parts of the body unrelated to the reproductive system such as the urinary system, or they may be present in the semen as a result of trauma, disease or infection to the penis, prepuce, urethra or accessory glands.

The significance and relationship of isolated abnormalities in the seminal plasma to the development of infertility in the dog is unknown.

The semen should be examined for the following abnormalities:

### Reduced overall volume

Prostatic secretion may be reduced due to chronic prostatic hyperplasia. However, as in other species, volume of semen is unlikely to be significantly correlated with fertility.

### Abnormal colour

Yellow or green discoloration indicates contamination with urine or inflammatory exudate. Brown discoloration is indicative of haemolysed blood and may be present in prostatic disease. Red colour indicates fresh blood and may be present due to trauma to the penis during ejaculation or collection of semen. It may also originate from the prostate and suggest prostatic disease. Both urine and fresh blood are spermicidal and may influence fertility.

**High white cell count**
White blood cell counts greater than 2000/ml indicate infection (see below).

**Infection**
The presence of micro-organisms should be interpreted together with the clinical signs before attaching too much significance to the finding. Infections can occur in the urinary tract, the epididymis, vas deferens or the prostate. The significance of bacterial infection in causing infertility is a matter of some debate but, in general it may cause oligozoospermia, teratozoospermia and asthenozoospermia, and may result in the production of anti-sperm antibodies.

Most normal dogs have a variety of microflora present in the prepuce and urethra. These include *Pasteurella, Streptocccus, Staphylococcus, Escherichia coli, Mycoplasma* and *Ureaplasma* species. These latter organisms may cause an opportunistic infection in the more distal portions of the reproductive tract. The presence of anaerobic bacteria in high numbers strongly indicates infection, while Gram-positive bacteria probably indicate distal urethral contamination.

In general, a total culture count of >10,000 bacteria per millilitre of ejaculate, particularly if accompanied by clinical signs, is significant and suggestive of disease in the urinary or reproductive tract.

Localization of the infection may be difficult. A positive diagnosis of reproductive tract infection may be made if a high white cell count in the semen, particularly the prostatic portion of the ejaculate, is accompanied by a concomitant negative urine culture taken by cystocentesis. Unfortunately, cross infections can occur between the tracts and complicate the diagnosis. Hence the importance of clinical signs.

Bacterial prostatitis may be acute or chronic in nature. Such conditions should be distinguished from benign prostatic hyperplasia, prostatic cysts and prostatic abscesses. Dogs with acute bacterial prostatitis are unlikely to present as infertility cases. Such dogs may be febrile, anorexic or depressed or show signs of caudal abdominal pain on digital rectal examination. Diagnosis is made on the basis of clinical, laboratory and ultrasonographic findings. Chronic bacterial prostatitis may be symptomless, although it may follow acute episodes of the condition. Affected dogs may show, or have a history of, haematuria, tenesmus, constipation or recurrent urinary tract infections. Diagnosis is made from the history, clinical signs and cytological evaluation and bacteriological culture of the prostatic fluid. Attempts should be made to distinguish prostatitis from urinary tract infection.

## INVESTIGATION OF INFERTILITY IN THE DOG

The infertile male dog presents special problems for the attendant clinician. Inadequate or insufficient investigation of such cases exacerbates the problem and may result in refuge being sought by the clinician in obscure hormonal regimens which have no rational basis. While costs, lack of facilities and expertise available are limiting factors, clinicians should strive to follow a logical sequence of investigation. This entails the collection of a full clinical history from the client and a complete clinical and reproductive tract examination.

### History
A complete previous and current medical and reproductive history should be obtained. This will enable the clinician to define the extent and duration of the problem. Information should be sought in relation to the breeding record and health status of the dog. Questions which may be posed include:

- How many bitches has the dog mated? To exclude individual bitch infertility, the number of bitches should be as large as possible
- Over what period did these matings take place? When was the last successful mating?
- How many mated bitches became pregnant and what was the average litter size? Were any defects noted in offspring?
- Was pregnancy diagnosis carried out on the bitches that did not whelp?
- Has there been any change noted in the conception rate, litter size or lengths of gestation in mated bitches?
- How was the time of previous matings decided? Was it based on bitch receptivity, or on veterinary assessment of the bitch, using appropriate clinical protocols, or was the date of mating decided by the stud and/or bitch owner using imprecise arbitrary methods based on fashion or breeder's preference? Was there a limitation on the number of matings to each oestrous bitch?
- What was the subsequent fertility of bitches that did not conceive? Did they later reproduce normally or develop uterine problems such as cystic hyperplasia which would have rendered them infertile at the time of mating?
- Has the breeding behaviour and libido been normal and unchanged? If not, were any medication or drugs used in an attempt to alleviate the problem?

- Has there been previous or present illness, accidents or injuries sustained, particularly in relation to any part of the reproductive tract?

## Clinical examination

A complete physical examination should be carried out to exclude systemic disease directly or indirectly affecting reproductive function. However, it is the examination of the reproductive system which is of primary importance. Some or all of the following steps and procedures may be carried out in suspected infertility cases. Although all are desirable, the circumstances of the case, the cost, and the facilities available often dictate the extent of investigation.

### Inspection and palpation of genital organs

The scrotum is examined for testicular content and for evidence of recent or previous injuries, dermatitis or adhesions to the underlying testicle.

The testicles, which should be freely movable within the scrotum, are examined for size, symmetry and consistency. Size should be approximately average for the breed, and one testicle should not be significantly larger than the other. Consistency should be 'plum-like', or similar to the consistency of the 'webbing' between a stretched thumb and index finger. Small, excessively hard, or soft 'flabby' testicles indicate testicular hypoplasia or degeneration. Enlargement or the presence of nodules indicates inflammatory or neoplastic changes.

The epididymides may be palpated at the dorsolateral aspect of the testicles. While areas of swelling, nodules or absence may indicate inflammation of the epididymis, granulomas or segmental aplasia, negative findings do not exclude epididymal pathology since significant non-palpable changes may have occurred.

To examine the prostate, the index finger of one hand is used per rectum to locate the symmetrically bilobed spongy structure, which is palpable cranial to the pelvis. The other hand also locates the prostate by abdominal palpation, and pushes the organ towards the rectally located index finger by trans-abdominal pressure.

The penis and prepuce should be the last of the genital organs to be examined before an attempt is made to collect semen, as manipulation may cause the dog to have an erection. They should be examined for evidence of trauma, infection, abnormal discharges, neoplasms or congenital defects. Particular attention should be paid to the ability of the dog to extrude the penis from the prepuce.

## Collection of semen

Collection should be carried out in quiet surroundings with the dog on a non-slip surface. The owner should be requested to bring a teaser bitch which is in oestrus. While not essential with every dog, it does make collection more certain, and ensures a good specimen. In some circumstances, so as not to cause disappointment, it is prudent to warn the owner against travelling long distances until a suitable bitch is available. In the absence of a suitable bitch, it may still be possible to collect semen, especially from experienced stud dogs, but younger less mature dogs, or dogs with poor libido, may present a problem. Some experienced studs are only interested when an oestrous bitch is presented.

Less effective stimulation may be afforded by the exposure of the dog to swabs impregnated with either the chemical pheromone *p*-hydroxybenzoic acid methyl ester, or the vaginal discharges of a previously available pro-oestrous or oestrous bitch. The swabs taken from the latter may be stored for future use at -20°C. At the appropriate time, one is thawed out and applied to the vulva of a substitute non-oestrous bitch prior to attempted semen collection.

The donor dog is stimulated by the bitch, if available, by allowing him to investigate, smell and lick the vulva. At the same time the operator massages the penis through the prepuce to cause an erection. Before full erection is attained, the prepuce is gently slipped back over the penis caudal to the engorged bulbus glandis. Moderate pressure, using both thumbs, is applied behind the bulbus glandis. If possible, touching the exposed glans penis should be avoided. The ejaculate is collected directly into a warm plastic, or glass, beaker, best held by an assistant. Alternatively, the dog may be allowed to ejaculate into a graduated glass tube, through a latex cone or special canine artificial vagina applied after extrusion of the erect penis. The collection vessel should be covered with an insulated jacket to avoid heat loss. Minimal use of lubricants and exposure of the semen to latex, is desirable.

The dog ejaculates in three fractions. The first is a clear pre-sperm fraction, during which the dog usually, but not always, makes vigorous pelvic thrusting movements. A dog that is slow to ejaculate may be encouraged by gently touching the exquisitely sensitive tip of the penis against the collection vessel. The second cloudy sperm-rich fraction follows, or accompanies the pelvic thrusts. Following this stage, the dog may attempt to turn around as happens during normal mating. To assist the dog, the penis can be rotated 180° in a caudal direction and the ejaculation allowed to proceed. The third fraction is the clear prostatic fraction. This is normally not collected with the semen sample, but a sample may be collected separately for culture. The third fraction is identified by observing individual clear drops being ejaculated every 1-2 seconds. Once this is observed, the dog has ejaculated all the spermatozoa that are going to be produced.

### Evaluation of semen

***Sperm motility:*** The beaker containing the semen sample should be quickly transferred to a heated water bath, or a drop applied to a warm clean glass slide, covered with a cover slip, and immediately assessed under the x100 lens of a microscope for forward progressive motility. A 1:1 dilution of semen may be made with 0.9% citrate- or phosphate-buffered saline at 37°C to allow assessment of motility in concentrated samples. In some ejaculates, if the first and second fractions are collected together, the nature of the first fraction appears to inhibit sperm motility, because following dilution, the progressive motility is markedly improved.

***Sperm concentration:*** Although a complete examination will include a total sperm count, it is usually obvious from the initial examination of the semen drop whether the sperm concentration is low. The concentration may be quite variable in any one sample depending on the amount of prostatic fraction included. Many investigators attempt to collect only the second fraction by waiting until the thrusting has finished and stopping once the third fraction is seen. However, for a variety of reasons, including inexperience, the first and second fraction may be collected together. Obviously, this will mean that the concentration of the ejaculate will vary according to the method used. For fertility assessment, the essential parameter is the total number of sperm in the ejaculate which is easily calculated by multiplying the number in the drop by the total volume. The same result will be obtained whichever collection method is used.

Precise counts can be carried out in practice using a 1/100 white blood cell dilution kit and Neubauer haemocytometer. A 20 μl aliquot of diluted semen (diluted according to kit directions) is loaded into the haemocytometer chamber. The number of cells in the central square millimetre x $10^6$ x semen volume is the number of spermatozoa per ejaculate. The daily sperm output may be determined by collecting daily for 5 days until the sperm numbers are stabilized, i.e. when the extragonadal reserves are depleted. In the normal fertile dog, this averages 400 x $10^6$ spermatozoa over 5 days.

***Sperm morphology:*** This may be evaluated after making a smear of the semen sample, similar to a blood smear. After drying, the smear is stained for 5 minutes using eosin-nigrosin, aniline blue or crystal violet stain and examined under x1000 magnification with oil immersion. At least 100 sperm cells should be evaluated for normality, or the presence of sperm defects.

***White cell count:*** The number of white blood cells (WBCs) per microlitre may be calculated by counting the number of WBCs in the four large corner squares of the haemocytometer and multiplying by 250. If infection is suspected, a differential white cell count should be performed.

| Parameter | Measurement |
|---|---|
| pH | 6.3–6.7 |
| Volume: | 1–30 ml |
| 1st fraction | 1–12 ml |
| 2nd fraction | 1–2 ml |
| 3rd fraction | up to 20 ml |
| Progressive motility | >70% |
| Normal sperm morphology | >80% |
| Abnormal sperm morphology | <20% |
| Sperm count per ejaculate | >200 x $10^6$ |
| WBCs | <2000/ml |
| Alkaline phosphatase (IU/l) | 5000–40,000 |

***Table 8.1:*** *Normal parameters for canine ejaculate.*

***Semen cultures:*** Cultures of total semen and prostatic portions of the ejaculate should be carried out, especially if ejaculate cytology suggests infection.

***Volume and pH of the semen sample:*** These should be determined.

***Measurement of semen alkaline phosphatase:*** ALP is present in normal dog semen and originates from the epididymides. Low concentration or absence indicates incomplete ejaculation or bilateral obstruction of the epididymides or vas deferentia.

The normal parameters for the canine ejaculate are shown in Table 8.1.

Minor deviations from normal do not mean that the dog is infertile. Sometimes, a less than satisfactory result in one parameter is often offset by a corresponding excellence in another, e.g. a dog with a semen sample with less than the normal total number of spermatozoa may be acceptable for at least reduced stud usage if the spermatozoa have excellent motility. In previously fertile dogs, or where specific diagnosis is unclear, repeated semen evaluation at a later date is prudent and essential. This is usually carried out some 2 months after the initial examination, which allows time for renewed spermatogenesis to take place.

**Endocrinological evaluation**

Since many hormones are secreted in an episodic manner, caution should be used in assessing single serum samples. Many of the conditions that might warrant such investigation are untreatable, so the cost of testing should be considered. Moreover, delays and extra expense may be incurred by non-specialized laboratories having to set up these assays specially.

The following protocols may be helpful in establishing the source of suspected endocrine dysfunction.

***Testosterone production:*** The most reliable method of assessing testosterone production is the stimulation test using human chorionic gonadotrophin (hCG) (44 IU/kg, i.m.) or gonadotrophin-releasing hormone (GnRH) (2.0 μg/kg, i.m.). Serum testosterone concentration should be measured before and 4 hours after the administration of hCG or before and one hour after the administration of GnRH. Normal post-stimulation testosterone values range from 3.7 to 7.5 ng/ml.

***Serum gonadotrophins:*** Because of the episodic nature of their secretion at least three resting samples at 20–30 minute intervals are recommended for luteinizing hormone (LH) and follicle stimulating hormone (FSH) estimations. Normal values range from 34 to 85 and 73 to 84 ng/ml, respectively. These assays may not be commercially available.

***Interpretation of endocrine values:*** Generally, low serum testosterone and LH concentrations indicate a possible hypothalamic, pituitary or testicular dysfunction. Persistent low LH values following challenge testing by GnRH, 2.0 μg/kg, given i.m., suggest pituitary or hypothalamic involvement. However, a rise in LH following such administration, indicates normal pituitary/hypothalamic function and localizes the problem to the testicles. Low testosterone and elevated LH indicate primary Leydig cell dysfunction. Concentrations of FSH may rise in the serum of dogs with gonadal dysfunction, probably reflecting reduced inhibin feedback from the abnormal gonads with the elevation correlating to the severity of the altered spermatogenesis.

In the absence of other changes, low LH levels should be interpreted cautiously since values are low in normal dogs. They may also be lowered by certain drug therapies, e.g. steroids, anaesthetics and sedatives.

**Radiography**

This procedure is carried out on infertile dogs suspected of having prostatic disease. Survey radiographs of the caudal abdomen may indicate enlargement or irregularity of the gland. The significance of the findings, in relation to the effect on fertility, will obviously depend on the overall history and age of the dog.

**Ultrasonography**

This procedure is principally used in evaluation of the prostate but is being used increasingly to investigate testicular pathology. The prostate and testicles are best evaluated in the sagittal and transverse planes using a 5 or preferably 7.5 MHz real-time sector scanner. An enema should be administered prior to scanning the prostate, to eliminate colonic contents which may mimic peripheral prostatic disease. Conditions such as cysts or abscesses are visualized clearly. Other less distinct but echogenically complex areas may indicate neoplasia or areas of infection within the gland. Ultrasonography of the testicles may reveal non-palpable neoplasms, abscesses or areas of cavitation.

**Testicular biopsy**

This procedure may be helpful with diagnosis and prognosis in valuable dogs with persistent azoospermic or oligozoospermic semen. It should not be carried out in fertile dogs, as it may lead to a decrease in semen quality due to the local inflammation induced by the procedure. Specimens can be obtained by either incision or needle biopsy.

***Incision biopsy:*** A small incision is made, under general anaesthesia, through the skin and tunica albuginea of the testicle, which is located and immobilized as for castration. The small wedge of tissue, containing a piece of the seminiferous tubules which normally bulge from the incision site, can be 'scooped' cleanly away by using a fine blade. The specimen should be placed in a preservative solution, such as Bouin's or formol saline prior to examination by a veterinary pathologist.

*Needle biopsy:* This procedure can be performed under deep sedation using a fine needle directed posteriorly and diagonally within the ventral hemisphere of the testicle. Atrophy of the testicle can occur following needle biopsy but usually the surrounding tissues remain normal.

## THERAPY AND MANAGEMENT OF INFERTILITY IN THE DOG

Many cases of infertility may be transient if the original cause of the condition did not have a severe effect, e.g. short febrile episodes, mild scrotal dermatitis, minimal physical and psychological stress, some drugs or toxins. Recovery from such insults, with return to normal or near normal fertility, may take as long as 2–6 months. This is because not only does spermatogenesis take approximately 62 days and epididymal transport 15 days but it may also require multiple generations of sperm to be formed in order to raise very low sperm counts. It is essential, therefore, that a number of examinations at intervals of 2 months are made before condemning a previously successful stud as sterile.

### Poor libido and difficulty in mating

Treatment is difficult and often unsuccessful (Figure 8.1). Management practices which may have contributed to the development of the condition should be investigated and corrected. Mating with an experienced gentle bitch should be encouraged in a suitable and unthreatening environment. A shy or timid breeder may be encouraged to perform if he observes an unaffected male mating a bitch. Patience, time and constant re-enforcement and mating with a quiet mature bitch may help to return to stud a dog that has suffered injury at an earlier mating. Diagnosis and successful treatment of underlying systemic diseases may result in a dog regain-

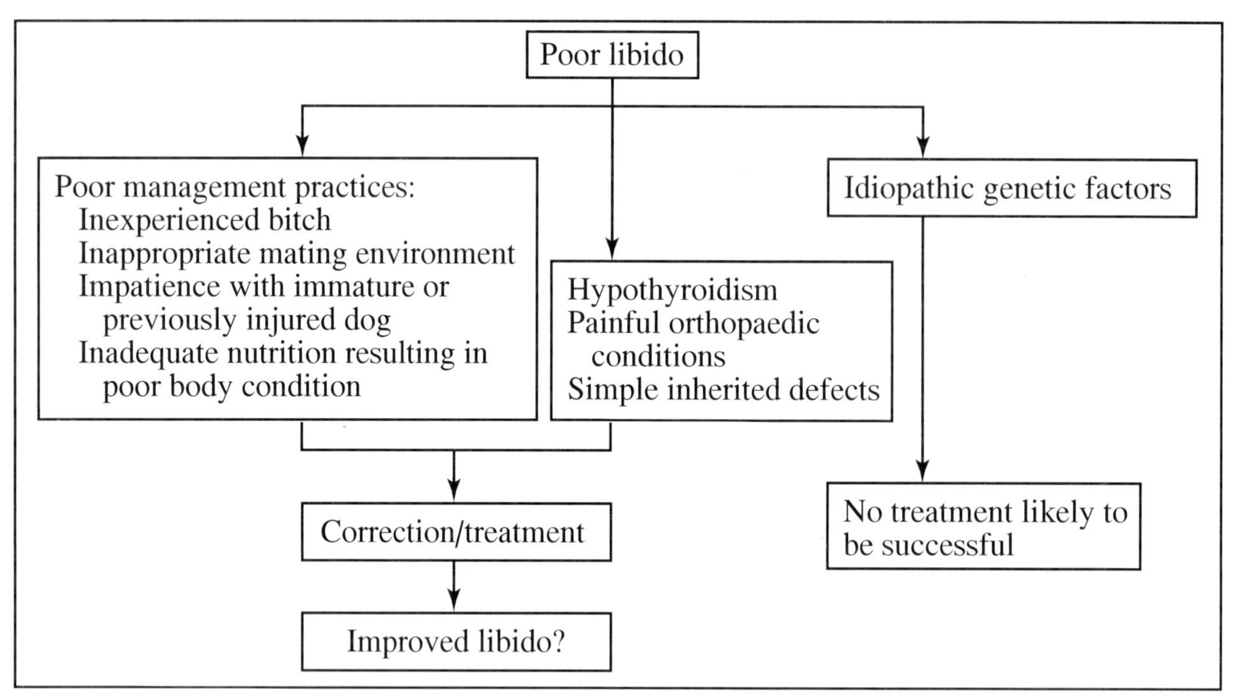


*Figure 8.1: Summary of therapy and management of dogs showing poor libido.*

ing normal libido. Thyroid gland function should be evaluated to exclude hypothyroidism as a factor. Non-steroidal anti-inflammatory drugs in dogs with painful orthopaedic conditions may allow limited use of such dogs. Similarly, simple correction of acquired or congenital defects of the genitalia, such as phimosis of the prepuce or persistent frenulum, may be successful. Stud dogs should be fed a digestible, high nutrient quality, low residue diet to ensure good body condition. Successful treatment of idiopathic or genetically based cases of poor libido is unlikely to be successful and treatment is unwarranted. Testosterone concentrations are not commonly low in affected dogs and treatment using testosterone is contraindicated. Its administration will cause negative feedback to the pituitary gland, causing decreased luteinizing hormone release followed by a decrease in endogenous testosterone production. Protracted use will also temporarily effect spermatogenesis and result in increased number of sperm abnormalities. The author has used GnRH 2.0 μg/kg, i.m., or hCG 500 IU s.c., twice weekly for 3 months, to stimulate Leydig cell function and thus enhance endogenous testosterone in situations where it is suspected that clinical requirements for testosterone are higher than normal. Such treatments are speculative and results equivocal. Semen may be collected from valuable dogs by appropriate methods and bitches artificially inseminated. However, such practices may favour perpetuation of the condition in offspring and are therefore considered unethical.

## Dogs with normal semen analysis

This is summarized in Figure 8.2.

Individual bitch problems and poor breeding management should be excluded by mating such dogs with a bitch of excellent past fertility. Changes occurring in the vaginal mucosa during pro-oestrus and oestrus should be observed in the bitch using a speculum; vaginal smears should be taken and progesterone concentration assessed to determine the optimal time for breeding (see other chapters). Pregnancy and litter size should be confirmed at 25–28 days post ovulation.

If the mating is unsuccessful, slides and formalin-fixed samples of semen should be sent to a specialized institution for evaluation and electron microscopy to identify sperm defects that are not obvious but may preclude fertility. Similarly, definitive diagnosis of chromosomal abnormalities may only be made by submission of a blood sample to a laboratory for karyotyping of peripheral blood lymphocytes.

Surgical intrauterine insemination with half the ejaculate placed at the tip of each horn at the appropriate time in oestrus may improve fertility in dogs with poor sperm longevity and in dogs where anti-sperm antibodies are being produced.

Immunosuppressive doses of glucocorticoids to suppress anti-sperm antibody production have been used in other species but are untested in the dog. Apart from other side-effects, such agents may decrease sperm production.

## Dogs with poor sperm and semen quality

The prognosis for recovery in such dogs is often poor (Figure 8.3). Treatment and management protocols are directed at reducing the infertility to the point where the stud dog may be used to a limited degree. Some treatments may not be practicable due to lack of facilities/expertise or because of expense. Alternatively, the client may wish to implement all of the recommendations described below.

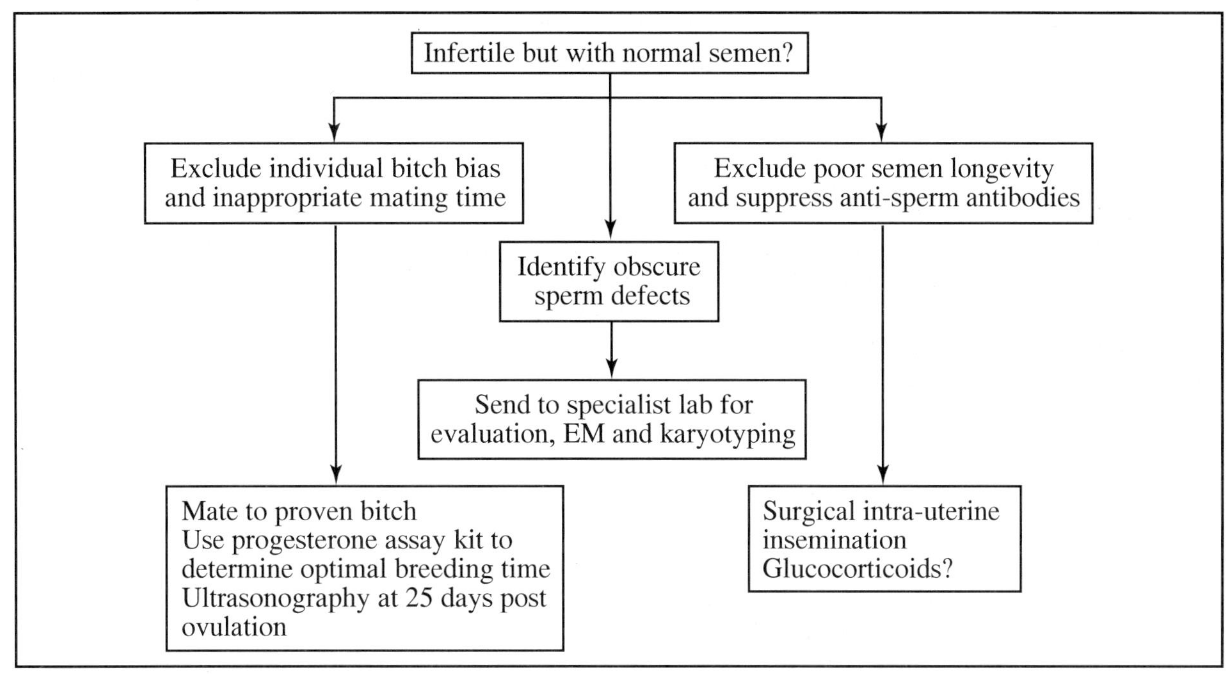


***Figure 8.2**: Summary of therapy and management of infertile dogs with apparently normal serum.*

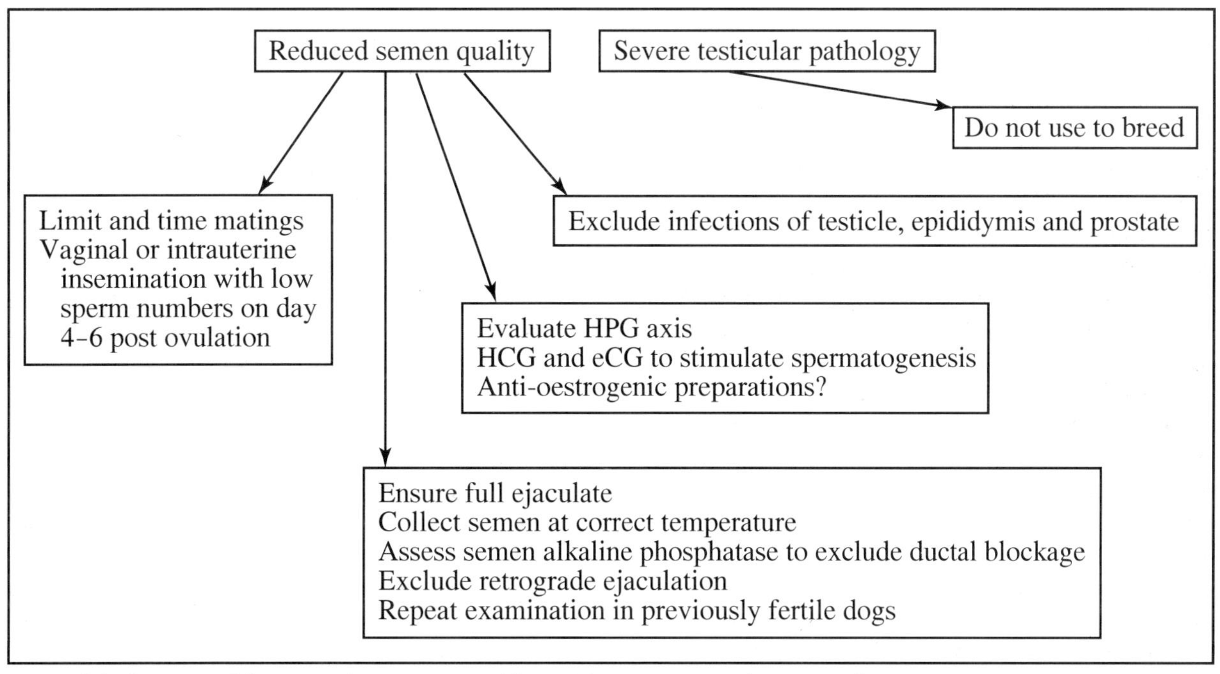


***Figure 8.3**: Summary of therapy and management of dogs with poor sperm and semen quality.*

**Azoospermia oligozoospermia and teratozoospermia**

1. Ensure a full ejaculate has been collected and repeat examinations, particularly in previously fertile dogs.
2. After collection, take a urine sample to eliminate the possibility that retrograde ejaculation into the bladder has occurred.
3. Assess the levels of alkaline phosphatase in several ejaculates to distinguish ductal blockage e.g. bilateral sperm granulomas (<10,000 IU/l) from incomplete ejaculation.
4. Eliminate such causes as immaturity, testicular hypoplasia and pathological factors resulting in testicular degeneration and age-related gonadal failure.
5. Exclude infections of the reproductive tract as a potential cause using the protocols described above. Treat chronic bacterial orchitis or epididymitis and chronic bacterial prostatitis (CBP)with long-term (4–12 weeks) antibiotics. The latter condition is recalcitrant to treatment because of difficulties of antibiotic penetration. Antibiotics with low protein binding and high lipid solubility should be selected,

e.g. chloramphenicol and fluoroquinolones. Determination of the pH of the prostatic fluid (usually neutral or slightly acidic) is important prior to treatment since weak base antibiotics such as erythromycin and trimethoprim-sulpha will achieve a higher prostatic concentration.

6. Evaluate the hormonal-pituitary-gonadal (HPG) axis with emphasis on Leydig cell function because low testosterone levels may lead to morphological defects. There is no effective treatment for dogs with primary testicular failure evidenced by low testosterone and elevated LH and/or FSH, or for dogs with normal testosterone and LH, but high FSH. Dogs with low testosterone and low gonadotrophins should be checked for pituitary neoplasms. Where gonadotrophic function is low, resulting in spermatogenetic dysfunction, a recommended protocol is to give hCG at 500 IU s.c. twice weekly to stimulate Leydig cell function, and equine chorionic gonadotrophin (eCG) at 20 IU/kg s.c. three times weekly to stimulate spermatogenesis. Treatment should be continued for approximately 3 months.
7. Semen evaluations should be carried out at least one week prior to mating as oligozoospermic dogs may be made azoospermic by repeated matings. Anecdotal success has been claimed for the use of anti-oestrogen preparations such as clomiphene and tamoxifen in oligozoospermic men. Some improvement in sperm quality has been achieved in a small number of dogs using the synthetic androgen mesterolone at the dosage rate of 0.75-1.5 mg/kg.
8. Time and limit matings carefully with oligozoospermic dogs. Appropriately timed vaginal AI or surgical insemination in selected bitches may achieve success with low sperm numbers. Where sperm numbers are $>100 \times 10^6$ per ejaculate, vaginal AI may be successful if performed on days 4, 5 and 6 post-LH peak, as determined by progesterone assay. If this fails, or if normal sperm numbers are low but greater than $20 \times 10^6$ per ejaculate, surgical AI may be performed on day 5 post LH peak.
9. Epididymal aspirates containing viable spermatozoa in cases of bilateral sperm granulomas have been surgically inseminated in other species but this procedure has not been reported in the dog.

### Asthenozoospermia

1. Exclude and treat reproductive tract infections as necessary.
2. Exclude incomplete ejaculation.
3. Exclude urine contamination of the sample by ensuring the dog urinates prior to collection.
4. Ensure semen samples are collected and evaluated at the correct temperature to exclude cold shock.
5. If no specific cause is identified, the dog should be sexually rested and re-examined in 2 months.

## INVESTIGATION AND CAUSES OF INFERTILITY IN THE CAT

A medical and reproductive history should be obtained. In the cat this tends to be obscure, however the intention is to define the extent and duration of the problem as in the dog.

A general clinical examination and specific reproductive tract examination should be performed.

The testicles should be spherical, smooth, firm, bilaterally symmetrical in size and shape and non-painful. Evidence of inflammatory or traumatic changes should be noted and treated using general principles. Feline infectious peritonitis has been associated with orchitis and the development of testicular degeneration in the cat. Cryptorchidism occurs with greater frequency in pure bred cats, and as in other species, is associated with aspermatogenesis. In severe forms of orchitis and in cryptorchidism, castration is indicated. Testicular hypoplasia with associated azoospermia due to chromosomal disorders such as 38,XY/57,XXY (tortoiseshell and white colour) and 38,XXY have been occasionally reported in the cat.

The penis is enclosed in a free prepuce and is directed backwards. The os penis is ungrooved. The cranial two-thirds of the penis contain 100-200 cornified papillae which are believed to stimulate the female during mating and by so doing increase the number of ovulations. Diphallus (double glans penis) and persistent frenulum have been reported. Urethral obstruction is quite common in the cat and may be associated with haematoma formation. Unsuccessful mating may result in a ring of hair being caught around the glans penis. This may be removed by gently extruding the penis.

### Collection and evaluation of semen

#### Artificial vagina

Up to 20% of cats can be trained in 2 weeks to use the artificial vagina (AV). This may be made by attaching a 1-2 ml rubber bulb pipette to a 3 x 44 mm test tube, cutting the end off the rubber bulb and inserting the adapted apparatus into a 60 ml polyethylene bottle filled with water at 52°C. The rolled end of the bulb is stretched over the rim of the bottle for fixation. A teaser queen may be necessary to stimulate the tom, who is further encouraged by firm pressure on the dorsal part of the pelvic region. On erection, the AV is slipped on to the penis and ejaculation takes place in 1-4 minutes.

| Parameter | Measurement |
|---|---|
| pH | 7.0-7.9 |
| Volume: | |
| Artificial vagina | 0.02-0.12 ml |
| Electroejaculation | 0.233 ml |
| Sperm count: | |
| Artificial vagina | 13-143 x $10^6$ |
| Electroejaculation | 28 x $10^6$ |
| Motility | >80(%) |
| Abnormal spermatozoa | <10(%) |

***Table 8.2:*** *Some normal parameters of cat semen.*

### Electroejaculation

It is necessary to give a general anaesthetic prior to this procedure. A combination of xylazine 2 mg/kg and ketamine hydrochloride 10 mg/kg or metetomidine 80 μg/kg with ketamine hydrochloride 5 mg/kg, may be used. Atipamamezole may be used to accelerate recovery following collection. A lubricated Teflon rectal probe 10 x 12 cm, with three longitudinal stainless steel electrodes 5 cm each in length, is inserted into the rectum to a depth of some 6 cm. Stimuli of 2-8 V and 5-220 mA are applied in a series of 60 stimuli with an interval of 2 seconds between each stimuli.

The methods used to evaluate semen are the same as in the male dog. Normal parameters are shown in Table 8.2.

## Other causes of infertility

### Loss of libido

New or unfamiliar surroundings, restriction to a cage or an aggressive queen may result in a temporary or permanent loss of libido. Malnutrition or obesity may have a similar effect. Removal of the cause may result in return to normality but treatment of a congenital poor libido is not advised, since the condition is hereditary.

### Failure of mating

This may occur in spite of mounting and arousal in cases where hair accumulates around the glans penis as described earlier.

### Retrograde ejaculation

Retrograde ejaculation has been reported, but there is no known treatment.

## REFERENCES


Axner E, Strom B, Linde-Forsberg C, Gustavsson I, Lindblad, K and Wallgren M (1996) Reproductive disorders in 10 domestic male cats. *Journal of Small Animal Practice* **37**, 394-401

Burke TJ (1986) *Small Animal Reproduction and Fertility*. Lea and Febiger, Philadelphia

Christiansen IBJ (1984) *Reproduction in the Dog and Cat*. Baillière Tindall, London

Ellington JE (1994) Diagnosis, treatment and management of poor fertility in the stud dog. *Seminars in Veterinary Medicine and Surgery (Small Animal)* **9**, 46-53

England GCW (1996) Reproductive biology in the male dog. *The Veterinary Annual* **36**, 187-201

Feldman EC and Nelson RW (1996) *Canine and Feline Endocrinology and Reproduction, 2nd edition.* WB Saunders, Philadelphia

Johnston GR, Feeney DA, River B, and Walter PA (1991) Diagnostic imaging of the male canine reproductive organs: methods and limitations. *Veterinary Clinics of North America Small Animal Practice* **21**, 553-589

Johnston SD (1991) Performing a complete canine semen evaluation in a small animal hospital. *Veterinary Clinics of North America Small Animal Practice* **21**, 545-551

Meyers-Wallen VN (1991) Clinical approach to infertile male dogs with sperm in the ejaculate. *Veterinary Clinics of North America: Small Animal Practice* **21**, 609-633

Olson PN (1991) Clinical approach for evaluating dogs with azoospermia or aspermia. *Veterinary Clinics of North America: Small Animal Practice* **21**, 591-608

Root MV and Johnston SD (1994) Basics for a complete reproductive examination of the male dog. *Seminars in Veterinary Medicine and Surgery (Small Animal)* **9**, 41-45

Sakamoto Y, Matsumoto T, Mizunoe Y, Haraoka M, Sakumota M and Kumazawa J (1995) Testicular injury induces cell-mediated autoimmune response to testis. *Journal of Urology* **153**, 1316-1320

Wallace MS (1992) Infertility in the male dog. *Problems in Veterinary Medicine* **4**, 531-544.

CHAPTER NINE

# Mating and Artificial Insemination in the Dog

*Wenche Farstad*

## INTRODUCTION

Examination of the bitch and stud should be scheduled for about 3 weeks before mating. The stud, if he is proven and has been able to sire healthy, viable puppies in recent previous matings, may not need a fertility examination. However, if his last fertile breedings date back more than a year, or he is older than 7 years, a fertility examination should be carried out. This clinical examination should include a general health check, semen sample examination, visual inspection of the penis and prepuce, and palpation of the prostate, the testicles and epididymides. The bitch should undergo a thorough health check, should be vaccinated and treated for internal parasites. Her vulva should be examined visually, and the vestibule and vagina should be examined by palpation for the presence of vaginal strings (remnants of the hymen) or strictures. During oestrus, vaginal smears and/or tests for plasma progesterone should be carried out to assess the optimal time for breeding. This chapter is written on the assumption that these checks have been adequately considered and correctly undertaken.

## THE PSYCHOLOGY OF MATING

Before attempting to perform artificial insemination (AI) when there appears to be a mating problem, it is important both for the dog owner and the veterinarian to consider the physiological and psychological mechanisms of normal mating. The psychology of mating is not always understood, even among experienced breeders. The female should be brought before the stud at his kennels rather than bringing the male to the bitch's premises; this is particularly important when presenting a young male for his first mate. It is also preferable for the debutant male to mate with an experienced bitch, and, similarly, the maiden bitch should be mated with an experienced male.

There are subtle interactions between the bitch and the dog, individual likes and dislikes, which may cause either a stud dog or a bitch to refuse mating with a particular individual. Alternatively, the social status of the dog or bitch, the dominance factor, may be the cause of the problem. Young or subordinate males trying to mate a dominant bitch may show submissive behaviour, resulting in abnormal orientation when attempting to mount or during intromission. A very dominant, although oestrous female may discourage a subordinate male entirely from trying to mount her, but will readily mate with a more dominant stud. When presented with a mating problem with no obvious physical explanation, and where the timing in oestrus has not been inappropriate, such psychological factors should be considered.

## MATING BEHAVIOUR OF THE BITCH

There are often signs of a bitch approaching oestrus several weeks before pro-oestrus. She may tease or mount other dogs, both males and females. She urinates more frequently, she may lose or increase her appetite or lose concentration when performing tasks, such as obedience training. To the experienced owner, such behavioural changes can be a fairly reliable sign of an approaching heat.

During pro-oestrus, the bitch is usually more interested in interacting with other dogs. She is, however, usually passive when approached by the male, or she may snap or avoid him if he tries to sniff at her hindquarters. In late pro-oestrus her behaviour is more playful, and she will sniff the male, allow him to sniff her, and may often initiate a playful chase. At the onset of oestrus, the bitch will show true oestrous behaviour, i.e. she will stand still, wagging her tail and deviating it to one side ('flagging'). Some females may present their vulva by elevating it when sniffed by the male, and often contractions may be observed in the perineal and rectal muscles. The bitch often lowers her back (lordosis) if the dog puts a paw on her back. If the male is somewhat timid, she may back up towards him, poke him in the side with her nose, paw on his back or even mount him. When she is mounted, she will stand quietly, with her hind legs placed wide, vulva elevated and tail flagging.

## MATING BEHAVIOUR OF THE MALE

Young males who exhibit mounting and thrusting behaviour towards their littermates from a very early age, i.e. 4–5 weeks old, are normal. These activities are important for the male to learn proper mating behaviour. Also, mounting other dogs or objects in the house, within reasonable limits, assists in the development of sexual performance. Such mating behaviour, however, should be firmly discouraged if it is directed towards humans, especially children.

Pheromones passed in the vaginal discharge and urine from the bitch may attract dogs from a considerable distance, and the dog's pursuit is persistent when following a bitch on heat. When housed with females on heat, a dog may refuse to eat or drink for several days. Some dogs are also extremely vocal and howl incessantly for days. Some may try to escape by breaking through doors or windows, digging through their pen or jumping fences; when prevented from escaping, dogs may show destructive behaviour. Dogs with prostatic disease, such as cystic hyperplasia of the prostate gland, may start to bleed from the penis due to emptying of blood-filled prostatic cysts during the peak of an adjacent bitch's oestrus. The dog should if possible be moved for the time the bitch is on heat.

When a male encounters a bitch on heat, his body language usually indicates playfulness with a degree of caution, depending on his social status. He approaches the bitch with his tail wagging and ears erect, and usually avoids staring at the bitch. He may stand next to her if allowed to approach, and lick her ears and mouth. If encouraged, he will sniff her, and put a paw on her back or rest his head on her back. If a bitch is receptive, he will usually try to mount her after only a short foreplay. If she is reluctant, he may initiate play by lowering his front, lying down or inviting her to chase him.

## NORMAL MATING

This consists of six stages:

1. Mounting
2. Thrusting and intromission — initial erection
3. Erection — swelling of the bulb — release of first fraction: clear prostatic fluid
4. Ejaculation — release of second fraction: white sperm-rich fraction — rotation
5. Tie — sperm transport — release of third fraction: clear prostatic fluid
6. Break of the tie and dismount.

### Mounting

If the dog is allowed to mount the bitch (Figure 9.1), he will usually grasp her around her flanks and push himself forward and initiate thrusting movements with his pelvis. An experienced male will usually orientate his penis, with controlled movements, to locate the vulvar orifice. Erection is only sufficient to enable intromission with the support of the os penis. Contact with the vulva initiates the involuntary powerful thrusting movements.

*Figure 9.1: Mounting.*

### Thrusting and intromission

When the penis is in the vulva, the thrusting movements become more powerful and deeper. To accomplish this, the male usually dances from side to side (treading) behind the bitch with one hind leg above the ground, and his body partly elevated above her.

### Erection and swelling of the bulb

During this phase full erection is accomplished. The completion of the erection involves elongation of the glans penis, the bulbus glandis remaining fastened to the os penis and the *pars longa glandis* sliding forward over the os penis. The bulbus glandis undergoes swelling, which is a prerequisite for the tie or lock of the penis in the vulva. Erection is stimulated by the sight, sound and smell of an oestrous female. Erection is due to impulses from the *nervi erigentes* composed of parasympathetic fibres from the pelvic and sacral nerves. These nerve impulses lead to dilation of the external and internal pudendal arteries to the cavernous body of the penis, due to contraction of the ischiourethral muscles, and venous flow is prevented. Blood retained in the sinuses of the cavernous tissue of the bulbus glandis causes the bulbus to swell. The contraction of the bulbospongiosus and ischiocavernosus muscles of the penis and vulvar constriction during the tie also aid in the intensity of the erection. During this phase the first fraction of the ejaculate, which ranges from 1 to 2 ml of clear prostatic fluid, is released.

### Ejaculation and rotation

#### Ejaculation

When the penis is fully erect, the second, or sperm-rich semen fraction (1–2 ml), which is normally white in colour, is released. The release of the sperm-rich fraction takes 1–2 minutes. Ejaculation is due to stimulation of the sympathetic nerves of the penis. The semen and the prostatic fluid are expelled by peristaltic contractions in the muscles surrounding the urethra, particularly the bulbocavernosus and ischiocavernosus muscles. The thrusting movements normally stop at the time when ejaculation starts.

### Rotation

The dog lifts one leg and turns around so as to assume a tail-to-tail position (Figure 9.2). The tip of the penis remains in its original position, directed towards the cervical os of the female. The 180° turn is made possible by a twist of the penis just behind the bulbus. The penis is very elastic in this area, and the twisting does not seem to cause any discomfort. The penile bone of the dog prevents occlusion of the urethral opening during the erection and twist. It has been suggested that the rotation causes occlusion of the emissary vein of the glans, thus preventing detumescence. The dog does not have to turn. Some dogs prefer to stay parallel to the bitch, and others prefer to rest on the bitch's back. This may be uncomfortable for the bitch if the dog is heavy, or if tie is of long duration, so gently lifting the dog off the bitch may ease her discomfort.

## The tie

During the tie, the dogs usually stand quietly in a tail-to-tail position (Figure 9.3), and the last of the three fractions of the ejaculate, 5–20 ml of prostatic fluid, is released. Rhythmic contractions in the dog's perineal muscles along with pumping movements of the tail are observed.

The duration of the tie varies from 10 to 45 minutes, but the author has observed ties lasting up to one hour. The length of the tie does not seem to have any influence on whether conception results. Conception can occur even without a tie as long as the second (sperm-rich) fraction has been ejaculated, usually accomplished just after the dog has ceased thrusting. Some claim that no tie may reduce litter size; however, this is not the experience of the author.

An experienced bitch will stand firmly and quietly through mounting, intromission and tie. Occasionally the bitch may try to walk away and the male is pulled along with her. Usually this causes no harm to bitch or stud, but it is wise to hold the bitch by the collar to keep her quiet.

***Figure 9.2:*** *Rotation.*

***Figure 9.3:*** *The tie.*

## Breaking of the tie

Before the tie is broken, the swelling of the bulbus goes down, allowing the partners to move apart quietly. The male usually licks his penis, which then retracts into the prepuce. Hairs from the abdomen and prepuce can be trapped between the penis and the preputial orifice during the retraction, or the skin on the outer layer of the prepuce may be rolled inward. This may cause pain to the dog, and he may then require some assistance. Normally, full retraction can be achieved by pressing inwards on the prepuce about 2–3 cm behind the orifice, causing eversion of the preputial orifice.

After mating the female will usually lick her vulva vigorously because there is some discharge of semen at this time. Some bitches may show great excitement by jumping and playing or rolling around on the ground. There is no need to try to restrain the bitch, or refuse to allow her to urinate at this time, because most of the semen has already been transported into the uterus and oviducts during the tie.

# ABNORMAL MATING

## Mating problems due to the bitch

Mating problems may be caused by the bitch or the dog. Attempting to mate a bitch too early in oestrus, although she may accept mounting and attempts of intromission, often leads to difficulties because the vulva is too oedematous. This results in failure to accomplish the tie because the male is unable to thrust the penis far enough inside the vagina. Often, trying again the next day may prove successful, particularly with inexperienced animals, where assisting the dog by holding his hindquarters in position while guiding the penis into the vulva can be helpful.

Some bitches which are normally very good natured and placid may panic during intromission and become very aggressive towards the dog or even try to bite the owner. Such bitches should be calmed with an appropriate tranquillizer to avoid injury to the dog or to the people handling the bitch. Artificial insemination may be considered in such cases, although Kennel Club regulations regarding the registration of litters born as a result of AI should be considered.

Constriction in the annular muscles of the vagina either from a ring stricture, or because of a rigid hymen, will not permit the entry of the penis to its full length. A persistent hymen is a problem in some maiden bitches, and breaking the hymen with a gloved finger may be possible. Surgical removal of the hymen can be carried out, but the possibility of scar tissue formation must be considered, as this may lead to problems at parturition. Artificial insemination is the best alternative in cases of vaginal stricture or persistent hymen, provided there is an opening large enough for the insemination equipment to pass through. According to the author's experience this type of vaginal stricture does not result in dystocia at parturition.

Occasionally, bitches experience extreme swelling of the vaginal mucosa during oestrus, which results in prolapse of this hyperplastic tissue. Vaginal hyperplasia does not permit intromission, and very often prohibits even the use of insemination equipment. Conservative treatment with mild astringent and disinfectant solutions is recommended, whilst occasionally surgical removal of the protruding mucosa is necessary. An uncomplicated vaginal hyperplasia will usually recede after oestrus; there is a tendency, however, for a bitch to experience vaginal hyperplasia at subsequent oestrus. Mating of bitches prone to vaginal hyperplasia is therefore not recommended.

### Mating problems due to the dog

One of the more common mating problems of the dog is caused by inexperience. A novice male may start thrusting towards the anus or below the vulva, or even in the air because he is inexperienced in the clasping technique. Young males may be so excited that the bulbus swells before intromission, making intromission impossible. During this time, prostatic fluid is released, and some clear discharge may drip from the urethral orifice. Usually the sperm-rich fraction has not been ejaculated at this time. Removing the dog from the bitch for half an hour may calm him enough for a second try. If this is not successful, AI is the best alternative.

In rare cases, young males are not able to mate a bitch due to phimosis, i.e. adhesion between the prepuce and the penis, or because of a stricture of the preputial orifice. Surgical correction is necessary in both cases.

Absence of libido can be due to a variety of factors, such as dislike of the partner or incorrect mating time (psychological), testicular hypoplasia, prostatic problems, urethral calculi (rare), balanoposthitis, epididymiditis, vertebral arthrosis (common in Dachshunds, Pekingese), lower back problems (spondylosis in Boxers) or hind leg problems (hip dysplasia, stifle disease). A clinical examination should establish the cause of lack of libido. Sometimes with psychologically based libido problems, introduction to another female may solve the problem. If the dog becomes excited enough, he may then mate the other bitch, or allow semen to be collected for AI. Prostatic disease, urinary calculi, epididymiditis and balanoposthitis must be treated. If the libido problem is due to hereditary disease which may affect the dog's offspring, the ethical implications should be discussed with the owner, and probably the animal should not be bred.

## SEMEN COLLECTION AND PRESERVATION

### Semen collection

According to the author's experience, the best conditions for semen collection are the presence of a teaser bitch in oestrus (preferably at the peak of oestrus, although any stage may work as long as the bitch is held firmly so as not to cause injury to the dog), a quiet room with as few people as possible present, and access to outside runs if the dog prefers to play with the bitch before collection. The semen quality improves, and the speed of the whole procedure is greatly facilitated in these circumstances.

The dog is allowed to mount the teaser bitch, and as soon as the penis protrudes from the prepuce, the prepuce is manually deflected back and the bulbus glandis is held in a firm grip. The enlarged bulbus is held in the collector's hand, and the collector's fingers constrict at the point of torsion, caudal to the bulbus (Figure 9.4).

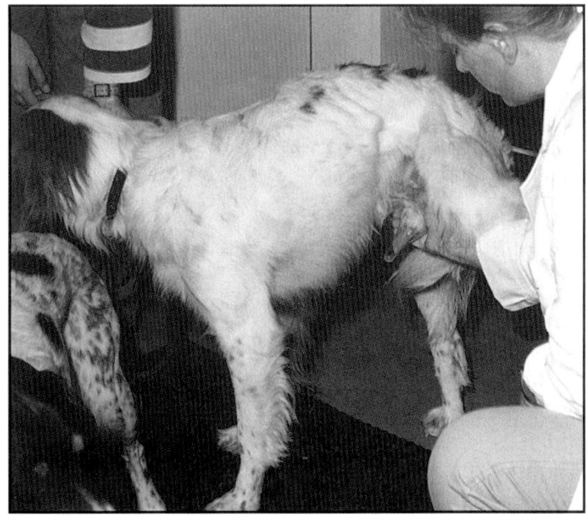

***Figure 9.4**: Semen collection using a teaser bitch. The prepuce is manually deflected back and the bulbus is held in a firm grip while the collector's fingers are constricting at the point of torsion. Semen may be collected directly into a disposable plastic funnel.*

Collection should not be attempted until full erection has occurred and the thrusting movements have ceased. The collecting device used by the author is a single-use, funnel-shaped plastic cup. If the dog is allowed to thrust his penis into this device, rupture of the small vessels on the surface of the corona may occur, and blood is mixed with the semen. Older dogs, especially, are prone to bleeding. It is not known whether blood is harmful to the spermatozoa, but the presence of blood makes semen evaluation extremely difficult. Where the ejaculate is well fractionated, only the sperm-rich fraction (second fraction) should be collected if the semen is to be frozen. Where the semen will be used immediately for artificial insemination, 1-2 ml of prostatic fluid can be allowed into the funnel, and the semen may be used directly for AI with no other extender.

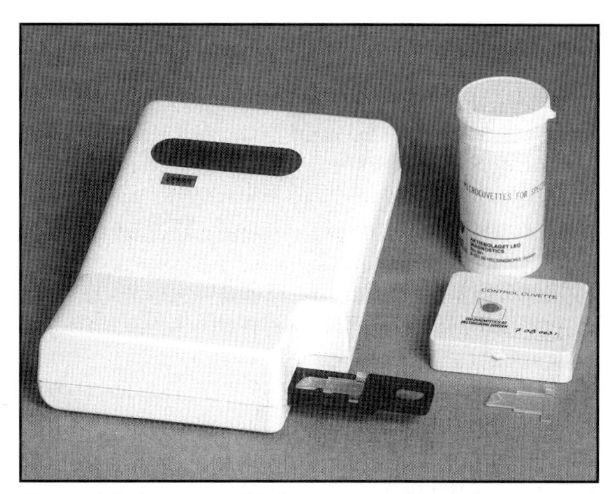

***Figure 9.5***: *Spermacue™ photometer calibrated to measure the density (spermatozoa/ml) in fresh undiluted dog and fox semen.*

## Semen examination

The semen should be examined for colour. Normal colour is from that of skimmed milk to slightly cream coloured, depending on density; never yellow (indicating contamination with urine or inflammatory cells) or red (blood tinged). If the colour is abnormal, the cause should be established by microscopic evaluation. Semen with leucocytes, erythrocytes or urine may be washed by centrifugation at 300-1000 rpm for 5 minutes in TRIS extender (Table 9.1) or saline with antibiotics, the supernatant discarded and the pellet resuspended in fresh extender with antibiotics. The semen may then be used for AI.

| Trishydroxymethyl aminomethane | 6.056 g |
|---|---|
| Citric acid | 3.400 g |
| Fructose | 2.500 g |
| Double-distilled water | 200 ml |
| Crystalline penicillin | 200 000 IU |
| Dihydrostreptomycin | 0.2 g |
| Add egg yolk just prior to dilution with water (20% v/v): 2 ml egg yolk to 8 ml of TRIS base + antibiotics | |

***Table 9.1***: *TRIS extender without glycerol for transport of fresh chilled semen. Buffer base with antibiotics.*

Assessment of semen density, i.e. spermatozoal concentration, can be made by using a counting chamber and a microscope or by using photometers calibrated for dog semen, such as the one in Figure 9.5 which is used in the author's laboratory. From the volume and concentration, the total number of spermatozoa is calculated. The total number of spermatozoa in dog semen varies from 200 to 1200 x $10^6$, with a second fraction concentration of 100-700 x $10^6$/ml.

Progressive motility (PM) is usually estimated by visual inspection with the microscope at x100 magnification. Progressive motility (number of spermatozoa with a forward progression) is estimated to the nearest 5% and values are normally from 75% to 90% in undiluted freshly collected semen. If the semen has been frozen, progressive motility post-thaw should have a minimum value of 50% if the semen is to be used for AI. The speed of forward progression (velocity) is also important. This can be subjectively scored on a scale from 0 to 5, where 0 denotes no forward progression (necrospermia) and 5 the most rapid forward movement. Normal fertile dog semen usually has a grade between 3 and 5, and semen used for cryopreservation should show fast forward progression, i.e. preferably grade 5 prior to freezing. Low temperature reduces sperm motility, and the use of a warming plate for the microscope or use of prewarmed slides is recommended, particularly when frozen-thawed semen is examined. Extenders, especially those containing glycerol, also depress motility temporarily; however, motility is usually restored in the semen after 2-3 minutes on the warming slide.

Recently, a test for membrane integrity and viability of canine spermatozoa, called the hypo-osmotic swelling test (HOST), has been evaluated. This test involves incubation of spermatozoa in hypo-osmotic media which results in swelling of the tail of spermatozoa. It has been found to correlate well with gross and progressive motility and viability, but not with concentration.

Morphology (M) is given as the percentage of spermatozoa with normal appearance and is evaluated at x200 magnification in a phase contrast microscope, or following staining with nigrosin-eosin, using light microscopy (see Chapter 8). Normal morphology of dog semen varies from 65% to 90%. The minimum percentage of morphologically normal spermatozoa in a dog ejaculate required to maintain normal fertility is not established, but usually values below 60% indicate some disturbance in testicular or epididymal function.

In pubertal dogs, up to 40% of the immature spermatozoa may have proximal cytoplasmic droplets. For semen intended for cryopreservation at least 75% of spermatozoa should be morphologically normal prior to freezing.

Evaluation of acrosome integrity is particularly relevant to evaluate the morphology of frozen-thawed spermatozoa. In the author's laboratory the Spermac™ stain is routinely used to evaluate the number of spermatozoa with intact acrosomes.

### Semen preservation

For artificial insemination with fresh semen immediately after collection, dilution of the semen is not necessary. However, if the semen is going to be transported with a delay in insemination of 2–3 hours or longer, it should be diluted with a semen extender and transported at 4°C, as the longevity of dog semen is considerably longer at 4°C than at 22°C. The author has used two extenders for transport of fresh semen with success (Tables 9.1 and 9.2).

In a recent experiment with TRIS-egg yolk, egg yolk-cream and egg yolk-milk extenders, the TRIS extender was found to be superior to the other two with respect to preserving dog semen at 4°C. In both extenders the egg yolk protects the sperm membrane against cold shock. Dilution of the sperm-rich fraction is usually approximately 1:6, depending on the initial semen concentration. It is important that the volume of diluent is larger than the initial semen volume, and if the semen collected is not well fractioned, centrifugation of the semen is beneficial to reduce the content of prostatic fluid in the semen before dilution.

Chilled transport is obtained by using a thermos flask partially filled with crushed ice. A plastic centrifugation tube (e.g. 10 ml, cell culture quality with screw top) may be used to transport the semen. This plastic tube should be protected by placing it into an insulating vial, such as the plastic vials used for transport of blood samples. The insulating vial with the tube should then be placed in the thermos flask on top of the layer of crushed ice and tissue paper wrapped around it. The semen should be carefully rewarmed to 30-35°C prior to insemination.

| | |
|---|---|
| High pasteurized cream 12% fat | 8 ml |
| 20% v/v egg yolk | 2 ml |

***Table 9.2**: The egg yolk–cream extender for fresh chilled semen.*

### Semen cryopreservation

A variety of freezing regimens, extenders and thawing protocols have been published in the literature. The author has worked mainly with the TRIS extender which has given good results for dog and fox semen over a number of years, yielding conception results of 67–80% in dogs and foxes following intrauterine deposition of 50–150 x $10^6$ spermatozoa per insemination, with two inseminations 24 hours apart.

## PROCESSING AND FREEZING OF DOG SEMEN

### Preparation and addition of extender

Glycerol is added to the TRIS base solution with antibiotics and this is used as the freezing solution (Table 9.1). The volume of glycerol added is either 12 ml (6% v/v) or 16 ml (8% v/v) depending on the freezing regimen, 6% for automatic freezing and 8% for manual freezing (see below). The distilled water content is adjusted accordingly to either 188 ml or 184 ml, respectively, instead of the original 200 ml.

The TRIS base solution with glycerol is mixed just prior to use with 20% egg yolk. The egg yolk is obtained from specific pathogen-free hens for export, or from consumer's eggs for domestic use. The yolk is separated from the egg white, rolled over a clean piece of paper tissue, gently broken and allowed to run into a funnel. Then the yolk is broken down by vigorous beating with a glass rod to minimize the size of the egg yolk globules. This reduces the frequency of attachment of sperm heads to the egg globules which makes semen evaluation more difficult. The egg yolk is then mixed with the TRIS buffer which is pre-warmed to 30°C. The mixture is stirred well, and further warmed to 35°C. Premixed ready-for-use TRIS-egg yolk extender can be frozen and kept frozen for 2 months prior to use. Gentle thawing in a water bath at 35°C is recommended before immediate use.

After the microscopic evaluation of the semen according to the criteria described above, the sperm-rich fraction of the ejaculate is diluted by dropwise addition of 35°C extender, until the desired concentration is reached. In the author's laboratory a total concentration of 100 x $10^6$ spermatozoa per millilitre is used routinely for freezing. However, when semen is of high quality, half that concentration, i.e. 50 x $10^6$/ml, may be used.

### Cooling and equilibration

After dilution, a sample is examined microscopically, and the extended semen is then poured into plastic centrifugation tubes which are placed into a beaker holding water at 35°C. The water beaker is then placed into a walk-in refrigerator at 4°C and left for 2 hours. During this time the semen is cooled to 4–5°C, and the cryoprotectant glycerol penetrates through the spermatozoal membrane.

### Semen packaging

The semen is stirred gently to remix after the 2 hours of equilibration and cooling, and is immediately placed into 0.5 ml plastic straws. This can be done by using commercially available filling devices or by suction through a latex tube with a mouthpiece. The semen must be filled in such a way that the powder between the two filter tips at the filter end of the straw is filled with liquid and solidifies. The straws are first filled half-way, and a small air bubble is allowed into the straw before the rest of the straw is filled to approximately 1 cm from the top. The air bubble prevents semen from being expelled due to the change of pressure within the straw during thawing. Then the straw is sealed using either ultrasound or commercially available sealing balls or sealing powders.

### Freezing regimens

There are principally two alternatives for freezing, either a manual, static protocol or an automatic, dynamic protocol, and the glycerol content added to the extender varies depending on which of the two freezing methods are used (see earlier for adjustment of water content in buffer base). The manual protocol involves the use of a polystyrene box (30 x 40 x 30 cm) with a removable metal rack placed 10 cm below the edge. The box is then filled with liquid nitrogen up to a level 4 cm below the rack.

The pre-filled straws, maximum 8–10 at a time, are placed horizontally on top of the rack, using forceps, and are left on the rack in the nitrogen vapour for 8 minutes. The forceps are then cooled in liquid nitrogen and applied on to the straws one by one to ensure complete crystallization (seeding). The straws are subsequently plunged into the liquid nitrogen. This freezing protocol is called static because there is a constant non-regulated flow of vaporized nitrogen during cooling and freezing.

Automatic freezing involves the use of freezing machines. It is called dynamic because vaporized nitrogen is let into the freezing chamber at variable speed by the pre-set freezing programme, allowing the cooling and freezing rates to be regulated. One such freezing programme using a Planer 10™ freezing machine was developed for fox semen by Hofmo (1988), and is used for dog semen in the author's laboratory when large quantities of semen are frozen. Straws are frozen horizontally on a rack with a removable lid, and several racks can be put into the freezing chamber. The freezing programme follows the regimen:

- -2°C/min from +4°C to -7°C
- -50°C/min from -7°C to -100°C
- -25°C/min from -100°C to -180°C

After the programme is completed, the whole rack is removed and placed directly into liquid nitrogen.

### Thawing

The straws are thawed in a water bath (thermos flask) at 70°C for 8 seconds. After thawing, the straw should be held vertically, filter tip down and sealed end up, then shaken to allow the air bubble in the middle of the straw to escape to the top. The straw is cut at the sealed end and a small drop of semen is placed on a slide on the warming plate for microscopic examination of post-thaw quality. Often, the spermatozoa need some time to start moving after thawing, but 2 minutes on the warming plate should ensure restoration of motility. Since other laboratories may use other freezing regimens, it is important to follow the instructions for thawing from the laboratory or company which has provided the frozen semen, since freezing and thawing regimens are closely connected.

## ARTIFICIAL INSEMINATION

### Artificial insemination with fresh semen

Artificial insemination with fresh semen is usually performed because there is a mating problem, the male and the female are some distance apart, one wants to inseminate more than one bitch with an ejaculate, or in order to prevent contact between bitch and dog for reasons of disease prevention.

### Artificial insemination with frozen semen

Artificial insemination with frozen semen is usually performed when the distance between bitch and stud is such that fresh semen cannot be sent, or because government regulations relating to import or export of semen prevent the movement of fresh semen between countries, or the dog is no longer available, but the semen has been frozen and kept in a semen bank. Both in Europe and the USA, semen banks are established either by the national kennel clubs, university institutions or private companies.

### Timing of the insemination

When fresh semen is used, the semen should be inseminated on the day of ovulation, with a second insemination 2 days later. When frozen semen is used, its longevity is reduced, and also the capacitation time is shorter because the freezing–thawing process influences the stability of the sperm acrosome and membranes. Therefore with frozen semen, insemination should be delayed until 1–2 days after ovulation, with a second insemination 24 hours after the first. Plasma or serum progesterone measurements provide a good indicator of the time of ovulation, and the progesterone concentration (measured by radioimmunoassay) should be approximately 30 nmol/l (10 ng/ml) on the first day and between 55 and 75 nmol/l (18–25 ng/ml) on the second day of insemination. Rapid enzyme-linked immunosorbent assay (ELISA) techniques are also available for the assessment of plasma progesterone in the practice laboratory. These may be used to give qualitative or quantitative assessment of the progesterone concentration. In foxes, it has been shown that two inseminations usually increase litter size; however conception rates may be just as high with one insemination if the timing is optimal.

## Insemination techniques

Fresh semen can be deposited into the vagina by using rigid plastic pipettes such as those used for intrauterine treatment of cows, but cut to an appropriate length to fit the bitch, with a 2-5 ml syringe attached (Figure 9.6). Syringes with rubber stoppers should be avoided, because some types of rubber can be toxic to spermatozoa.

The volume of the inseminate should not exceed 3 ml for large breeds and 2 ml for smaller breeds of dog, in order to reduce semen backflow. The plastic catheter should be inserted steeply upwards over the brim of the pelvis while the vulva is lifted upwards with the operator's index finger. When the catheter is in the vagina, the caudal end of the catheter is elevated, and should follow the dorsal wall of the vagina to the vaginal fornix avoiding inserting the catheter into the urinary bladder. The urethral orifice is located at the ventral wall of the pelvic brim. The insertion of the catheter should be supported with the operator's other hand palpating the cervix through the abdominal wall. The semen should be deposited slowly, and the bitch's hindquarters should be elevated for 10 min after deposition of the semen. Feathering the bitch around the vulvar area is considered beneficial for sperm transport because it stimulates contractions of the uterus.

In the author's laboratory a specially designed intrauterine catheter is used for both fresh and frozen semen (Figure 9.6). This equipment consists of a nylon guiding speculum and a metal catheter on to which a syringe is attached. The technique involves insertion of the guiding tube covering the metal catheter into the vagina. The guiding catheter is inserted into the vaginal fornix, and the metal catheter is pushed forward to the cervical os. The cervix is held by the operator's thumb and index finger across the abdomen of the bitch, and the metal catheter is inserted through the cervical canal after the cervical os is located with the tip of the catheter (Figure 9.7). The cervical canal is straight with little folding and quite open during oestrus, and passage through the canal is usually easy. Transcervical catheterization is done while the bitch is standing on a table and most commonly carried out without any sedation of the bitch (Figure 9.8). This method requires training, but once learned, the procedure takes only a minute or two and causes no discomfort to the bitch. When intrauterine insemination is used, it is not necessary to elevate the bitch's hindquarters after deposition of the semen.

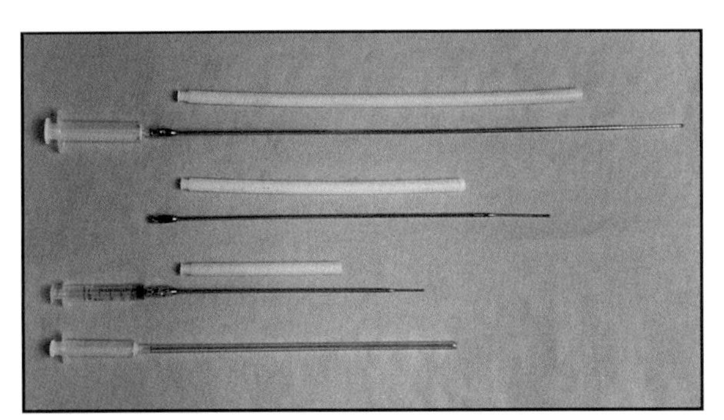

***Figure 9.6**: Three different sizes of catheter are shown with nylon guiding tubes intended for transcervical intrauterine artificial insemination in dogs. At the bottom is a simple plastic catheter for intrauterine treatment in cows that has been cut to an appropriate length for intravaginal insemination in the bitch.*

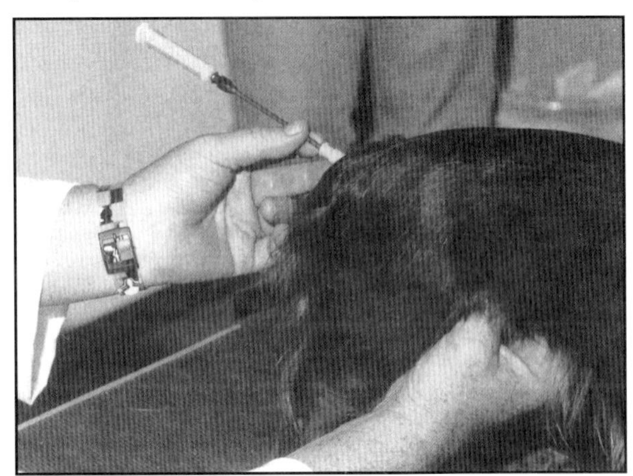

***Figure 9.7**: Artificial insemination in the bitch using the intrauterine catheter. The guiding tube is inserted over the metal catheter to protect the catheter from vaginal contamination and the vaginal mucosa from the tip of the catheter. The tube also aids in stretching the vagina sufficiently to locate the cervix through the abdomen. The cervix is fixed in a horizontal position, and the tip of the catheter is inserted through the cervix and into the uterus.*

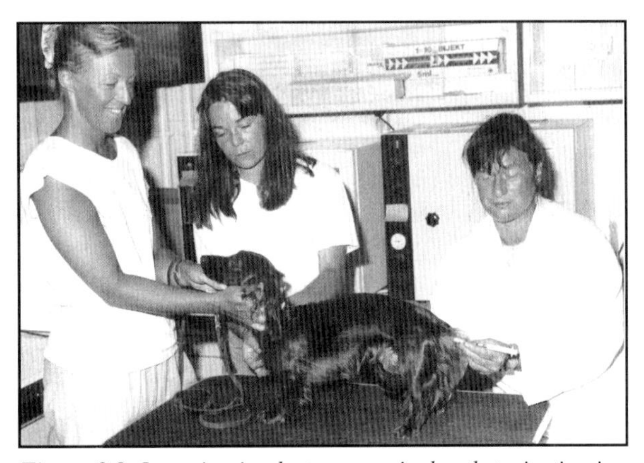

***Figure 9.8**: Insemination by transcervical catheterization is done while the bitch is standing on a table. In oestrous bitches there is usually no need to use sedation, and the procedure once learned, is performed in less than 5 minutes.*

The recommended insemination dose for frozen semen when inseminated transcervically into the uterus, is 100 x $10^6$ total spermatozoa per insemination, but litters can certainly be obtained at sperm numbers as low as 35-40 x $10^6$ spermatozoa per insemination provided the timing is correct and the post-thaw semen quality is high. The 100 x $10^6$/ml dose is based on the author's results with field trials with fox semen in which litter size was also an important parameter, and personal experience with dog semen. When surgical insemination is performed by laparoscopic exposure of the uterus, fewer spermatozoa may be required, but to the author's knowledge, no controlled experiments have so far been conducted to verify this.

Although pregnancies can be obtained from frozen semen with vaginal deposition of the semen, the results are consistently better with intrauterine deposition. Intrauterine insemination also seems to require fewer spermatozoa in the insemination dose.

When catheterization is not possible, or when the operator is not proficient in the catheterization technique, surgical laparoscopic insemination or insemination by endoscopic visualization of the cervical os is recommended. Surgical intrauterine insemination is the most common insemination technique for frozen semen in the USA and Canada. In most European countries, however, ethical concerns have led to the vaginal, endoscopic or catheterization approach.

## REFERENCES AND FURTHER READING


Andersen K (1975) Insemination with frozen dog semen based on a new insemination technique. *Zuchthygiene* **10**, 1

Evans HE and deLahunta A (1971) *Miller's Guide to the Dissection of the Dog. Revised reprint.* WB Saunders Company, Philadelphia

Farstad W (1984a) Bitch fertility after natural mating and after artificial insemination with fresh or frozen semen. *Journal of Small Animal Practice* **25**, 561-565

Farstad W (1984b) The correlation between a cyclus coefficient based on cytological indices in the vaginal smear and circulating progesterone in oestrous bitches. *Zuchthygiene* **19**, 211-217

Farstad W (1992) The optimum time for artificial insemination of blue fox vixens (*Alopex lagopus*) with frozen—thawed semen from silver foxes (*Vulpes vulpes*). *Theriogenology* **38**, 853-865

Farstad W (1996) Semen cryopreservation in dogs and foxes. Proceedings, XIII International Congress on Animal Reproduction and Artificial Insemination, Sydney. *Animal Reproductive Science* **42**, 1-4, 251-260

Farstad W and Andersen Berg K (1989) Factors influencing the success rate of artificial insemination in the dog. *Journal of Reproduction and Fertility* **39**, 289-292

Hofmo PO (1988) *Studies on Cryopreservation of Fox Spermatozoa and Evaluation of the Fertilizing Capacity of Frozen-thawed Silver Fox Spermatozoa.* PhD thesis, Norwegian College of Veterinary Medicine, Oslo

Kieffer JP (1992) Accouplement dans l'espece canine (mating in the canine species) In: *Les Indispensables de l 'Animal de Compagnie. Reproduction*, ed. C Dumon and A Fontbonne, pp. 67-73. P.M.C.A.C., Paris

Kumi-Diaka J and Badtram G (1994) Effect of storage on sperm membrane integrity and other functional characteristics of canine spermatozoa: In vitro bioassay for canine semen. *Theriogenology* **41,** 1355-1366

Laing JA, Brinley Morgan WJ and Wagner WC (1988) *Fertility and Infertility in Veterinary Practice, 4th edn*, pp. 10-12. Balliere-Tindall, London

Linde-Forsberg C (1995) Artificial insemination with fresh, chilled extended and frozen-thawed semen in the dog. *Seminars in Veterinary Medicine and Surgery (Small Animal)* **1**, 48-58

Nöthling JO and Volkman DH (1993) Effect of addition of autologous prostatic fluid on the fertility of frozen-thawed dog semen after intravaginal insemination. *Journal of Reproduction and Fertility* **47**, 325-327

Oettlé E (1986) Using a new acrosome stain to evaluate sperm morphology. *Veterinary Medicine* **3**, 263-266

Roberts SJ (1971) *Veterinary Obstetrics and Genital Diseases (Theriogenology)*, pp. 609, 620. Edwards Brothers Inc., Ann Arbor Michigan, and Ithaca, New York

Rodriguez-Gil JE, Montserrat A and Rigau T (1994) Effects of hypoosmotic incubation on acrosome and tail structure on canine spermatozoa. *Theriogenology* **42**, 815-829

Rota A, Ström B and Linde-Forsberg C (1995) Effects of seminal plasma and three extenders on canine semen stored at 4°C. *Theriogenology* **44**, 885-887

Silva LDM, Onclin K, Snaps F and Verstegen J (1995) Laparoscopic intrauterine insemination in the bitch. *Theriogenology* **43**, 615-623

Wilson M (1993) Non-surgical artificial insemination in bitches using frozen semen. *Journal of Reproduction and Fertility* **47**, 307-311

CHAPTER TEN

# Mating and Artificial Insemination in Domestic Cats

*Eva Axner*

## INTRODUCTION

Breeding animals should be free from defects and have a good temperament. Only females which have had uncomplicated births and have good maternal behaviour should be used for repeated breeding. Soundness and temperament should never be compromised, and pure-bred cats must also be good representatives of their breed as described in the breed standard. The goal should be to produce even litters with healthy kittens free from defects, rather than a few show-winning cats, if their littermates suffer from defects or are not typical of the breed. Inbreeding can, in the short term, give excellent-looking animals but will, in the long term, be detrimental and should be avoided, especially in numerically small breeds.

It is important to understand the physiology behind reproductive behaviour and the mechanisms of induced ovulation, in order to be able to differentiate between problems due to management and true reproductive problems. Numerous contagious diseases cause problems in catteries and efforts must be made to avoid their spread. Both the female and the male should be examined before mating for signs of disease and to exclude the presence of defects that could be hereditary. It is also advisable to test the animals for feline leukaemia virus (FeLV) and feline immunodeficiency virus (FIV). Cats which are to be mated should be vaccinated against feline herpes, feline calicivirus and feline panleucopenia and should also be treated for both external and internal parasites. Andrological and gynaecological examinations are not routinely performed before mating, but the male's testicles can be easily palpated and inspection of the penis is also easy to perform. If there is a history of reproductive failure, a more thorough examination should be carried out, including assessment of semen quality.

## MATING

### Oestrous behaviour of the queen

Oestrous behaviour is induced by rising concentrations of oestradiol produced by the growing ovarian follicles. Pro-oestrus is brief, lasting 1–2 days, and is not always displayed. The queen calls, rubs against objects and people, becomes more restless, but will not allow the male to mount her. When the queen is in oestrus, she calls with a repeated monotone, howling, rubs her head, rolls and crouches to the floor by lowering her chest and elevating her pelvis. A scant clear vaginal discharge can often be seen. Lordosis, treading of the hindlegs and lateral deviation of the tail, can be displayed spontaneously or induced by grasping the cat by the neck and stroking the base of the tail or the perineal area (see Figure 10.1). It is also common for the queen to lose her appetite and urinate more frequently. In some females oestrous behaviour is, however, less distinct or perhaps absent despite normal waves of active follicles and high concentrations of oestradiol. Vaginal smears or blood samples for oestradiol assay can be used to confirm oestrus in these individuals. The intensity of oestrous behaviour is breed related, with oriental breeds showing more intense behaviour than Persians.

### Mating behaviour

The male is attracted to the female by her vocalizing, odours and behaviour. Before mounting he grasps the female by the scruff of the neck with his teeth and will often tread with his hindlegs on the female's pelvis making the queen respond with a more pronounced standing reflex. Intromission only takes a few seconds (3–30 seconds), with the semen being deposited in the

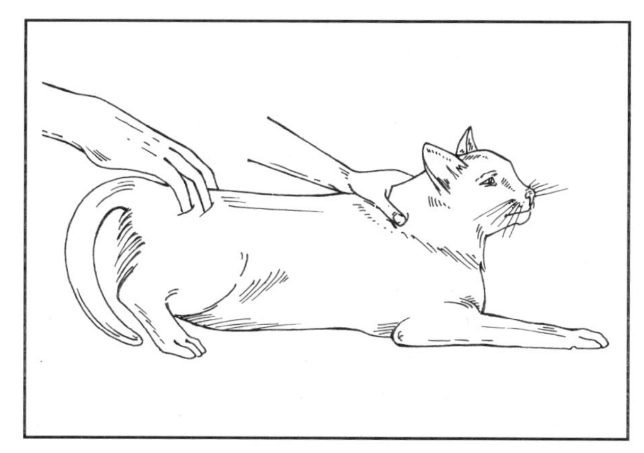

*Figure 10.1: Lordosis and treading with the hindlegs can be induced by grabbing the female by the scruff of the neck and stroking the base of the tail or the perineal area.*

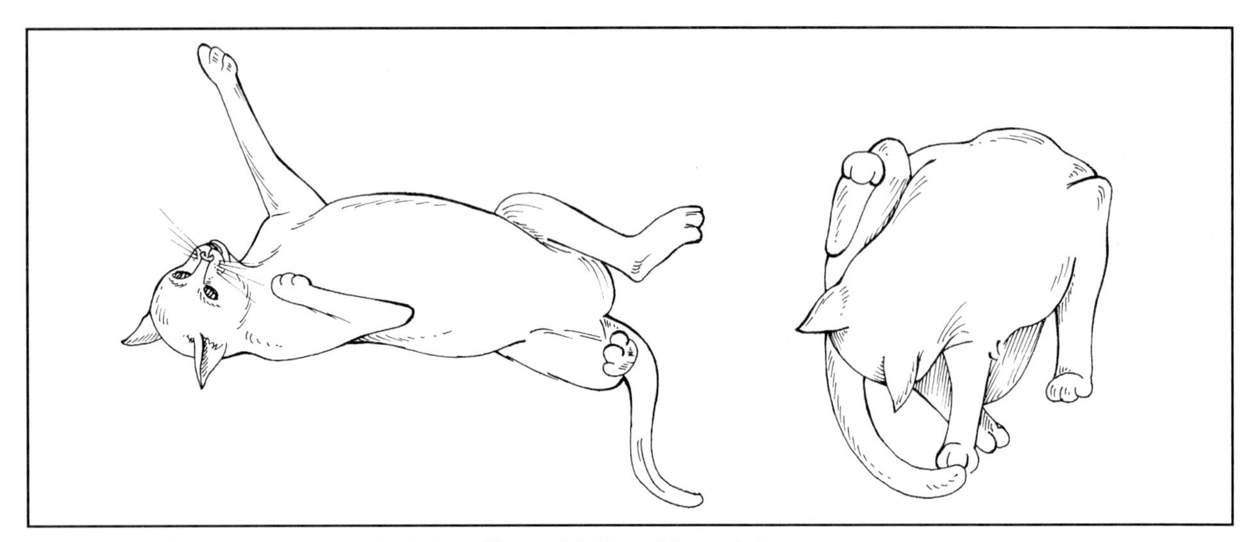

***Figure 10.2**: The post-coital reaction includes rolling and licking of the genitals.*

posterior vagina. The female emits a characteristic yowl, which increases in magnitude to a sharp cry when the male withdraws. An experienced male will quickly retreat and keep a safe distance from the queen, as she will strike out at him if he gets too close. The so-called 'copulation cry' and the rejection of the male is part of the female's post-coital reaction, which also includes intense rolling and licking of the genitals (Figure 10.2). The queen often allows remounting within 10 minutes but it may take more than an hour. If no post-coital reaction is seen, there has probably not been intromission. The post-coital reaction is thought to be caused by vaginal stimulation by the penile spines.

## Ovulation

The copulation stimuli trigger a release of gonadotrophin releasing hormone (GnRH) from the hypothalamus which in turn causes pituitary release of luteinizing hormone (LH) within minutes of copulation. When the concentration of LH is high enough, ovulation will be induced. Reproductive behaviour in cats is characterized by multiple matings and a single mating is often not sufficient to release LH and allow ovulation. Some females require at least four copulations to release enough LH. The beginning of oestrus does not always coincide with the ability to release LH in response to coitus, and not all queens will ovulate when mated on the first or second day in oestrus, probably because the oocytes are not fully matured. The oocytes are ovulated in an all-or-none fashion, which means that when there is a sufficient amount of LH, all mature oocytes will be ovulated. Ovulation occurs 26–29 hours after a mating where sufficient LH has been released. New data suggest that spontaneous ovulation is more common (35% of 20 queens in one investigation) than previously believed. A blood sample for analysis of progesterone from the queen one week after mating will show if ovulation has occurred. A basal concentration of progesterone indicates that there has been no mating, too few intromissions or that the queen was mated too early or too late in oestrus. Levels of progesterone over 15 nmol/l (5 ng/ml) indicate that the queen has ovulated. A prolonged interoestrus interval also indicates that ovulation has occurred, although not all females have regular cycles.

## Management of mating

Mating is more likely to be successful if the male is in his home environment, and therefore the female should usually be brought to the male. If the male is calm and experienced and the female is very frightened, it can, however, be better to take the male to the female. The animals may need some time to get used to each other and the new surroundings, and low lighting often makes them feel safer. If the female is inexperienced and frightened and rejects the male, she can be held near him, grabbed by the neck and stroked over the back to stimulate the mating position and to encourage the male to mount her. The animals should not be separated before the third day of oestrus, and at least four matings with a proper post-coital reaction should be observed to make sure that ovulation will occur. An inexperienced male should be allowed to mate an experienced and good natured queen the first time he is used, and an inexperienced female should be matched with an experienced male. Some older queens will, however, refuse a young and inexperienced male. Some problems encountered during mating are shown in Table 10.1.

# SEMEN COLLECTION AND SEMEN EVALUATION

## Semen collection

The two most common methods of semen collection from a male cat are by artificial vagina or by electroejaculation. Semen collection by vaginal lavage after mating has also been suggested. In addition

| Problem | Action |
|---|---|
| Inexperienced female rejects the male due to fear | If given sufficient time, mating usually occurs |
| Female still in pro-oestrus rejects male | Wait until female is in oestrus |
| Poor libido of male due to bad experience (aggressive females) | Try another quieter female |
| Aversion to one particular male or female | Change partner or use AI |
| Persistent frenulum allowing normal mounting but failure of intromission due to deviation of the penis or pain | Surgical correction of the persistent frenulum |
| Penile hair rings preventing intromission | Check the penis and remove hair rings |
| Dental problems interfering with the neck grip | Examine animals prior to mating |
| Clinical or sub-clinical disease causing decreased libido and/or testicular degeneration | Check for evidence of disease and treat the underlying disease |
| Congenital poor libido without other problems | Often hereditary, do not use for breeding |

***Table 10.1:*** *Mating problems and solutions.*

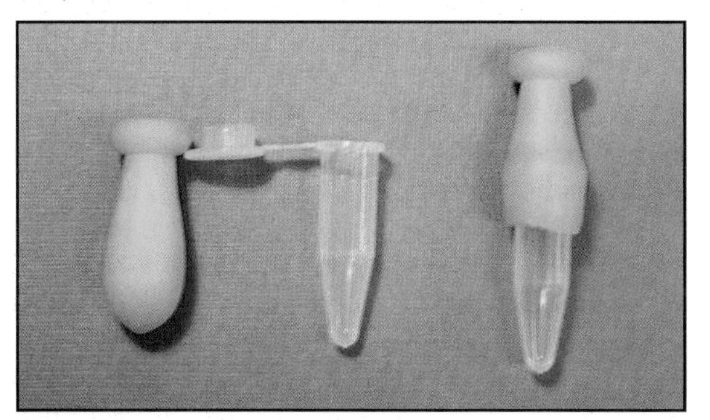

***Figure 10.3:*** *An artificial vagina for semen collection is made from a cut rubber bulb for a Pasteur pipette and a small test tube.*

spermatozoa can be collected from the cauda epididymidis after castration.

## Artificial vagina

The artificial vagina is made from a rubber bulb for a Pasteur pipette and a small test tube (Figure 10.3). The tom cat is allowed to mount a queen in heat and the artificial vagina is held in a position to facilitate intromission by the tom (Figure 10.4). The disadvantages of this method are that a teaser female is required and, to accustom the male to the procedure, a 2–3-week training period is usually necessary. This training is only successful in about two-thirds of males. Collection with an artificial vagina can be useful when working with a colony of laboratory animals, but is not usually practical in a clinical situation.

***Figure 10.4:*** *Semen collection by an artificial vagina. The male is allowed to mount an oestrous female and the artificial vagina is held in a position to facilitate intromission by the tom.*

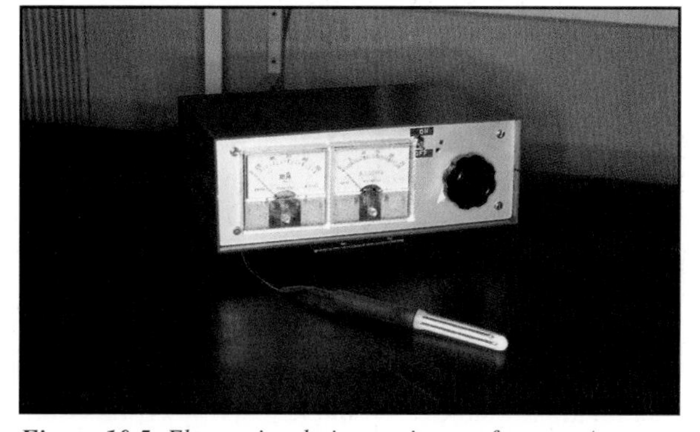

***Figure 10.5:*** *Electroejaculation equipment for cats. An electric stimulator and a rectal probe with three electrodes.*

## Electroejaculation

Electroejaculation can be used without prior training, but requires general anaesthesia to prevent discomfort to the cat. A rectal probe (1 x 12 cm) connected to an electric stimulator is used (Figure 10.5). The probe is made of non-toxic plastic and has three electrodes (1.5 mm x 5 cm) mounted longitudinally; the two outer electrodes are linked together, and the central one is of opposite polarity. It is important that the electrodes are tightly mounted to the probe, to prevent rectal mucosa from becoming trapped between the probe and the electrodes. Our laboratory uses a custom-made 50 Hz

| **Set** | **1** | | | **2** | | | **3** | |
|---|---|---|---|---|---|---|---|---|
| **Number of stimuli** | 10 | 10 | 10 | 10 | 10 | 10 | 10 | 10 |
| **Voltage** | 2 | 3 | 4 | 3 | 4 | 5 | 4 | 5 |

***Table 10.2:*** *Electroejaculation technique.*

sine-wave electroejaculator with a transformer which makes it capable of delivering a continuous range of voltages between 0 and 30 V. The unit is connected to a 220 V source. Voltage and ampage are monitored with a voltmeter and an ammeter.

Following induction of a surgical plane of anaesthesia, the probe is lubricated and carefully inserted approximately 7–9 cm into the rectum, with the electrodes directed ventrally. The penis is extruded by applying a gentle pressure at its base, and a small test tube is placed over the extruded penis. A total of 80 stimuli of between 2 and 5 V are delivered for each ejaculate (Table 10.2). The total sequence is divided into three series of 30 stimuli (10 at 2, 3 and 4 V), 30 stimuli (10 at 3, 4 and 5 V) and 20 stimuli (10 at 4 and 5 V). The cat is rested for 2–3 minutes between each series. The stimuli are administered for approximately 1 second from 0 V to the desired voltage, 2–3 seconds at the desired voltage and an abrupt return to 0 V for 2–3 seconds. Other sequences, with different numbers of stimuli and voltages have also been used. For each stimulus the cat responds with rigid extension of the hindlegs, indicating that the electrical stimulus has been adequate. A lack of extension at 2 V or more usually indicates improper positioning of the electrodes or interference by faeces. A proportion of the spermatozoa will be lost into the urinary bladder. Retrograde flow of spermatozoa to the urinary bladder is a normal component in the ejaculatory process of the domestic cat, but is perhaps exaggerated by the use of ($\alpha_2$-adrenoceptor agonists (xylazine and medetomidine) for sedation.

## Vaginal lavage after mating

This method is not very practical in a clinical situation, since few cats can be expected to try to mate within reasonable time after having been brought to the strange surroundings of a clinic. Other disadvantages are that lavage of the vagina requires sedation of the female and that the vaginal secretions and lavage fluid may affect semen quality. Vaginal lavage with saline at 37°C offers, however, a possibility to collect a semen sample if collection with an artificial vagina is unsuccessful and electroejaculation for some reason is not possible.

## Semen collection from the epididymis

Spermatozoa have also been collected from the cauda epididymidis after castration. The epididymal duct is flushed with a medium or spermatozoa are released by mincing the cauda. Collection from the epididymidis is only a practical method for collection of spermatozoa

| | |
|---|---|
| Colour | White: high semen concentration<br>Translucent: low semen concentration<br>Yellow: urine contamination, detrimental to semen, occurring more frequently if voltages ≥8 V are used for electroejaculation |
| Volume | Measured by a variable micropipette. Small volumes can be extended with saline or a buffer before further processing. Examine sperm morphology before mixing, as differences in osmolality can affect the proportion of abnormal sperm tails |
| Motility evaluation | Use phase contrast, magnification x100–400. Sperm velocity is assessed from 0 to 5, where 0 is used for immotile spermatozoa and 5 is used for spermatozoa with a rapid, steady forward progression |
| Sperm concentration | Measured using a counting chamber (e.g. Bürker chamber) under a microscope. Total number of spermatozoa is calculated from the concentration and the volume |
| Evaluation of the proportion of abnormal sperm heads | A stained smear (carbol-fuchsine or nigrosin–eosin) is examined under a light microscope at x1000. Abnormal head-forms include pear-shaped, narrow at the base, abnormal contour, undeveloped, detached and abnormal, narrow and variable size |
| Evaluation of other sperm abnormalities | A drop of formol-saline fixed spermatozoa is examined under a phase-contrast microscope at x1000. Abnormal sperm forms include proximal droplets, distal droplets, loose heads, acrosomal defects, acrosomal abnormalities, mid-piece abnormalities, single bent tails, coiled tails, double bent tails. In addition the proportion of spermatozoa devoid of these defects is estimated (assumed to be normal) |
| Evaluation of other cells in the ejaculate | Presence of leucocytes, spermiogenic cells and degenerative epithelial cells is evaluated on a stained (Papanicolaou) smear |

***Table 10.3:*** *Evaluation of semen. All material that comes into contact with the semen should be prewarmed to 37°C to avoid cold shock of the spermatozoa.*

for research or for preservation of semen from rare non-domestic felids killed by accident or which have died in captivity.

### Normal semen quality

Ejaculate volumes are small (<0.01-0.77 ml) but are generally larger when semen is collected by electroejaculation than when collected using an artificial vagina, due to overstimulation of the accessory sex glands with the former method. The total number of spermatozoa per ejaculate has been reported to be between 3 x $10^6$ and 153 x $10^6$; this is usually higher when semen is collected by artificial vagina than by electroejaculation. Sperm motility is highly variable and may be related to the duration of sexual abstinence. Osmolality of fresh semen collected with an artificial vagina is approximately 320 mOsm/kg and increases with storage time. The semen is rich in alkaline phosphatase and has a pH between 6.6 and 8.77. Males with more than 60% normal spermatozoa are classified as normozoospermic whilst males with less than 40% normal spermatozoa are classified as teratozoospermic. The relationship between semen quality and fertility has, however, not yet been established for the domestic cat; males with less than 40% normal spermatozoa may still have good fertility under natural conditions. If two semen samples are collected by electroejaculation within a short period of time, the second sample will generally have better motility and a higher proportion of normal spermatozoa. A study of 15 male cats of different breeds electroejaculated twice during the same anaesthesia showed means of normal spermatozoa of 40.9% in the first ejaculate and 54.6% in the second, with the first sample having a higher concentration of spermatozoa. More than one semen sample must therefore always be collected for evaluation of fertility.

## SEMEN PRESERVATION AND ARTIFICIAL INSEMINATION

### Storage of semen

Semen can be stored for a short time (24–48 hours) by extending it in a buffer and by chilling it. For long-time storage cryopreservation is necessary. Most studies have used *in vitro* methods to test the quality of the semen after storage and there are very few reports of pregnancy results after artificial insemination with stored cat semen.

### Chilled extended semen

The quality of chilled cat semen has only been tested by *in vitro* methods and the pregnancy results still remain to be investigated. A TesT-buffer (Table 10.4) with 20% egg yolk and 5% glycerol has been used for storing cat semen at 4–5°C. The motility declines and the percentage of intact acrosomes decreases during storage. Increasing amounts of egg yolk or sugars are detrimental to the motility after storage, indicating that a simple TesT-buffer without sugars or egg yolk could be a suitable extender.

| | |
|---|---|
| *N*-Trishydroxymethyl-methyl-2-aminomethane-sulphonic acid (Tes) | 11.2 g in 150 ml distilled water* |
| Trishydroxymethyl-aminomethane (Tris) | 2.9 g in 75 ml distilled water* |
| Penicillin B | 1000 IU/ml |
| Streptomycin | 1 mg/ml |

***Table 10.4:*** *TesT buffer. *Tes is titrated against Tris to pH 7.4. Egg yolk and glycerol are added when this buffer is to be used for cryopreservation of semen (this is not necessary when the semen is going to be chilled).*

| | |
|---|---|
| De-ionized water | 76 ml |
| Lactose | 11 g |
| Glycerol | 4 ml |
| Streptomycin sulphate | 1000 µg/ml |
| Penicillin G | 1000 IU/ml |
| Egg yolk | 20 ml |

***Table 10.5:*** *Egg yolk–lactose buffer.*

### Cryopreservation of semen

Cat semen can be frozen in an egg yolk–lactose extender (Table 10.5). Freezing in pellets and in straws gives comparable results regarding post-thaw sperm motility index, percentages of intact acrosomes and zona pellucida binding for this extender. Fresh semen is diluted 1:3 in Ham's F10 medium (with 5% fetal calf serum), centrifuged and resuspended in the extender. The extended semen is filled in 0.25 ml straws, equilibrated for 10 minutes at 22°C and manually lowered into $N_2$ vapour, held at -10°C for 60 seconds and then frozen at -40°C/min to -100°C before the straws are plunged into the liquid nitrogen. The straws are thawed for 10 seconds in air followed by 20 seconds in a 37°C water bath before the semen is mixed with Ham's F10 medium. Other buffers that have been used are TesT and Tris buffers. For semen extended in a TesT buffer with 20% egg yolk and 5% glycerol freezing in straws is superior to freezing in pellets. Post-thaw motility and the proportion of intact acrosomes decreases after freezing and thawing.

### Use of artificial insemination

Artificial insemination in the domestic cat has mostly been used for research purposes, where the domestic cat is often used as a model for wild felids. The techniques for artificial insemination and storage of semen could also, however, be a valuable tool in cat breeding programmes. In many breeds of cat there are only a limited number of breeding animals. This problem is accentuated by the fact that many male cats are castrated because their behaviour (e.g. spraying) makes them less suitable to keep indoors. Few individuals in a breed, and imbalance between the number of intact females and males, will sooner or later lead to inbreeding degeneration and the appearance of defects and diseases. Freezing of spermatozoa would allow shipping of semen world-wide for exchange of genetic material and would also make it possible to use a male's semen for breeding even after castration or death. Artificial inseminations would also considerably decrease the risk of spreading disease.

## Method of artificial insemination

### Induction of ovulation

In the absence of the mating stimuli ovulation can be induced by an injection of human chorionic gonadotrophin (hCG), which has potent LH bioactivity. Induction of ovulation is more successful on the third day of oestrus than on the first or second day. When 100 IU hCG is administered intramuscularly on the third day of oestrus, most queens will ovulate; excessive doses can result in ovarian hyperstimulation and degenerated oocytes. Anaesthesia can inhibit ovulation if administered after the induction of ovulation but before the ovulation. An alternative to hCG is to give 25 μg gonadotrophin releasing hormone (GnRH) intramuscularly.

### Artificial insemination with fresh semen

Pregnancies have been reported for both vaginal and intrauterine artificial insemination using fresh cat semen. When semen extended in 0.1 ml saline is deposited in the vagina both on the day of injection of hCG and at the time of ovulation 24 hours later, the pregnancy rate is higher than if a single insemination is performed on the day of hCG administration. Surgical intrauterine insemination gives the best results when carried out 31-50 hours after hCG administration. Fertilization can take place up to 49 hours after induction of ovulation. Tranquillizers are usually not necessary to perform vaginal inseminations.

| | |
|---|---|
| Induction of ovulation | 100 IU hCG on day 3 of oestrus |
| Ovulation | 26-29 hours after injection of hCG |
| Fertilization | Can take place up to 49 hours after injection of hCG |
| Anaesthesia | Can inhibit ovulation |
| Combined length of the vagina and vestibulum | 45-50 mm |

***Table 10.6:*** *Artificial insemination regimen.*

The domestic queen has a vestibulum 1-2 cm long and a narrow vagina 2.5-3.0 cm long. A 3.5 French tom cat catheter can be used for vaginal insemination. This is inserted 45-50 mm into the vagina as far as the cervix, where the semen is deposited (Figure 10.6). The queen should be held with elevated hindquarters for 10 minutes after the insemination. At least 5 x $10^6$ spermatozoa should be introduced when fresh semen is deposited in the vagina.

### Artificial insemination with frozen semen

Live young can be produced by artificial insemination with frozen-thawed cat semen. A pregnancy rate of around 10% can be expected after vaginal insemination in natural and hormone-induced oestrous queens. Poor fertility after artificial inseminations with frozen-thawed cat semen can be due to damage to the acrosomes after thawing even if the post-thaw motility is good. Timing of insemination and site of semen deposition are also important factors; as in the bitch, intrauterine inseminations give much better pregnancy results than vaginal inseminations, particularly when frozen-thawed semen is used. Surgical intrauterine insemination is not allowed in all countries, due to animal welfare considerations. A method for trans-cervical intrauterine insemination would probably greatly improve pregnancy results in feline artificial insemination.

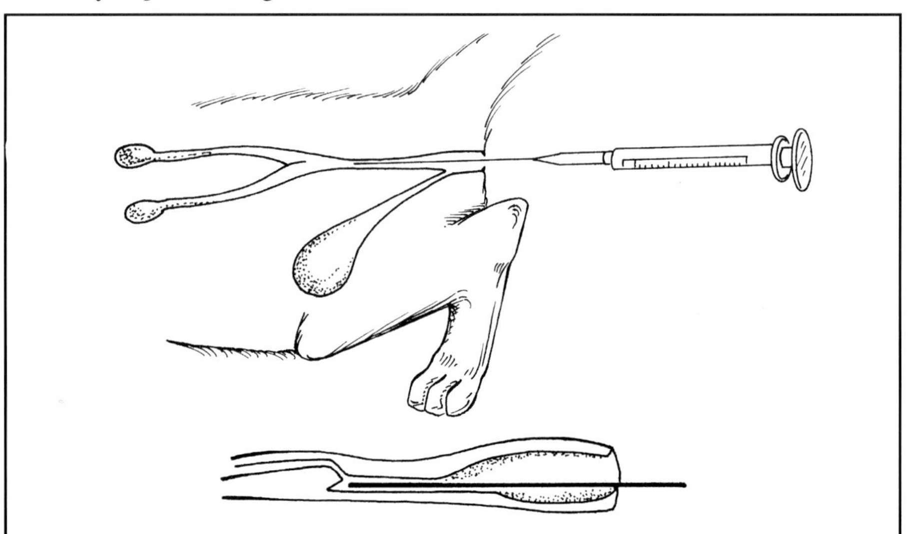

***Figure 10.6:*** *Vaginal insemination in the cat. The semen is deposited in the cranial vagina near the cervix.*

Semen collection, artificial insemination and preservation of cat semen are not widely used in practical cat breeding but may well become important tools in the not too distant future.

## REFERENCES AND FURTHER READING


Axnér E, Ström B and Linde-Forsberg C (1997) Sperm morphology is better in the second ejaculate than in the first in domestic cats electroejaculated twice during the same period of anesthesia. *Theriogenology*, **47,** 929–934

Axnér E, Ström B, Linde-Forsberg C, Gustavsson I, Lindblad K and Wallgren M (1996) Reproductive disorders in 10 domestic male cats. *Journal of Small Animal Practice* **37**, 394–401

Chakraborty PK, Wildt DE and Seager SWJ (1979) Serum luteinizing hormone and ovulatory response to luteinizing hormone-releasing hormone in the estrous and anestrous domestic cat. *Laboratory Animal Science* **29**, 338–344

Dooley MP and Pineda MH (1986) Effect of method of collection on seminal characteristics of the domestic cat. *American Journal of Veterinary Research* **47**, 286–292

Dooley MP, Pineda MH, Hopper JG and Hsu WH (1991) Retrograde flow of spermatozoa into the urinary bladder of cats during electroejaculation, collection of semen with an artificial vagina, and mating. *American Journal of Veterinary Research* **52**, 687–691

Glover TT and Watson PF (1985) The effect of buffer osmolality on the survival of cat (Felis catus) spermatozoa at 5°C. *Theriogenology*, **24**, 449–456

Glover TT and Watson PF (1987) The effects of egg yolk, the low density lipoprotein fraction of egg yolk, and three monosaccharides on the survival of cat (Felis catus) spermatozoa stored at 5°C. *Animal Reproduction Science* **13**, 229–237

Linde-Forsberg C (1990) Achieving pregnancy by using frozen canine chilled extended semen. *Veterinary Clinics of North America – Small Animal Practice* **21**, 467–485

Platz CC and Seager SWJ (1978) Semen collection by electroejaculation in the domestic cat. *Journal of the American Veterinary Association* **173,** 1353–1355

Platz CC, Wildt DE and Seager SWJ (1978) Pregnancy in the domestic cat after artificial insemination with previously frozen spermatozoa. *Journal of Reproduction and Fertility* **52**, 279–282

Pope CE, Turner JL, Quatman SP and Dresser BL (1991) Semen storage in the domestic felid: a comparison of cryopreservation methods and storage temperatures. *Biology of Reproduction* **44**, 257, 50

Sojka NJ, Jennings LL and Hamner CE (1970) Artificial insemination in the cat (Felis catus L.). *Laboratory Animal Care* **20**, 198–204

Watson PF and Glover TE (1993) Vaginal anatomy of the domestic cat (Felis catus) in relation to copulation and artificial insemination. *Journal of Reproduction and Fertility* Supplement **47**, 355–359

Wood TC, Swanson WF, Davis RM, Anderson JE and Wildt DE (1993) Functionality of sperm from normo- versus teratospermic domestic cats cryopreserved in pellets or straw containers. *Theriogenology* **39**, 342

CHAPTER ELEVEN

# Pregnancy Diagnosis, Abnormalities of Pregnancy and Pregnancy Termination

*Gary C.W. England*

## INTRODUCTION

In the majority of domestic species, pregnancy interrupts normal cyclical activity by increasing the length of the luteal phase and delaying the return to oestrus. The physiology of the bitch is unlike other species, in that the luteal phase is similar in length for both pregnancy and non-pregnancy. This fundamental difference in reproductive physiology is the reason why pseudopregnancy is a common and normal event in the bitch, and why endocrinological methods of pregnancy diagnosis and methods of pregnancy termination are dissimilar to many other domestic mammals.

In the queen, the physiology of pregnancy is similar to that of the bitch, although the endocrinology of non-pregnancy differs considerably.

## ENDOCRINOLOGY OF PREGNANCY

### Bitch

Following ovulation and the formation of the corpora lutea, there is little difference in progesterone concentration between pregnant and non-pregnant bitches. Progesterone concentration is, on average, slightly higher in pregnant bitches and the progesterone plateau is broader during pregnancy, whilst the luteal phase is slightly longer in non-pregnancy (Figure 11.1). Oestradiol concentration does not differ between pregnant and non-pregnant bitches, and oestradiol concentrations tend to increase consistently during the luteal phase. There may, however, be pregnancy-specific increases in oestrone and oestrone sulphate concentrations. Whilst one study demonstrated an increase in urine total oestrogen concentration during pregnancy, there have been no subsequent investigations, and the value of oestrogen measurement as a method of pregnancy diagnosis remains to be elucidated.

The pregnant luteal phase lasts approximately 63 days from ovulation to parturition, and the non-pregnant phase lasts approximately 66 days. During the second half of the luteal phase the progesterone concentration starts to decline and there is a concomitant increase in plasma prolactin concentration. Prolactin is secreted by the pituitary and is luteotrophic in nature; by supporting the corpora lutea, prolactin tends to maintain an elevated progesterone concentration. Inhibition of prolactin secretion, or antagonism of its actions, results in an abrupt termination of the luteal

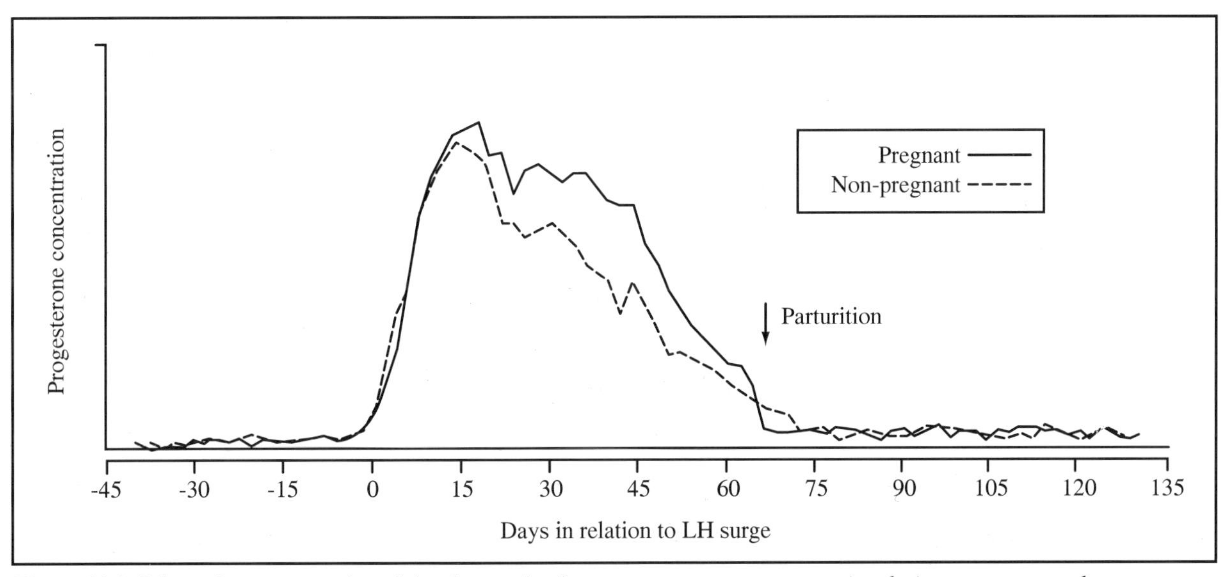


***Figure 11.1**: Schematic representation of the changes in plasma progesterone concentration during pregnancy and non-pregnancy in the bitch.*

phase. Prolactin concentration has been shown to be four times greater in pregnant bitches compared with non-pregnant bitches, however in some non-pregnant individuals prolactin concentrations are high. Luteinizing hormone (LH) is another luteotrophic agent in the bitch and its inhibition may result in luteolysis. The LH secretion pattern has not been characterized throughout the pregnant or non-pregnant luteal phase. Interestingly, the corpora lutea appear to be almost autonomous during the first 15 days after ovulation, and at this time have relatively little luteotrophic support.

Relaxin is the only known pregnancy-specific hormone in the bitch. It is present in plasma 25 days after ovulation and peak values are reached at approximately day 50.

In addition to these hormonal changes associated with pregnancy, a variety of other physiological changes have been observed in the bitch. For example, there is a pregnancy-specific increase in blood volume which contributes to a normochromic, normocytic anaemia, that is associated with a significant reduction in the percentage packed cell volume. Additionally, pregnancy-specific increases in plasma fibrinogen and other acute phase proteins are found from approximately day 20 onwards.

## Queen

In most queens, ovulation is not spontaneous and must be induced by coitus or artificial stimulation; however, in a small number of queens, ovulation may occur without an obvious stimulus. In the majority, copulation produces a rapid pituitary-mediated release of LH. Once LH concentration exceeds a threshold value, ovulation occurs.

Usually multiple copulations within a short time period are required to produce an LH surge of sufficient magnitude to induce ovulation. Progesterone concentration remains basal until after the mating-induced LH surge, and increases after ovulation; peak values are reached one month after mating. In pregnant queens progesterone concentration declines slowly until day 60, and remains relatively low for the last week before declining abruptly at parturition; the duration of pregnancy is usually 64–68 days from mating. In queens that have been mated but do not become pregnant, progesterone concentration is initially identical to that of early pregnancy but then rapidly returns to basal values, with the luteal phase lasting between 30 and 45 days before they return to oestrus (Figure 11.2). These queens are said to be pseudopregnant, although the only clinical sign is an absence of oestrus. Non-ovulating (non-mated or inadequately mated) queens do not have a luteal phase and return to oestrus at an interval of approximately 21 days. Throughout pregnancy the primary source of progesterone is the ovary, and there does not appear to be a significant contribution from the placenta.

Oestradiol concentration is elevated during the last week of pregnancy, and prolactin concentration is elevated in the last third of pregnancy and throughout lactation. Prolactin has a luteotrophic action similar to that in the bitch. There are no significant changes in prolactin concentration during pseudopregnancy.

Relaxin is present in the plasma from approximately day 25, and, similar to the situation in the bitch, it increases to peak values at approximately day 50, and declines after parturition.

# DIAGNOSIS OF PREGNANCY

The peculiarities of the oestrous cycle of the bitch and queen mean that endocrinological methods of pregnancy diagnosis cannot simply be adapted from other species. However, imaging technologies such as diagnostic B-mode ultrasonography and radiography are equally applicable to all domestic species.

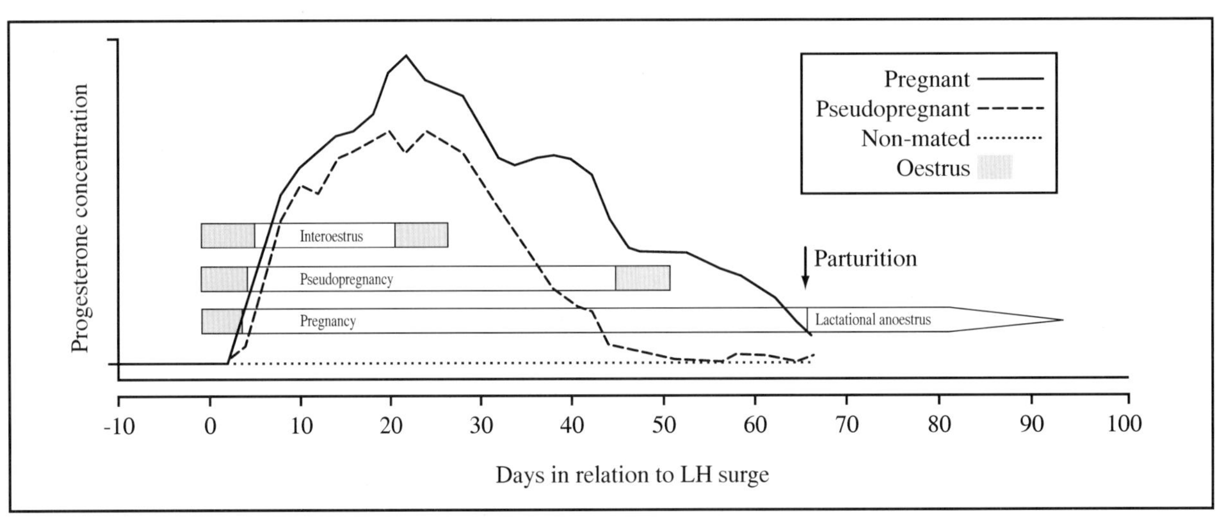


*Figure 11.2: Schematic representation of the changes in plasma progesterone concentration during pregnancy, pseudopregnancy and after non-ovulation in the queen. Oestrus is longer for non-mated, non-ovulating individuals. A variable period of anoestrus follows pregnancy. In all cases the return to oestrus may be influenced by season of the year.*

The diagnosis of early pregnancy in the bitch can be confused because of the difference between the 'actual' and the 'apparent' length of pregnancy. The onset of parturition occurs over a narrow spread of 3 days when considered in relation to the time of ovulation. However, matings that occur before ovulation may be fertile should spermatozoa reside within the reproductive tract 'waiting' for ovulation. The result of an 'early' mating produces an apparently long pregnancy length. Conversely, after ovulation oocytes are not immediately fertilizable and must undergo further maturation which takes 2 days. Oocytes may then remain fertilizable for at least 3 further days. A 'late' mating, for example, on the last day that eggs are fertile (5 days after ovulation) would result in an apparently short pregnancy length. This information is relevant since bitches may be examined during early pregnancy. If performed in relation to an 'early' mating this may result in a false negative diagnosis, since the examination may be undertaken before a positive diagnosis could be made. These problems do not occur in the cat, which is an induced ovulator, where in most cases pregnancy length can be estimated from the mating date.

### Absence of oestrus

In many species the absence of oestrus at 21 days can be used as an indicator of pregnancy. In the bitch, however, the interoestrous interval is identical in pregnant and non-pregnant cycles, whilst in the queen, the absence of a return to oestrus indicates only that ovulation has occurred; it is not specific for pregnancy (Figure 11.2). A failure to return to oestrus after day 45 in the queen may be used as a positive sign for pregnancy, although by this time the pregnant queen is usually easily identified by changes in her physical appearance. Some domestic cats occasionally mate spontaneously during pregnancy.

### Behavioural changes

Both pregnant and non-pregnant bitches may exhibit behavioural changes typical of pregnancy. These are usually associated with an increase in plasma prolactin concentration and are not specific for pregnancy. Food intake usually increases by approximately 50% in the second half of pregnancy; however, it is not uncommon for pregnant bitches to have a brief period of reduced appetite approximately 3–4 weeks after mating.

In the queen, behavioural changes usually only develop during late pregnancy, by which time the non-pregnant queen will have returned to oestrus, and pregnant queens can be easily identified by their physical appearance.

### Physical changes

Pregnant bitches commonly develop a slight mucoid vulval discharge approximately one month after mating, although this may also be seen in non-pregnant animals. The teats begin to become pink and erect. Body weight begins to increase from day 35 onwards, and may increase by to up to 50% of normal. Abdominal swelling may be noted from day 40 and this may progress to abdominal distension from day 50 onwards. However, these changes may not be obvious in primigravida or bitches with small litters. Mammary gland enlargement is usually obvious from day 40, at which time serous fluid can be expressed from the glands. Colostrum may be present in the teats in the last 7 days of pregnancy. Mammary changes may vary considerably between primigravida and multigravida, and care should be taken when assessing changes in mammary size and secretion, since similar features are common in pseudopregnant bitches.

In the queen, body weight and the appearance of the abdomen may alter during pregnancy although the changes are less obvious than in the bitch, and may not be noticed until after day 50. An increase in the size and the degree of reddening of the nipples is commonly seen from day 21 onwards. Mammary enlargement occurs from day 58 and colostrum may be present in the teats in the last 7 days of pregnancy. Mammary enlargement and the secretion of milk is not normally a feature of pseudopregnancy in the cat, and therefore these signs may be used as indicators of pregnancy.

### Abdominal palpation

The technique of abdominal palpation can be highly accurate, but may be difficult in obese or nervous animals, and may be inaccurate if the bitch was mated early and pregnancy is not as advanced as anticipated. The palpation features are similar in bitches and queens, although in the queen palpation may be technically easier.

The optimum time for the diagnosis of pregnancy is approximately one month after mating, at which time an experienced clinician should achieve an accuracy of nearly 90%. Before this time the conceptuses are small tense swellings within the uterine horns and may not be detected because of their small size. From days 26 to 30, the conceptuses are spherical in outline, and may vary between 15 and 30 mm in diameter. They are tense fluid-filled structures and can be readily palpated in a relaxed bitch. From day 35, the conceptuses become elongated, enlarged and tend to lose their tenseness. They may be less easy to palpate at this time. After day 45, the uterine horns tend to fold upon themselves, resulting in the caudal portion of each horn being positioned against the ventral abdominal wall, and the cranial portion of the same horn being positioned dorsally. After day 55, the fetuses can often be identified, especially if the forequarters of the bitch are elevated and the uterus is manipulated caudally towards the pelvis.

It is difficult to count accurately the number of conceptuses by palpation, except when performing an examination at approximately day 28 in a relaxed and thin bitch.

## Identification of fetal heart beats

In late pregnancy, it is possible to auscultate fetal heart beats in both the bitch and queen, using a stethoscope. Fetal hearts may also be detected by recording a fetal ECG. Both of these are diagnostic of pregnancy. Fetal heart beats are not difficult to detect since the heart rate is usually more than twice that of the dam.

## Radiography

In both the bitch and the queen, uterine enlargement can be detected from day 30. At this stage the enlarged uterus can be readily identified in the caudal abdomen, originating dorsal to the bladder and ventral to the rectum; it frequently produces cranial displacement of the small intestine. However, the early pregnant uterus has only soft tissue opacity and it cannot be differentiated from other causes of uterine enlargement, such as pyometra which occurs at the same stage of the oestrous cycle. Pregnancy diagnosis is not possible until after day 45 (day 40 in the queen) when mineralization of the fetal skeleton is detectable radiographically. Progressive mineralization results in an increasing number of bones that can be identified. It is unlikely that the fetuses will be damaged by ionizing radiation after day 45; however sedation or anaesthesia of the dam may be required and is a potential risk. In late pregnancy, the number of puppies can be reliably estimated by counting the number of fetal skulls.

## Endocrine tests

Plasma concentrations of progesterone are not useful for the diagnosis of pregnancy in the bitch and, whilst there may be pregnancy-specific changes in plasma and urine oestrogen concentration, these have not been adequately evaluated to provide a clinically useful pregnancy test.

There is, however, a significant elevation of plasma prolactin in pregnant bitches compared with non-pregnant bitches, and it is possible that prolactin assays may, in the future, become useful as methods of pregnancy diagnosis.

Measurement of the hormone relaxin is diagnostic of pregnancy; however, at present there is no available commercial assay for relaxin concentration.

In the queen, plasma progesterone concentration is elevated in both pregnancy and pseudopregnancy; therefore measurement of this hormone before day 45 is not diagnostic of pregnancy. Whilst a pregnancy-specific increase of plasma relaxin concentration occurs from day 25, assay of relaxin concentration is not possible in the queen.

## Acute phase proteins

An acute phase response occurs in pregnant bitches at approximately the time of implantation. This response appears to be unique to the bitch, and measurement of fibrinogen, C-reactive protein, or other acute phase proteins are sensitive markers for pregnancy. The initial rise in these proteins occurs from day 20 onwards with a peak at approximately day 40.

These methods appear to be reliable, although false positive diagnoses may result from inflammatory conditions (such as pyometra, which occurs at the same stage of the oestrous cycle). The rise in fibrinogen concentration is now the basis of a commercial pregnancy test.

## Ultrasound examination

Diagnostic B-mode ultrasonography can be used for early pregnancy diagnosis in both the bitch and the queen. Ultrasonography is a non-invasive imaging modality which is safe both for the operator and the patient. The most accurate time to perform an examination is generally one month after the last mating.

### Technique

In both species, the homogeneous uterus can be identified dorsal to the bladder (Figure 11.3). In the bitch, conceptuses may be first imaged from 15 days after ovulation, at which time they appear as spherical anechoic structures approximately 2 mm in diameter. The anechoic fluid is the yolk sac, and it is not until the yolk sac is filled with a sufficient volume of fluid that the pregnancy can first be imaged. Uterine enlargement occurs during the luteal phase, whether the bitch is pregnant or non-pregnant. During early pregnancy the embryo is located adjacent to the uterine wall and is not imaged. The conceptus rapidly increases in size and may lose its spherical outline, becoming oblate in appearance. From day 20 after ovulation the conceptus is approximately 7 mm in diameter and 15 mm in length and the embryo can be imaged. The presence of the embryonic heart beat can be detected from approximately 22 days after ovulation. The fetal membranes are complicated and may appear confusing ultrasonographically, although the developing allantois initially appears as a nearly spherical structure

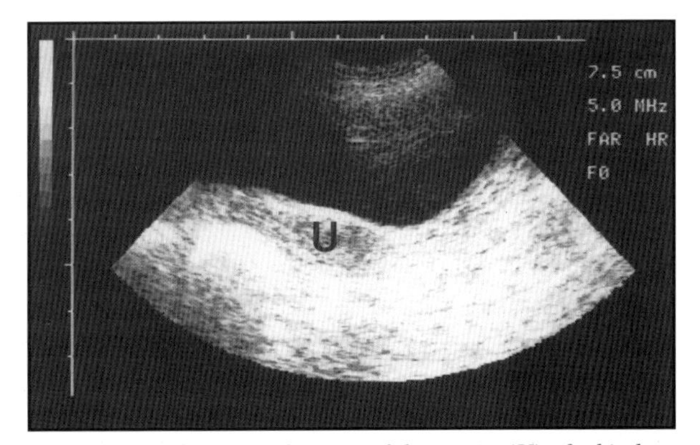


***Figure 11.3**: Ultrasound image of the uterus (U) of a bitch positioned dorsal to the anechoic bladder. In this image, the bitch was examined in the standing position with the transducer placed on the ventral abdomen. 5.0 MHz transducer, scale in cm.*

within the conceptus which subsequently increases in size and surrounds the yolk sac. A third fluid-filled sac, the amnion, may be noted later in pregnancy since it is surrounds, and initially is in close apposition to, the fetus (Figure 11.4).

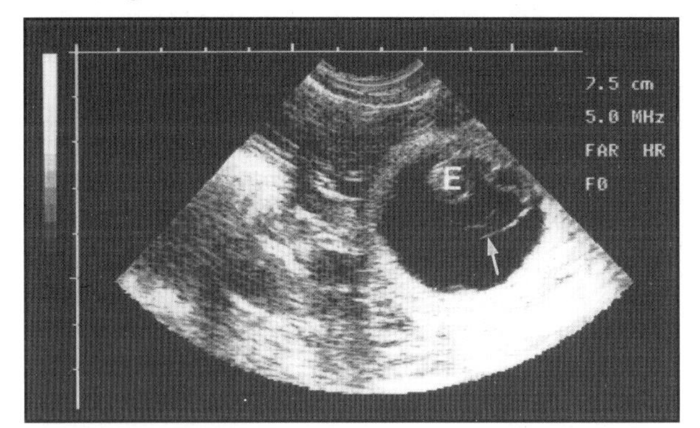


***Figure 11.4**: Ultrasound image of a single conceptus, demonstrating the embryo (E), embryonic membranes (arrow), and surrounding anechoic allantoic fluid. 5.0 MHz transducer, scale in cm.*

The most rapid growth of the fetus occurs between days 32 and 55, and during this time the limb buds become apparent and there is clear differentiation of the head, trunk and abdomen (Figure 11.5). The zonary placenta can usually be easily identified from this stage of pregnancy onwards in both the bitch and queen (Figure 11.6). The fetal skeleton becomes evident from 40 days onwards when fetal bone appears hyperechoic, and casts acoustic shadows. The heart can now be easily identified as the hyperechoic valves can be seen to move. The large arteries and veins can be seen cranially and caudally (Figure 11.7). Lung tissue surrounding the heart is hyperechoic with respect to the liver, and the region of the forming diaphragm can easily be identified. From 45 days onwards, it is possible to identify the fluid-filled (anechoic) stomach, and a few days later the bladder can be imaged.

In late pregnancy, the head, spinal column and ribs produce intense reflections and become more easily identifiable. In the last 20 days of gestation, the kidneys can be seen and in late pregnancy small intestine may be detected.

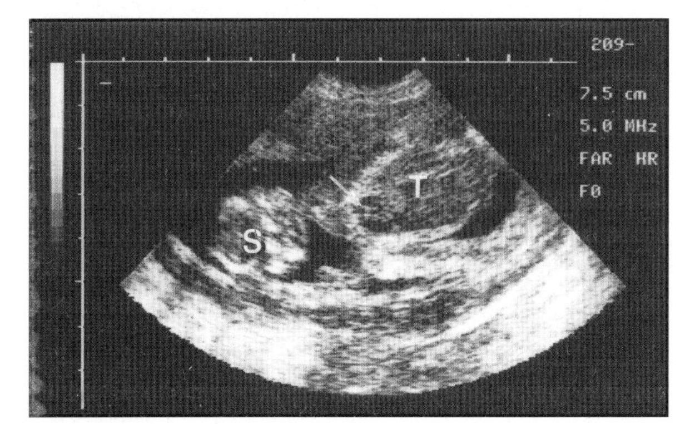


***Figure 11.5**: Ultrasound image of a fetus demonstrating the fetal skull (S) and trunk (T). The heart (arrowed) can be clearly identified. 5.0 MHz transducer, scale in cm.*

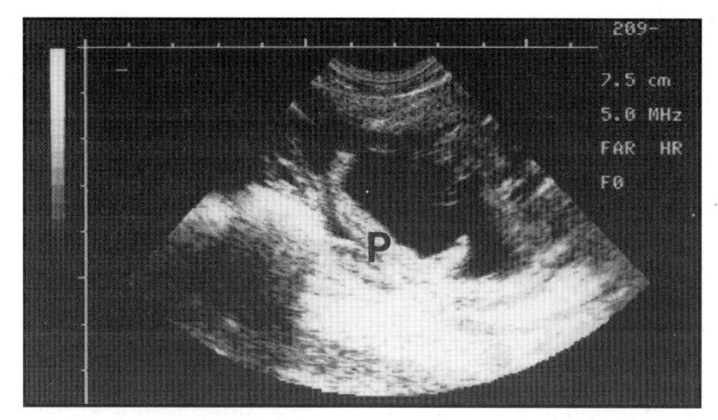


***Figure 11.6**: Ultrasound image of a conceptus demonstrating the C-shaped zonary placenta (P) seen in longitudinal section. 5.0 MHz transducer, scale in cm.*

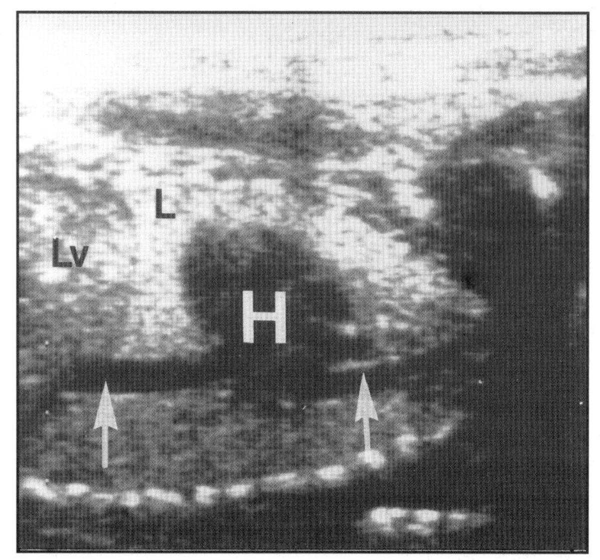


***Figure 11.7**: Ultrasound image of a fetal thorax, demonstrating the heart (H), liver, (Lv), lungs (L) and major vessels (arrowed). 7.5 MHz transducer, scale in cm.*

A positive diagnosis of pregnancy can be made earlier in the queen, and it is possible to identify the fluid-filled conceptus from 11 days after mating. At this time the conceptus appears similar to that of the dog. Embryonic tissue can be imaged from day 14, and cardiac motion is often detected one day later. Fetal structures appear similar to those described for the dog. Whilst embryonic and fetal development may appear to occur earlier in the queen than the bitch, when events are related to the fertilization period, they are similar.

In several domestic species and in humans, ultrasonography is used to determine the gestational age. This may also be of value in bitches with multiple or uncertain mating times. Conventionally this is achieved using measures of fetal size; however, these have only been established for a few breeds of dog, and therefore an alternative approach is to use the time that specific organs can be imaged using ultrasound. For example, the fetal bladder is usually only imaged during the last 20 days of gestation.

### Accuracy

It is unlikely that bitches or queens examined in late pregnancy would be incorrectly diagnosed as non-pregnant. However, in the bitch, early examinations may be inaccurate, especially if conducted in relation to the time of mating and not the time of ovulation (the latter is often not known by the owner). This may result in a false negative diagnosis, other causes of which include overlooking a conceptus, or acoustic artefacts produced by gas or faecal material hiding a conceptus. False positive diagnoses may be the result of confusion of empty loops of small intestine with early pregnancy. Fetal resorption may also produce a disparity between the number of conceptuses imaged and the number of offspring born.

The accuracy of detecting absolute fetal number is poor, and the greatest accuracy is at the first examination one month after mating. The lowest accuracy is in late pregnancy. Most commonly the number of fetuses is underestimated and the accuracy is reduced in large litters.

## DISORDERS OF PREGNANCY

Pregnancy is a normal physiological state and it is often necessary to reassure owners of pregnant animals that normal exercise and minor injuries will not be deleterious.

### Bitch

#### Normal physiological changes

A number of normal physiological events occur during pregnancy which may be mistakenly thought to be abnormal. In the bitch, the most common is the development of a normochromic, normocytic anaemia, associated with a significant reduction in the percentage packed cell volume. It is also common for pregnant bitches to have a reduced appetite and to demonstrate a small volume of mucoid vulvar discharge at approximately one month after mating.

#### Conception failure

The absence of pregnancy despite an apparently normal mating is a frequent cause for concern for many breeders. In these cases it is not uncommon for the cause to be attributed to an 'infection', and veterinary surgeons are often persuaded to culture specimens collected from the vagina and/or prepuce. The isolation of bacteria should not, however, lead to a diagnosis of infectious infertility since in more than 99% of cases the organisms are commensal bacteria. The 'treatment' of normal commensal bacteria is unnecessary, inappropriate and may lead to the establishment of resistant bacteria, yeast or fungal organisms.

The most common reason for conception failure is mating at an inappropriate time. The fertilization period of the bitch is between 2 and 5 days after ovulation (see Chapter 1), and in some cases the problem occurs because the behavioural signs of oestrus do not always correlate with the changes in peripheral plasma hormones. However, most commonly, the problem relates to a misunderstanding of the normal reproductive physiology, and the owner's desire to have their bitch mated a certain number of days after the onset of pro-oestrus. Bitches may ovulate as early as day 5 or as late as day 30 after the onset of pro-oestrus; therefore, the concept of 'counting days' is wholly inappropriate. A variety of methods can be used to detect the optimal time of mating and these are discussed in Chapter 1.

#### Embryonic resorption

The true incidence of embryonic resorption is unknown; however, owners frequently complain about bitches which appear to have been pregnant but do not subsequently whelp. This observation is unreliable, although some scientific studies have shown a disparity between the number of puppies born and the number of ovulated oocytes, and a recent ultrasound study demonstrated that resorption of one or two conceptuses occurs in up to 10% of pregnancies. In these cases the pregnancy continues and normal puppies are born at term. These cases clearly differ from those reported by breeders when the entire litter is resorbed; the incidence of the latter is unknown. There are many potential causes of resorption, including embryonic abnormalities, abnormal maternal environment and infectious agents (see below). However, the most likely cause of an isolated resorption is either abnormal development of the embryo, or competition for uterine space. Treatment at the time of a resorption is not warranted unless the bitch becomes systemically unwell.

#### Fetal abortion

Expulsion is the commonest sequel to fetal death in the bitch. Mummification is rare and the majority of bitches are presented with a dark red vulval discharge; the aborted material may have been eaten by the bitch and be unavailable for inspection. Fetal abortion may have several causes, including fetal defects, abnormal maternal environment, infectious agents and trauma. In cases of infectious abortion, the bitch is usually systemically ill. Fetal death can be confirmed using real-time ultrasonography, when loss of fetal fluid and absence of heart beats are noted. Radiography may be useful for the detection of fetal or uterine gas, overlapping of the cranial bones and abnormal fetal posture, all of which are signs of fetal death. Treatment aimed at preventing an ongoing abortion is inappropriate, and it is best to encourage expulsion, possibly by using ecbolic agents such as oxytocin (5–10 IU), combined with parental fluid therapy and antimicrobial preparations. Progesterone should not be administered during an abortion since it encourages closure of the cervix and may result in endotoxaemia.

The cause of an abortion may be established by serological examination of the dam and bacterial isolation from the fetal membranes and stomach; however, in many cases non-specific bacteria, including *Escherichia coli*, streptococci, *Proteus* and *Pseudomonas*, are identified.

### Infectious agents

**Brucella canis:** Abortion caused by *Brucella canis* occurs most commonly between days 45 and 55 of pregnancy; however, there may be early fetal resorption, or the birth of stillborn or, more rarely, weak puppies. *B. canis* can be transmitted in several ways including contact with aborted fetal or placental tissue, contact with the vaginal discharge of infected bitches, venereal transmission and congenital infection. The most common method of infection is venereal.

Between 1.5 and 6.6% of dogs in the USA have antibodies diagnostic of brucellosis; however, the organism is not present in the UK. The isolation of the bacterium from blood or aborted tissue is diagnostic of the disease. However, there may be prolonged periods when the bitch is not bacteraemic, so that a negative blood culture does not rule out infection. Fortunately, diagnosis using the plate agglutination test for screening and tube agglutination for confirmation is not difficult, titres of 1:200 or greater being diagnostic of infection.

**Toxoplasma gondii:** Infection with *Toxoplasma gondii* is an uncommon cause of abortion in the bitch. The public health consequences of *Toxoplasma* infection should be considered whenever it is diagnosed.

*T. gondii* may also cause premature birth, stillbirth and neonatal deaths. Surviving infected puppies may carry the infection.

***Canine herpes virus:*** The outcome following exposure to canine herpes virus depends largely upon the time of infection. Abortion generally follows infection in mid-pregnancy, whilst infection during early pregnancy may result in fetal death and subsequent mummification, and infection during late pregnancy results in premature birth. It appears that infection of the pregnant bitch results in the production of placental lesions and direct transmission to the fetus. Infected placentas are macroscopically underdeveloped and possess small greyish-white foci which are characterized by focal necrosis and the presence of eosinophilic intranuclear inclusion bodies.

The virus has also been recovered from vesicular lesions on the genitalia of bitches. Frequently, these lesions develop during pro-oestrus, suggesting that venereal transmission is important in adult dogs. Recrudescent canine herpes with virus shedding may be stimulated by the stress of pregnancy and parturition.

***Canine distemper virus:*** Experimental exposure of pregnant bitches to canine distemper virus produces either clinical illness in the bitch with subsequent abortion, or subclinical infection of the bitch and the birth of clinically affected puppies. This provides evidence for transplacental transmission, although the frequency of this under natural conditions is unknown.

***Canine adenovirus:*** Infection with canine adenovirus during pregnancy can result in the birth of dead puppies or weak puppies which die a few days after parturition. In most cases, however, the virus is ingested by the neonate and causes subsequent mortality. Carrier bitches occur and may therefore act as a source of infection for puppies.

### Habitual abortion

Habitual abortion is an uncommon clinical condition in the bitch. In many alleged cases, pregnancy has never been confirmed by a reliable method, and most are probably non-pregnant bitches that are repeatedly mated at an inappropriate time.

Where repeated abortion has been adequately documented there is often an abnormal uterine environment (such as cystic endometrial hyperplasia), or other disease of the uterus or placenta present. However, some workers claim that habitual abortion is the result of progesterone deficiency due to poor luteal function. In the bitch the corpora lutea are the principal source of progesterone throughout gestation, and very low concentrations of progesterone (6 nmol/l; 2 ng/ml) are required to support the pregnancy. Measurement of plasma progesterone at the time of an abortion often reveals that concentrations are basal; however, this is likely to be the result of the abortion rather than its cause. It is not appropriate to supplement bitches with progesterone to prevent habitual abortion unless a persistently low plasma progesterone concentration has been documented. Progesterone or progestogen supplementation during pregnancy may produce masculinized female puppies or cryptorchid male puppies, and may possibly impair or delay parturition, resulting in fetal death. Progestogen therapy should be restricted to those cases in which a true luteal insufficiency has been diagnosed, and then only short acting progestogens administered orally should be used.

### Diabetes mellitus

During the pregnant and non-pregnant luteal phase, there is increased secretion of growth hormone. This results in peripheral insulin antagonism, resulting in a transient reversible type II diabetes (although the clinical signs may be severe and the changes are not always reversible). In non-pregnant animals, ovariohysterectomy is useful, since it reduces the period of instability associated with the luteal phase. In pregnant animals, the diabetes may be severe, and the sensitivity to insulin may be reduced as early as 35 days after ovulation. These animals can be difficult to manage, and often ovariohysterectomy after parturition is the best option to allow stabilization of the bitch and prevent further breeding.

### Hypocalcaemia

Hypocalcaemia (eclampsia, puerperal tetany) occurs most frequently during late pregnancy or in early lactation. It is seen more commonly in small to medium sized bitches. The early clinical signs are restlessness, panting, increased salivation and a stiff gait which progress to muscle fasciculations and pyrexia. If untreated, tetany and death result. The aetiology of hypocalcaemia is probably related to calcium loss in the milk, combined with poor dietary availability. Slow intravenous administration of a 10% calcium gluconate solution to effect (5–10 ml) is usually curative. During administration, heart rate and rhythm should be carefully monitored.

After recovery, further supplementation may be given subcutaneously and orally to prevent recurrence. Hypocalcaemia may be prevented using oral calcium supplementation in the last few days of pregnancy and during lactation. In several species, excessive oral calcium administration has been shown to reduce the intestinal absorption of calcium and inhibit the secretion of parathyroid hormone. This does not, however, appear to be a clinically significant problem in the bitch.

### Pregnancy hypoglycaemia

Low blood glucose concentration and ketosis may occur during pregnancy in the bitch. The condition is uncommon, but may result in weakness that progresses to coma and death. The diagnosis of the condition is based upon the clinical signs and the measurement of plasma glucose concentration. Following the intravenous administration of glucose, these cases respond rapidly.

Some bitches with hypoglycaemia may be wrongly diagnosed as being hypocalcaemic, based upon the clinical signs. These animals may temporarily respond to the intravenous administration of calcium gluconate solution, but relapse. This should raise the suspicion of another cause.

The aetiology of pregnancy hypoglycaemia is uncertain, as during progesterone dominance glucose concentration is normally elevated.

## Queen

### Normal physiological changes

Progressive enlargement and reddening of the nipples is a common and normal feature of pregnancy in the queen. A small volume of mucous material may be present at the vulva, although this is not observed as frequently as in the pregnant bitch.

### Conception failure

Conception failure in the queen is most commonly due to a low plasma LH concentration following mating. In these queens, LH fails to exceed the threshold concentration necessary to stimulate ovulation. Such cases are characterized by a return to oestrus every 3 weeks (Figure 11.2). Queens which ovulate following mating but do not become pregnant, and those which spontaneously ovulate, have an interoestrus interval of between 30 and 45 days. An alternative method of diagnosing ovulation failure is to measure plasma progesterone concentration 7–10 days after mating. A low progesterone concentration demonstrates an absence of luteal tissue and therefore a failure of ovulation.

Multiple matings within a short period of time are required to ensure an adequate release of LH; alternatively, human chorionic gonadotrophin (500 IU) can be administered on day one of oestrus. This mimics an LH surge and results in ovulation.

### Embryonic resorption and fetal abortion

Isolated embryonic resorption with continuation of the pregnancy may be recognized in the queen at routine ultrasound examination. The incidence of resorption is unknown; however, it probably has a similar aetiology to that in the bitch.

Other causes of resorption and abortion include environmental, nutritional, genetic and infectious factors.

### Infectious agents

***Feline leukaemia virus:*** Feline leukaemia virus (FeLV) may cause a variety of reproductive tract diseases including embryonic resorption; fetal abortion and the birth of permanently infected kittens also occur. The virus appears to cross the placenta; however, the actual abortion may be the result of secondary bacterial infection following FeLV-induced immunosuppression.

FeLV may be suspected clinically on the basis of tumour development or signs of other FeLV-related disease, including a history of reproductive failure or deaths in young cats. FeLV infection can be diagnosed by virus isolation (virus detected in plasma), immunofluorescence (detection of virus protein p27 within neutrophils), or enzyme-linked immunosorbent assay (ELISA; detection of virus protein p27 in plasma).

A cat that is positive by immunofluorescence for p27 is likely to be excreting FeLV; most of these cats remain positive for life. A cat that is negative by immunofluorescence for p27 has no detectable infected blood cells. A positive ELISA result for p27 demonstrates circulating p27, and most of these cats are excreting FeLV. As with a negative immunofluorescence result, a negative ELISA result demonstrates no detectable p27, but does not exclude the possibility that the virus is being incubated.

Several studies have found some cats that are positive by ELISA but negative by immunofluorescence, or by virus isolation. It appears that these cats do not give birth to infected kittens and do not excrete the virus.

Owners should be encouraged to test queens before mating, and discouraged from breeding from affected queens, since their offspring are born persistently infected. Infected kittens usually develop an FeLV-related disease soon after birth. Vaccination is available to provide protection against FeLV and its related diseases.

***Feline herpes virus:*** Abortion during the fifth or sixth week of pregnancy may follow infection with feline herpes virus I. In experimental studies both uterine and placental lesions have been demonstrated, although in the clinical situation abortion may be the result of a non-specific reaction to the infection, such as pyrexia. The virus is normally transmitted via the respiratory tract, and approximately 80% of cats exposed remain chronically infected. Accurate diagnosis can be difficult, although intranuclear inclusion bodies within nasal or conjunctival smears can be detected by indirect immunofluorescent techniques. The virus may be confirmed by isolation in tissue culture. Herpes virus abortion is normally diagnosed on the basis of the clinical signs and isolation of the virus. All breeding animals should be vaccinated, since this provides solid immunity against infection.

***Feline panleucopenia virus:*** Whilst abortion is a relatively common complication of feline panleucopenia virus infection, in many cases the clinical signs are stillbirths, neonatal deaths and the birth of kittens with cerebellar hypoplasia. The virus passes across the placenta and the outcome depends upon the time of infection; infection in early pregnancy results in fetal abortion, whilst infection later in pregnancy results in cerebellar hypoplasia and stillbirths. The virus is transmitted by direct contact with saliva, faeces and urine, and diagnosis is made on the basis of the clinical signs, histopathological findings, virus isolation and a rising antibody titre. Serological diagnosis can be made by virus neutralization or ELISA assay with paired serum samples. All breeding animals should be vaccinated.

***Feline infectious peritonitis virus:*** Abortion during the last 2 weeks of pregnancy is not uncommon following infection with feline infectious peritonitis virus. The virus may also cause endometritis, stillbirths, and chronic upper respiratory tract disease and fading kitten syndrome.

At the time of abortion, queens are not always ill and the abortion may be unnoticed by the owner. Diagnosis is made by serological and pathological investigation. Antibody can be detected by indirect immunofluorescence or ELISA; however, whilst very high titres may be diagnostic, moderate titres may be present in some normal cats. In many cases a diagnosis is only confirmed at post-mortem examination, when mesothelial hyperplasia and granulomatous infiltrates with focal necrosis are found.

**Toxoplasma:** Abortion may rarely be the result of *Toxoplasma gondii* infection. Some kittens are, however, born congenitally infected with the protozoan. The role of this organism in abortion can be demonstrated using serological screening.

**Chlamydia:** The feline strain of *Chlamydia psittaci* may cause abortion in the queen. The aetiology of the abortion and the mode of transmission has not been elucidated, although the organism may be found within the reproductive tract of aborting queens. Direct isolation and the demonstration of high antibody titres may be diagnostic of infection, although these findings should be interpreted with care, since the organism may be opportunistic.

### Habitual abortion

There is only anecdotal evidence to suggest that habitual abortion is a clinical problem in the queen. As in the bitch, habitual abortion has been suggested to be the result of a low plasma progesterone concentration. However, the minimum plasma progesterone concentration required to maintain pregnancy is 3–6 nmol/l (1–2 ng/ml), and in the author's experience values this low are uncommon, even immediately prior to an abortion. Most cases of habitual abortion are the result of an abnormal uterine environment such as endometrial hyperplasia. The administration of progesterone or progestogens to 'prevent' habitual abortion would therefore appear to be unwarranted, and these agents should be used with care because they may produce masculinized female kittens and cryptorchid male kittens. Depot progestogens may also impair or delay parturition. Progestogens should be used only in those cases in which a true luteal insufficiency has been diagnosed.

### Hypocalcaemia

Hypocalcaemia (eclampsia, puerperal tetany) is rare in cats, but it may be seen during late pregnancy or in early lactation. The aetiology, clinical signs and treatment are similar to those seen in the bitch. The condition can be treated by the slow intravenous administration of a 10% calcium gluconate solution to effect (2–5 ml). During administration heart rate and rhythm should be carefully monitored.

## UNWANTED MATING AND PREGNANCY TERMINATION

Unwanted mating in the bitch and queen is a common clinical problem in veterinary practice. Bitches are frequently presented during oestrus soon after an owner has observed a mating, or is suspicious that a mating has occurred. In the queen, mating is frequently not observed and the animal is often presented when the owner is concerned over the development of a pregnancy.

There are several options that may be considered in the animal that has had an unwanted mating, including those which prevent or interfere with implantation, those which alter the normal endocrine environment and induce resorption or abortion, and those which are directly embryotoxic. It is noteworthy that of the agents discussed below, oestradiol benzoate is the only product licensed in the UK for treating bitches for unwanted mating, and no products are licensed for use in the cat.

## Bitch

In the bitch, the treatment chosen following an unwanted mating will depend upon the time of presentation to the veterinary clinic in relation to the presumed mating, and the likelihood of the animal becoming pregnant. In some cases, the risk of a pregnancy is low and it may be prudent to wait until a diagnosis of pregnancy is made before instituting treatment. In other cases the animal may only be presented when there is concern that a pregnancy has been established.

### Bitches presented immediately following a mating

There are several options when a bitch is presented 1-4 days after mating:

- If the bitch is not required for breeding, an ovariohysterectomy may be performed after the end of oestrus
- If the risk of pregnancy is high, the bitch may be treated in an attempt to prevent implantation
- If the risk of pregnancy is low, it may be prudent to wait until a positive diagnosis of pregnancy can be made (requires ultrasound examination, measurement of acute phase proteins, or accurate palpation at approximately 28 days after mating) before initiating therapy to terminate pregnancy.

The risk of a pregnancy can be determined by the use of vaginal cytology and measurement of plasma concentrations of progesterone.

***Vaginal cytology:*** Vaginal cytology can be performed to assess the stage of the oestrous cycle (see Chapter 1) and to attempt to identify spermatozoa within the vagina. Normal vaginal cytological techniques are however not reliable for detecting spermatozoa within the vagina when the interval between mating and examination is more than 24 hours. After this time, a modified technique is required. This involves placing a cotton swab moistened with saline into the vagina for one minute, and then incubating the tip with 0.5 ml physiological saline in a small test-tube for 10 minutes. The swab should then be squeezed dry, and the saline centrifuged at 2000 rpm for 10 minutes. Spermatozoal heads can be identified by microscopic examination of the sediment after staining with Diff-Quik® as for vaginal cytology (see Chapters 3 and 4). Using this technique, spermatozoa can be found in 100% of samples in which mating occurred within the previous 24 hours, and in 75% of samples in which mating occurred within the previous 48 hours.

***Plasma progesterone concentration:*** Several ELISA techniques are commercially available for the measurement of plasma progesterone concentration. These assays give either qualitative or quantitative results within a short time (45-60 minutes) of sample collection. In the bitch, plasma progesterone concentration is low during anoestrus and pro-oestrus. Concentrations increase when preovulatory follicular luteinization occurs, and there is a rapid rise in progesterone concentration after ovulation. Basal progesterone concentrations (<6 nmol/l; <2 ng/ml) generally indicate a low risk of conception, elevated concentrations (>32nmol/l; >10 ng/ml) indicate a high risk of conception, and intermediate concentrations have a medium risk. Progesterone concentrations remain elevated after the fertile period, and values should be interpreted in the light of the clinical history and assessment of vaginal cytology.

### Bitches presented several days after mating

There are several options for bitches that are presented more than 5 days after a mating including:

- Treatment with agents that will induce luteolysis (and therefore resorption) regardless of whether the animal is pregnant
- Diagnosis of pregnancy at approximately 28 days after mating, and, if positive, termination of pregnancy either by altering the endocrine environment or using agents with direct toxic effects upon the embryo.

### Prevention of implantation

***Oestrogens:*** Oestrogens are widely used to prevent unwanted pregnancies. It is thought that they act by altering the transport time of zygotes, and interfering with implantation. In some bitches given oestrogens during oestrus, a subsequent early luteolysis has been noted, and this effect may be mediated either directly upon luteal progesterone production or by inhibition of LH release which is luteotrophic.

Diethylstilboestrol, oestradiol benzoate, oestradiol cypionate and mestranol have been widely used to prevent implantation. Only oestradiol benzoate is licensed for the treatment of unwanted mating in the UK. This is administered within 4 days of mating, or at a low dose (0.01 mg/kg), using a repeated administration regimen, at 3 and 5 days (and possibly also 7 days) after mating. The low dose regimen has been advocated in an attempt to reduce the possibility of adverse effects.

Oestrogens may result in the development of pyometra, since they potentiate the stimulatory effects of progesterone on the uterus and they cause cervical relaxation, thus allowing vaginal bacteria to enter the uterus. Oestrogens may also produce dose-related bone marrow suppression, and the toxic dose for an individual may lie within the therapeutic range. Other less severe effects include alopecia, skin hyperpigmentation and mammary and vulval enlargement.

***Tamoxifen:*** Tamoxifen is an anti-oestrogen which can be used to prevent or terminate pregnancy in the bitch, although the exact mechanism of action is unknown. It may interfere with zygote transport and/or implantation. Relatively high doses of the drug given twice daily orally during pro-oestrus, oestrus, or early metoestrus (dioestrus) are efficacious. A high incidence of pathological changes of the bitch's reproductive tract is induced by tamoxifen, including ovarian cysts and endometritis, and the compound is therefore of little value for potential breeding animals.

**Termination of pregnancy**

Progesterone is the principal hormone of pregnancy, and is produced throughout gestation by the corpora lutea. Therapies directed against progesterone are therefore useful for pregnancy termination and may produce either resorption (treatment early in pregnancy) or abortion (treatment later in pregnancy). Strategies for pregnancy termination include inducing lysis of the corpora lutea, removing support for the corpora lutea (by inhibiting LH and prolactin release which are the primary luteotrophic factors), or by blocking the action of progesterone. A further method of pregnancy termination is the use of specific embryotoxic agents.

***Prostaglandins:*** Prostaglandins cause lysis of the corpora lutea and reduce plasma progesterone concentration. In addition, they stimulate uterine contraction. Canine corpora lutea are, however, more resistant to the action of prostaglandin than those of other species, and repeated therapy for several days is necessary to induce complete luteolysis. Prostaglandins have a variety of adverse effects in the bitch including, salivation, vomiting, pyrexia, hyperpnoea, ataxia and diarrhoea; high doses may be lethal. In general therefore low doses of dinoprost (150 μg/kg) given twice daily for 5 days are most suitable since this regimen minimizes the adverse effects.

Prostaglandins may be used as early as 5 days after the onset of metoestrus (dioestrus) to induce luteolysis. They may therefore be used in bitches in which a diagnosis of pregnancy has not yet been made. However, more common treatment regimens utilize twice-daily administration from 23 days after ovulation onwards for approximately 5 days.

***Prolactin antagonists:*** Prolactin is a principle luteotrophic agent in the bitch. Prolactin antagonists such as bromocriptine and cabergoline reduce plasma progesterone concentration, and are especially efficacious when administered during the second half of the luteal phase.

Bromocriptine is less specific than cabergoline and produces some adverse effects, of which the most common is vomiting. When administered daily for 7 days after day 45, bromocriptine (20 μg/kg) can be used to terminate pregnancy, although the efficacy of the regimen is low. Cabergoline (5 μg/kg), however, can be used successfully when given daily for 7 days after day 40, and treatment even as early as day 30 is efficacious in the majority of bitches. Treatment before day 40 usually results in resorption, whilst later treatment results in expulsion.

***Combination of prostaglandin and prolactin antagonists:*** The combination of cabergoline and the synthetic analogue of prostaglandin $PGF_{2\alpha}$, cloprostenol, induces pregnancy termination from day 23 after ovulation. Cabergoline (5 μg/kg) can be administered orally daily, and cloprostenol (5 μg/kg) subcutaneously every other day. This regimen reduces the adverse effects of prostaglandin therapy alone, and increases the efficacy of prolactin antagonists. When bitches are treated for approximately 9 days, 100% will resorb, and there are generally no adverse effects.

***Progesterone synthesis inhibitors:*** Progesterone production can be blocked by hydroxysteroid dehydrogenase isomerase enzyme inhibitors, which prevent the conversion of pregnenolone to progesterone. Pregnenolone is biologically inactive, and therefore progesterone support for the pregnancy is lost, resulting in resorption or abortion.

Epostane is one such enzyme inhibitor that has no intrinsic oestrogenic, androgenic or progestogenic activity. When administered subcutaneously at the onset of metoestrus (dioestrus), epostane will prevent or terminate pregnancy. This route of administration is associated with a high incidence of injection site abscessation. However epostane given in high doses (5.0 mg/kg) orally for 7 days will successfully terminate pregnancy without apparent adverse effects. Epostane is not available in the UK.

***Progesterone antagonists:*** Mifepristone (RU486) is an orally administered progesterone and glucocorticoid receptor antagonist. After the administration of RU486, peripheral progesterone concentration is unaltered but its action is blocked. When administered from one month after ovulation fetal death occurs within 5 days of treatment. Bitches can be treated earlier in pregnancy, although the treatment period must be increased. Pregnancy loss is characterized by resorption during early pregnancy or abortion during late pregnancy. Adverse effects have not been noted with this treatment. A similar progesterone antagonist, aglepristone (RU46534) has recently been marketed for subcutaneous use in bitches in France. The product is administered on two occasions at 5.0 mg/kg separated by 3 days, and has been found to be efficacious and to produce minimal adverse effects.

***GnRH agonists and antagonists:*** Specific LH antagonists are not available; however, the continuous administration of gonadotrophin releasing hormone (GnRH), or GnRH agonists causes a down-regulation effect which may be used to withdraw gonadotrophin support of the corpora lutea. This results in a decline in progesterone concentration and resorption or abortion.

Similarly, the daily administration of GnRH antagonists successfully suppresses luteal function.

The agent most widely tested is detirelix, which is a GnRH antagonist. The daily administration of detirelix produces abortion or resorption depending upon the stage of pregnancy; however, this agent is not available in the UK. Administration very early in the pregnancy is not efficacious even when the dose is increased. This reflects the relative independence of the corpora lutea from gonadotrophin support during early pregnancy, and attempts to overcome this have been made by combining the GnRH antagonists with prostaglandin. The action of these two agents appear to be synergistic, and they may be useful for the termination of pregnancy as early as day 4 of metoestrus (dioestrus), although reported success rates are only 80%.

***Corticosteroids:*** In many species, glucocorticoids are administered during late pregnancy to induce abortion. Glucocorticoid administration results in the production of oestrogen and prostaglandin by the fetoplacental unit.

Single doses of glucocorticoids are not efficacious in the bitch, although dexamethasone, when administered twice daily for 10 days, commencing one month after ovulation, produces abortion. The mechanism of action is uncertain; however, the repeated administration and high doses make the method impractical for clinical use.

***Embryotoxic agents:*** Several novel embryotoxic agents have been evaluated in the bitch. These drugs are often phenyltriazole isoindole and isoquinoline compounds. They are most effective when given around the time of implantation and often only a single administration is required. One agent, lotrifen, has a slow release from depot injection and may be efficacious when administered any time during the first 15 days after mating. Success rates have been shown to be up to 90%; however, some bitches may deliver weak puppies which rapidly die. Many of these agents have toxic side effects, including vomiting, diarrhoea, weight loss, pyrexia, lethargy, purulent meningitis and immunosuppression. For this reason these products are not presently available for clinical use.

## Queen

Many owners are unaware when their queen is in oestrus, and frequently there is little evidence of mating. Therefore few queens are presented immediately following an unwanted mating. It is more common for queens to be presented when there are clinical signs of a pregnancy. The physiology and endocrinology of pregnancy are similar in the queen and the bitch, and whilst it is likely that similar regimens can be used to prevent implantation or to terminate pregnancy, many treatments have not been adequately investigated in this species.

### Queens presented immediately following a mating

It is uncommon for owners to be suspicious that a queen has been mated unless this has been witnessed. Even in the case of an observed mating, it is difficult to determine the risk of pregnancy. There are therefore several treatment options if the queen is presented 1-4 days after mating, including the following:

- If the queen is not required for breeding, an ovariohysterectomy may be performed immediately, or one week after oestrus, when the queen will either be interoestrus or in early pregnancy
- The queen may be treated in an attempt to prevent implantation
- The queen may be examined for an early pregnancy diagnosis at 3 weeks, and, if positive, therapy to terminate the pregnancy can be initiated.

### Queens presented several days after mating

There are several options for queens that are presented more than 5 days after a mating, including:

- Treatment with agents that will induce luteolysis (and therefore resorption) regardless of whether the animal is pregnant
- Diagnosis of pregnancy at approximately 21 days after mating, and, if positive, termination of pregnancy.
- If the queen is not required for breeding, an ovariohysterectomy may be performed as described above.

### Prevention of implantation

***Oestrogens:*** There is limited information concerning the use of oestrogens to prevent implantation in queens. Diethylstilboestrol (0.5 mg/cat) has been shown to be efficacious when given 1-2 days post mating, and in the USA oestradiol cypionate (0.25 mg/cat) is widely used in a similar regimen. Oestradiol benzoate has been reported anecdotally to be a suitable treatment. None of these agents is licensed for this purpose.

Oestrogens have adverse effects in queens that are similar to those reported in the bitch.

***Progestogens:*** There is limited evidence to show that several progestogens, including megoestrol acetate and delmadinone acetate, may be used to prevent implantation if given on the day of mating. The mechanism of action has not been elucidated and these agents are not licensed for this purpose in the UK.

**Termination of pregnancy**

In the queen, the primary source of progesterone throughout pregnancy is the ovary. Pregnancy termination can therefore be achieved by influencing progesterone production or its action.

The principal luteotrophic agent in the queen appears to be prolactin, similar to the situation in the bitch.

***Prostaglandins:*** Prostaglandin $F_{2\alpha}$ (dinoprost; 0.4 mg/kg) can be used to induce abortion when administered daily for 5 days after day 33. The mechanism of action is related to lysis of the corpora lutea and uterine spasmogenic activity. Similar to the bitch, repeated therapy is required, since the corpora lutea appear to be relatively insensitive to the action of prostaglandin. Adverse effects are frequently noted following prostaglandin administration and include nausea, vomiting and diarrhoea.

***Prolactin antagonists:*** Prolactin is the principle luteotrophic agent in the queen and inhibition of prolactin secretion results in the demise of the corpora lutea. Prolactin antagonists such as bromocriptine and cabergoline may therefore be used to terminate pregnancy. Bromocriptine is less specific and produces more adverse effects than cabergoline. Cabergoline (5 μg/kg daily for 7 days) can be useful for pregnancy termination as early as day 30, and when treated at this time, embryonic resorption is the common outcome. Some early treatments are, however, not successful and treatment should be delayed until after day 40, when efficacy is 100%. Later treatments are associated with fetal abortion.

***Combination of prostaglandin and prolactin antagonists:*** The combination of cabergoline (5 μg/kg daily for 10 days) and the synthetic analogue of prostaglandin $PGF_{2\alpha}$, cloprostenol (5 μg/kg every other day for 10 days), may be used to induce pregnancy termination from day 30 after mating. This regimen allows the dose and frequency of prostaglandin administration to be reduced, and increases the efficacy of cabergoline.

In all cases, fetal abortion results, however the queens remain well and returns to normal fertility.

***Other agents:*** There are no studies of the use of progesterone synthesis inhibitors or progesterone antagonists in the queen; however, it is likely that these agents, and others such as GnRH agonists and antagonists, would terminate pregnancy. Similarly, embryotoxic agents have not been evaluated.

## CONCLUSION

Recent studies have improved the understanding of normal reproductive physiology of the bitch and queen, and have allowed major developments in the areas of pregnancy detection, treatment of diseases during pregnancy, and pregnancy termination. These areas of research are extremely important, considering the polarity associated with dog reproduction; on the one hand attempting to increase productivity for show and research animals, and on the other, aiming to prevent reproduction because of the considerable pet overpopulation problems.

## FURTHER READING

Concannon PW (1991) Reproduction in the dog and cat. In: *Reproduction in Domestic Animals, 4th edn*, ed. PT Cupps, pp. 517-554. Academic Press, San Diego

Concannon PW, England GCW, Verstegen JP and Russell HA (1993) Fertility and infertility in dogs cats and other carnivores. Proceedings of the Second International Symposium on Canine and Feline Reproduction. *Journal of Reproduction and Fertility*, Supplement 47

Concannon PW, England GCW, Rijnberk A, Verstegen JP and Doberska C (1997) Reproduction in dogs, cats and exotic carnivores. *Journal of Reproduction and Fertility*, Supplement 51

England GCW (1995) Small animal reproductive ultrasonography. In: *Veterinary Ultrasonography*, ed. PJ Goddard, pp. 55—85. CAB International, London

CHAPTER TWELVE

# Parturition

*Catharina Linde-Forsberg and Annelie Eneroth*

## NORMAL PARTURITION

### Duration of pregnancy

Apparent gestation length in the bitch averages 63 days, with a variation from 56 to 72 days if calculated from the day of the first mating to parturition. This surprisingly large variation is due to the long behavioural oestrous period of the bitch. Actual gestation length determined endocrinologically is much less variable, parturition occurring 65 ± 1 days from the preovulatory LH peak, i.e. 63 ± 1 days from the day of ovulation.

In the queen, ovulation occurs in response to mating, between 24 and 48 hours after coitus. Because of repeated coitus, parturition can be expected between 63 days after the first mating to 63 days (plus 24–48 hours) after the last mating. Feline gestation periods of 56–71 days have been reported, but the average ranges between 63 and 67 days.

Gestation length both in the dog and cat has been reported to be shorter for larger litter sizes, but this remains uncertain. Breed differences in gestation length, although not well documented, have been postulated for both dogs and cats.

### Litter size

The litter size in dogs varies, ranging from as few as one puppy in the miniature breeds to more than 15 in some of the giant breeds. It is smaller in the young bitch, increases up to 3-4 years of age, and decreases again as the bitch gets older. A litter size of only one or two puppies predisposes to dystocia because of insufficient uterine stimulation and large puppy size, 'the single-puppy syndrome'. This can be seen in dog breeds of all sizes. Breeders of the miniature breeds tend to accept small litters, but should be encouraged to breed for litter sizes of at least three to four puppies to avoid these complications.

In cats, litter size may vary between one and nine kittens, with an average-size litter being 3.5–4.6 kittens. For primiparous females, it has been found that litter size is smaller, averaging 2.8 kittens. Small litter size appears not to predispose to inertia in cats.

### Early fetal loss

The true incidence of early embryonic or fetal death (before 45 days of pregnancy) and spontaneous abortion in the bitch and queen is unknown, and may be difficult to determine since the condition may pass unnoticed. The female may consume aborted fetuses, or resorption of conceptuses may occur until day 45 of pregnancy without noticeable signs. In pigs the embryonic wastage is reported to be 30–50%.

### Perinatal loss

Based on a number of surveys, puppy losses up to weaning age appear to range between 10 and 30% and average around 12% in the dog. More than 65% of puppy mortality occurs at parturition and during the first week of life; few puppies die after 3 weeks of age. Perinatal death is 12.8% by weaning age in the cat. Inbreeding is said to be associated with a high incidence of fetal and neonatal mortality. An apparently short gestation length (less than 61 or 62 days) appears to increase the risk of mortality for kittens.

### Physiology of parturition

An understanding of the course and control of normal parturition (eutocia) is necessary for the diagnosis and treatment of abnormal parturition (dystocia). Exact mechanisms which allow parturition to commence and proceed normally in the bitch and queen are still unknown. Studies of canine and feline parturition and extrapolations from other species provide information on the physiological and endocrinological changes important for normal parturition.

Stress produced by the reduction of the nutritional supply by the placenta to the fetus stimulates the fetal hypothalamic–pituitary–adrenal axis; this results in the release of adrenocorticosteroid hormone and is thought to be the trigger for parturition. An increase in fetal and maternal cortisol is believed to stimulate the release of prostaglandin $F_{2\alpha}$, which is luteolytic, from the feto-placental tissue, resulting in a decline in plasma progesterone concentration. Increased levels of cortisol and of prostaglandin $F_{2\alpha}$-metabolite have been measured in the prepartum bitch. Withdrawal of the progesterone blockade of pregnancy is a prerequisite for the normal course of canine and feline parturition; females given long-acting progesterone during pregnancy fail to deliver. In correlation with the gradual decrease in plasma progesterone concentration during the last 7 days before whelping, there is a progressive qualitative change in uterine electrical activity, and a significant increase in uterine activity occurs during the last 24 hours before parturition with the final fall in plasma progesterone concentration. The change in the oestrogen:progesterone ratio is probably a major cause of placental separation and cervical dilation, although in the dog oestrogens have not been unambiguously shown to increase before parturition as they do in many other species. In the queen, a slight rise in oestradiol just before parturition has been observed. Oestrogens sensitize the myometrium to oxytocin, which in turn initiates strong contractions in the uterus when not under the influence of progesterone. Sensory receptors within the cervix and vagina are stimulated by the distension created by the fetus and the fluid-filled fetal membranes. This afferent stimulation is conveyed to the hypothalamus and results in the release of oxytocin. Afferents also participate in a spinal reflex arch with efferent stimulation of the abdominal musculature to produce abdominal straining. Relaxin causes the pelvic soft tissues and genital tract to relax which facilitates fetal passage. In the pregnant bitch, this hormone is produced by the ovary and the placenta; it rises gradually over the last two-thirds of pregnancy. Prolactin, the hormone responsible for lactation, begins to increase 3–4 weeks following ovulation and surges dramatically with the abrupt decline in serum progesterone just before parturition.

## Signs of impending parturition

### Bitch

Relaxation of the pelvic and abdominal musculature is a consistent but subtle indicator of impending parturition. The clinically most important sign is the drop in rectal temperature (Figure 12.1) caused by the final abrupt decrease in progesterone concentration. In the last week before parturition, the rectal temperature of the bitch fluctuates, and drops sharply approximately 8–24 hours before parturition and 10–14 hours after the concentration of progesterone in peripheral plasma has declined to less than 6 nmol/l (2ng/ml). This drop in rectal temperature is individual but also to a certain extent seems to depend on body size. Thus, in miniature-breed bitches it can fall to 35°C, in medium-sized bitches to around 36°C, whereas it seldom falls below 37°C in bitches of the giant breeds. This difference is probably an effect of the surface area:body volume ratio. Several days before parturition, the bitch may become restless, seek seclusion or be excessively attentive, and may refuse all food. The bitch may show nesting behaviour 12–24 hours before parturition, concomitant with the increasing frequency and force of uterine contractions, and shivering, attempting to increase the body temperature.

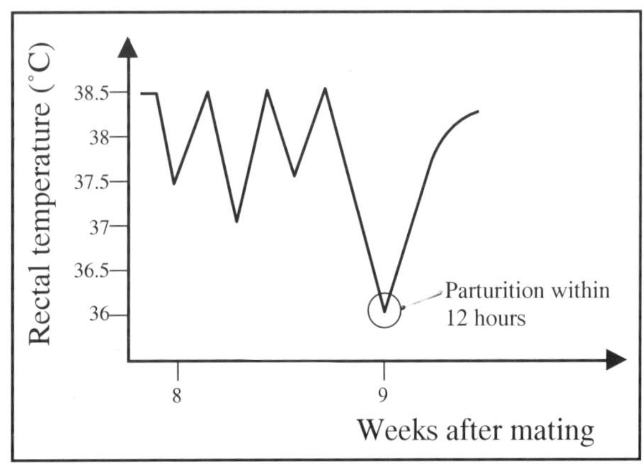


*Figure 12.1: The best sign of impending parturition in the bitch is the marked drop in rectal temperature. During the last week of pregnancy, the temperature will fluctuate because release of prostaglandins causes a transient fall in peripheral plasma progesterone concentration, and progesterone is thermoregulatory. During first stage labour, the drop is more pronounced, and the bitch should be in second stage labour within 12 hours after reaching the lowest temperature, and before temperature has returned to normal.*

### Queen

A decrease in rectal temperature during the preceding 12 hours or in the first stage of labour may be noted in the queen, although this sign is not as reliable as in the bitch. During the ninth week, the queen is generally less active. The primiparous queen especially may become more anxious and apprehensive during the last 2 days of pregnancy, searching for a place in which to deliver. Some queens will refuse all food from 12–24 hours before delivery, while others may not become anorexic at all and may even feed during parturition.

In primiparous bitches and queens, lactation may be established less than 24 hours before parturition, while after several pregnancies, colostrum can be detected as early as 1 week prepartum.

## Stages of parturition

Parturition is divided into three stages, with the last two stages being repeated for each puppy or kitten delivered.

### First stage

The duration of the first stage is normally between 6 and 12 hours. It may last for up to 36 hours, especially in the nervous primiparous animal, but for this to be considered normal, the rectal temperature must remain low. Vaginal relaxation and dilation of the cervix occur during this stage. Intermittent uterine contractions, with no signs of abdominal straining, are present. The dam appears uncomfortable, occasionally glancing at her abdomen, and the restless behaviour becomes more intense. Panting, tearing up and rearranging of bedding, shivering and occasional vomiting may be seen in the bitch. The queen may vocalize, turn around in circles and wash herself constantly. Some dams show no behavioural evidence of first stage labour. The weak uterine contractions increase both in frequency and intensity towards the end of the first stage.

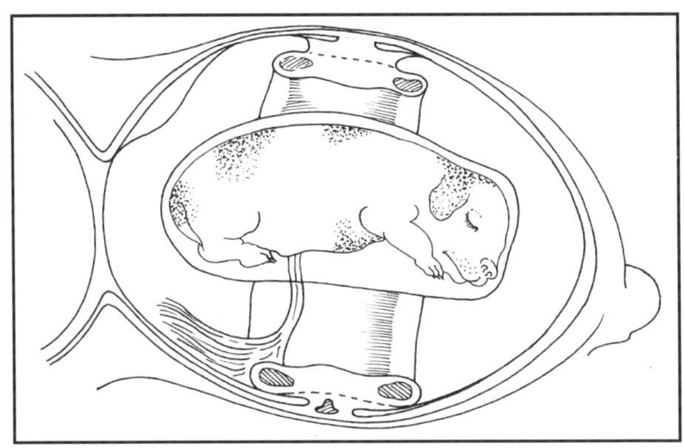

*Figure 12.2: Schematic picture of the fetus and the fetal membranes in the dog.*

During pregnancy the orientation of the fetuses within the uterus is 50% directed caudally and 50% cranially, but this changes during first stage labour as the fetus rotates on its long axis, extending its head, neck and limbs to attain normal birth position, resulting in 60% of puppies and kittens being born in anterior and 40% in posterior presentation. The fluid-filled fetal membranes are pushed ahead of the fetus by the uterine propulsive efforts and dilate the cervix.

## Second stage

The duration of the second stage is usually 3–12 hours, in rare cases up to 24 hours. At the onset of second stage labour, the rectal temperature rises and quickly returns to normal or slightly above normal. The first fetus engages in the pelvic inlet, and the subsequent intense, expulsive uterine contractions are accompanied by abdominal straining. On entering the birth canal, the allantochorionic membrane may rupture and a discharge of some clear fluid may be noted. Covered by the amniotic membrane, the first fetus is usually delivered within one hour after onset of second stage labour in the cat and within 4 hours in the dog. Normally, the bitch or queen will break the membrane, lick the neonate intensively and sever the umbilical cord. At times the dam will need some assistance to open the fetal membranes to allow the newborn to breathe, and sometimes the airways will have to be emptied of fetal fluids (see the chapter on the neonate). The umbilicus should be clamped with a pair of forceps and cut with a blunt pair of scissors to minimize haemorrhage from the fetal vessels, leaving about 1 cm of the umbilicus. In cases of continuing haemorrhage the umbilicus should be ligated (see Chapter 14).

***Diagnosing second stage labour:*** It is of utmost importance that the clinician is able to determine whether the dam is in the second stage or still in first stage labour. Inexperienced breeders tend to get nervous during first stage, not fully understanding the function of this preparatory stage of parturition during which recurring uterine contractions, the softening of the birth canal and opening of the cervix take place.

There are three signs which indicate that the bitch or queen has entered into second stage labour:

- The passing of fetal fluids (first water bag bursts)
- Visible abdominal straining
- The rectal temperature returning to normal level.

If one or more of these signs has been observed the female is in second stage labour.

In normal labour, the female may show weak and infrequent straining for up to 2 and at the most 4 hours before giving birth to the first fetus. If the female is showing strong, frequent straining without producing a puppy or kitten, this indicates the presence of some obstruction and she should not be left for more than 20–30 minutes before seeking veterinary advice.

The dam should be examined if:

- She has a greenish/red-brownish discharge but no puppy/kitten is born within 2–4 hours
- Fetal fluid was passed more than 2–3 hours ago but nothing more has happened
- The dam has had weak, irregular straining for more than 2–4 hours
- The dam has had strong, regular straining for more than 20–30 minutes
- More than 2–4 hours have passed since the birth of the last puppy/kitten and more remain
- The dam has been in second stage labour for more than 12 hours.

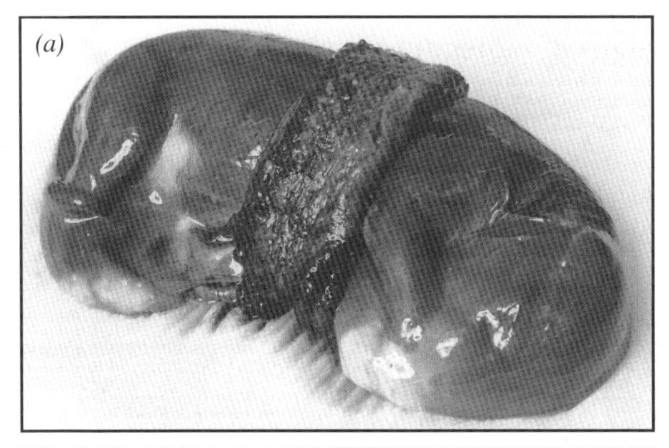

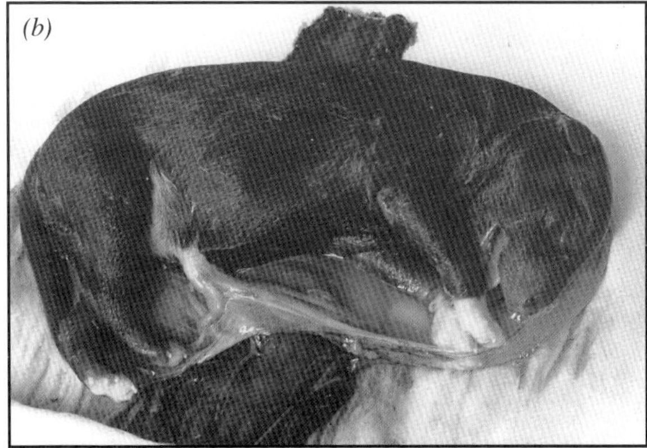

***Figure 12.3:*** *(a) A puppy in intact fetal membranes, following delivery by Caesarian operation. (b) The fetal membranes have been opened and the puppy is gasping for air.*

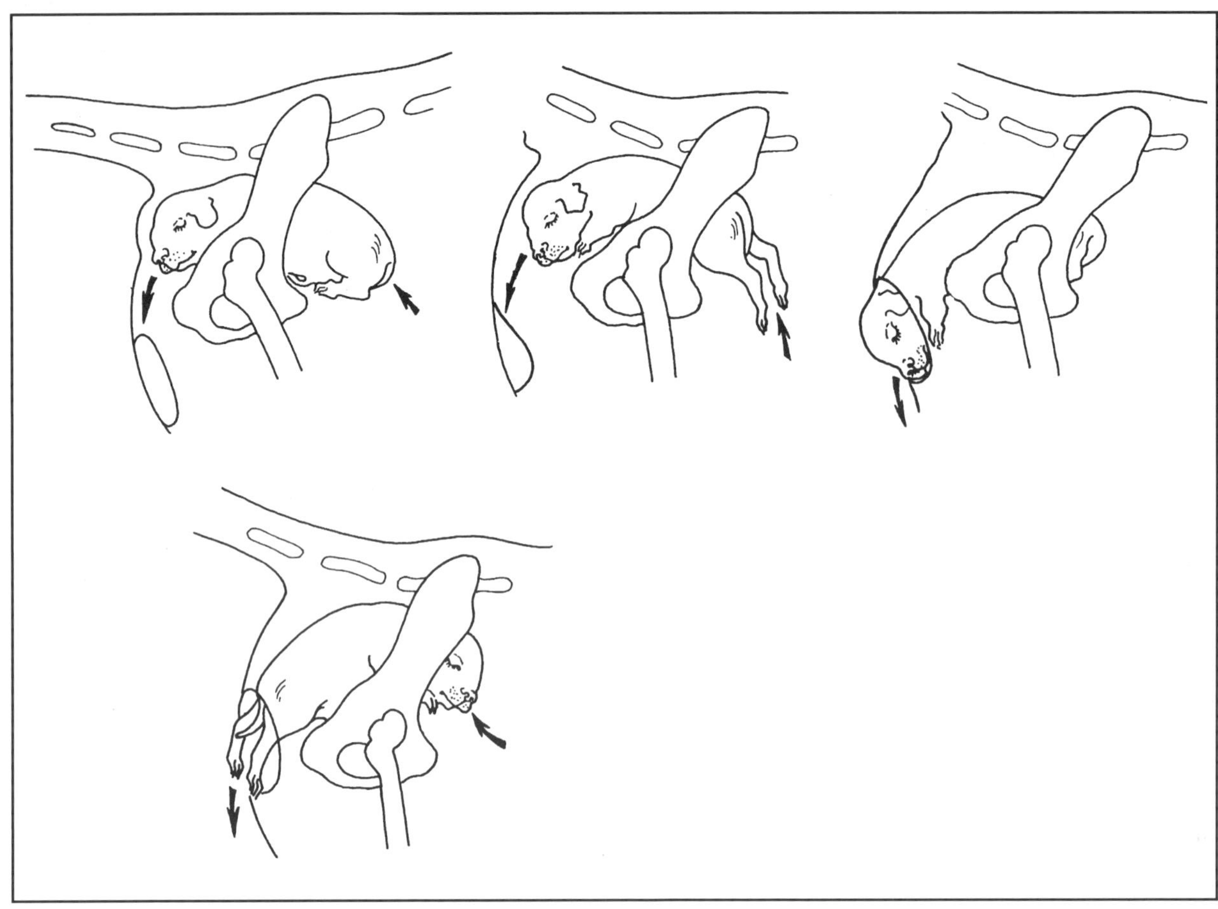

*Figure 12.4: Normal birth of a puppy in anterior and posterior presentation.*

### Third stage

The third stage of parturition, expulsion of the placenta and shortening of the uterine horns, usually follows within 15 minutes of the delivery of each fetus. Two or three fetuses may, however, be born before the passage of their placentas occurs. The dam should be discouraged from eating more than one or two of the placentas because she may develop diarrhoea and vomiting. Aspiration pneumonia due to placental vomitus could be life threatening. Lochia, i.e. the post-partum discharge of fetal fluids and placental remains, will be seen for up to 3 weeks or more, being most profuse during the first week. The discharge is greenish in colour in the bitch, and red-brown in the queen. Uterine involution is completed after 12–15 weeks in the bitch.

The dam should be examined if:

- All placentas have not been passed within 4–6 hours (although placental numbers may be difficult to determine because of the dam eating them)
- The lochia are putrid and/or foul smelling
- There is continuing genital haemorrhage
- The rectal temperature is higher than 39.5°C
- The general condition of the dam is affected
- The general condition of the puppies/kittens is affected.

### Interval between births

Expulsion of the first fetus usually takes the longest. The interval between births in normal uncomplicated parturition is from 5–120 minutes. In almost 80% of cases, the fetuses are delivered alternately from the two uterine horns. When giving birth to a large litter, a bitch or a queen may stop straining and rest for more than 2 hours between the delivery of two consecutive fetuses. The second stage straining will resume, followed again by the third stage, until all the fetuses are born.

### Completion of parturition

Parturition is usually completed within 6 hours after the onset of second stage labour, but it may last up to 12 hours. It should never be allowed to last for more than 24 hours because of the risks involved both for the dam and the fetuses.

## DYSTOCIA

### Definition

Dystocia is defined as difficult birth or the inability to expel the fetus through the birth canal without assistance.

### Frequency

Dystocia is a frequent problem in both the dog and the

cat. The true incidence of dystocia in the bitch is probably around 5% overall, but it may amount to almost 100% in some breeds of dog, especially of the achondroplastic type and those selected for large heads. The risk of dystocia appears to be significantly higher in pedigree cats than in cross-bred cats. The litter prevalence of dystocia in the dolichocephalic breeds (Siamese-type and Cornish Rex) appears to be around 10%, in the brachycephalic breeds (Persian, British Shorthair and Devon Rex) it appears to be around 7%, and in the mesocephalic breeds (mixed breeds, Abyssinian, Burmese and Manx) it is as low as around 2%.

## Clinical assessment

When a case of dystocia is presented, an accurate history and a thorough clinical examination of the patient are important prerequisites for proper management. The three criteria for being in second stage labour, namely passage of fetal fluids, visible abdominal straining and temperature returned to normal, should be assessed. An evaluation of the female's general health status should be made and signs of any adverse effects of parturition noted. Observation should be made of behaviour, character and frequency of straining, examination of the vulva and perineum noting colour and amount of vaginal discharge. Mammary gland development, including congestion, distension, size and presence of milk, should be evaluated. Palpation of the abdomen, roughly estimating the number of fetuses and degree of distension of the uterus, should be carried out. Digital examination of the vagina using aseptic technique should be undertaken to detect obstructions and determine the presence and presentation of any fetus in the pelvic canal (Figure 12.5). In most bitches, it is not possible to reach the cervix during first stage, but an assessment of the degree of dilation and tone of the vagina may give some indication of the status of the cervix and the tone of the uterus. Pronounced tone of the anterior vagina may indicate satisfactory muscular activity in the uterus, whereas flaccidity may indicate uterine inertia. The character of the vaginal fluids also will indicate whether the cervix is closed, with the production of a fluid which is scant and sticky creating a certain resistance to the introduction of a finger, or open when fetal fluids lubricate the vagina making the exploration easy. When the cervix is closed the vaginal walls also fit quite tightly around the exploring finger, whereas with an open cervix the cranial vagina appears more open.

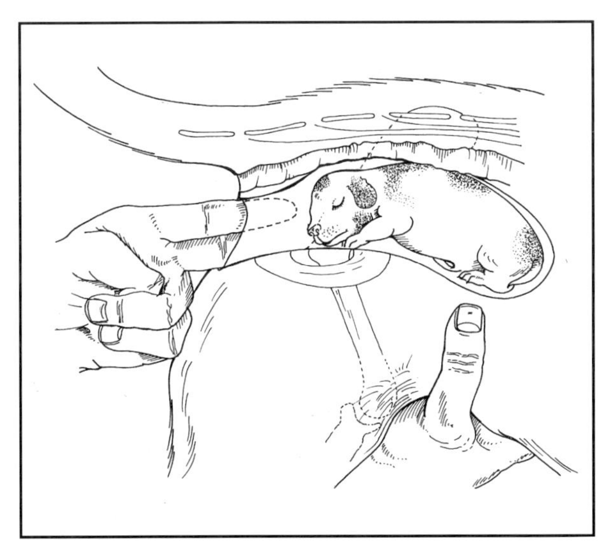

*Figure 12.5: Assessing fetal disposition in the bitch in second stage labour. Redrawn after Shille (1983).*

Radiographic examination is valuable to assess gross abnormalities of the maternal pelvis, number and location of fetuses, to estimate their size, detect congenital defects, or signs of fetal death. In the dead fetus, intra-fetal gas will appear 6 hours after death and can be detected in the radiograph, while overlapping of cranial bones and collapse of the spinal column will not be seen until 48 hours have passed after the death of the fetus. Ultrasound examination will determine fetal viability or distress, normal heart-rate being 180-240 per minute, decreasing in the compromised fetus.

## Diagnosis

The range of normal variations observed in dogs and cats at parturition makes recognition of dystocia difficult, especially for the inexperienced observer. The following criteria for dystocia may assist in making the diagnosis:

- The rectal temperature has been down and returned to normal with no signs of labour
- Green vulvar discharge (dog), red-brown discharge (cat) but no fetuses delivered. (These discharges emanate from the marginal haematoma of the placentas and indicate that at least one placenta is beginning to become separated from the maternal blood supply. They are normal once birth is underway)
- Passing of fetal fluids 2-3 hours before and no signs of labour
- Absence of labour for more than 2 hours or weak, infrequent labour for more than 2-4 hours.
- Strong and persistent non-productive labour for more than 20-30 minutes
- Evidence of an obvious cause of dystocia, e.g. pelvic fracture or a fetus stuck in the birth canal and partially visible
- Signs of toxaemia (disturbed general condition, general oedema, shock) when parturition should be occurring.

## Maternal causes of dystocia

Traditionally, dystocia is classified as being of either maternal or of fetal origin, or a combination of both (Table 12.1).

| Cause | Bitch (%) | Queen (%) |
|---|---|---|
| Maternal: | 75.3 | 67.1 |
| Primary complete inertia | 48.9 | 36.8 |
| Primary partial inertia | 23.1 | 22.6 |
| Birth canal too narrow | 1.1 | 5.2 |
| Uterine torsion | 1.1 | – |
| Uterine prolapse | – | 0.6 |
| Uterine strangulation | – | 0.6 |
| Hydrallantois | 0.5 | – |
| Vaginal septum formation | 0.5 | – |
| Fetal: | 24.7 | 29.7 |
| Malpresentations | 15.4 | 15.5 |
| Malformations | 1.6 | 7.7 |
| Fetal oversize | 6.6 | 1.9 |
| Fetal death | 1.1 | 1.1 |

***Table 12.1**: Causes of dystocia in bitches and queens.*

### Uterine inertia

Uterine inertia is by far the most common cause of dystocia in dogs and cats. It is classified into primary and secondary inertia.

In primary inertia the uterus may fail to respond to the fetal signals because there is only one or two puppies, and thus insufficient stimulation to initiate labour (the single-puppy syndrome) or because of overstretching of the myometrium by large litters, excessive fetal fluids or oversized fetuses. Other causes of primary inertia may be an inherited predisposition, nutritional imbalance, fatty infiltration of the myometrium, age-related changes, deficiency in neuro-endocrine regulation, or systemic disease in the dam. *Primary complete uterine inertia* is the failure of the uterus to begin labour at full term. *Primary partial uterine inertia* occurs when there is enough uterine activity to initiate parturition but insufficient to complete a normal birth of all fetuses in the absence of an obstruction.

Secondary uterine inertia is always due to exhaustion of the uterine myometrium caused by obstruction of the birth canal and should be distinguished from primary inertia.

***Management:*** In cases of primary uterine inertia, the owners should initially be instructed to try to induce straining by actively exercising the dam, for instance by running around the house or up the stairs with her. A considerable number of puppies and kittens are born in the car on the way to the veterinary surgery. Most of these would probably have been delivered in the calm and quiet of home had the owners tried to induce straining themselves, thus giving the puppies and kittens a better start in life and possibly also resulting in the whole litter being born without further intervention. Another means of induction of straining in the female with insufficient labour is by feathering of the dorsal vaginal wall (Figure 12.6). Feathering is accomplished by inserting one or two fingers into the vagina of the queen or bitch, respectively, and to push or 'walk' with them against the dorsal vaginal wall thus inducing an episode of straining (the Ferguson reflex). Feathering can also be effective in initiating labour after correction of the position or posture of a fetus.

Nervous voluntary inhibition of labour due to psychological stress may occur, mainly in the nervous primiparous animal. Reassurance by the owner or administration of a low dose tranquillizer may remove the inhibition. Once the first fetus is born, parturition will usually proceed normally.

The female with complete primary uterine inertia is often presented with a bright and alert appearance, a normal rectal temperature, and no evidence of labour. The cervix is often dilated, and vaginal exploration easy to perform due to the presence of fetal fluids, but the fetus may be out of reach because of the flaccid uterine wall. Before initiation of medical treatment of uterine inertia any obstruction of the birth canal must be excluded.

Calcium solutions and oxytocin are the drugs of choice in cases of uterine inertia. Oxytocin has a direct action on the rate of calcium influx into the myometrial cell, which is essential for myometrial contraction. Many do not respond to oxytocin alone but require prior administration of a calcium solution. Therefore, some 10 minutes before the administration of oxytocin, 10% calcium gluconate, 0.5–1.5 ml/kg body weight (dog 2–20 ml and cat 2–5 ml) should be given by slow i.v. infusion (1 ml/min) with careful monitoring of the heart rate. Bitches of some of the smaller breeds may be especially prone to hypoglycaemia, particularly after prolonged straining. In such cases, a dilute (10–20%) glucose solution can be added to the infusion or given i.v. in doses of 5–20 ml. The recom-

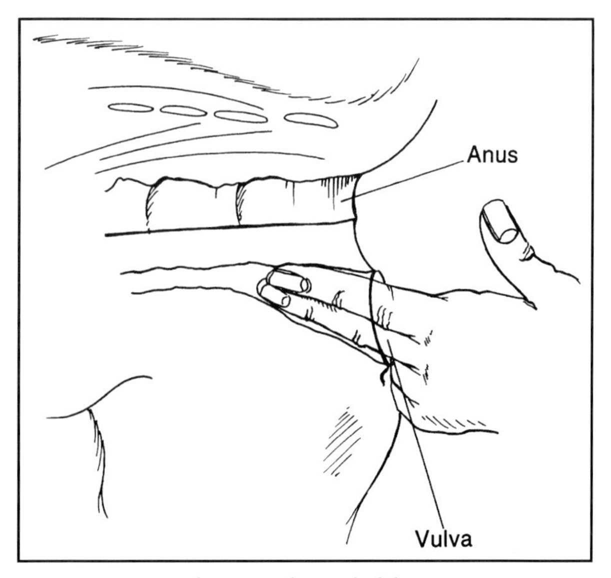


***Figure 12.6**: 'Feathering'. The roof of the vagina is 'tickled' with the fingers to stimulate contractions.*

mended dose of oxytocin for the bitch is 1-5 IU given i.v. or 2.5-10 IU i.m. and for the queen 0.5 IU given i.v. or i.m., which can be repeated at 30-minute intervals. The total maximum dose should never exceed 3 IU in the cat. The response to treatment will be reduced with each repeated administration. Higher doses than recommended or too frequent administration may result in prolonged contracture of the myometrium, preventing fetal expulsion and impeding utero-placental blood flow. The disadvantages of oxytocin include a tendency to cause premature induction of placental separation and cervical closure. If there is no response to treatment after a second administration of oxytocin the dam should be delivered without further delay, either with the aid of obstetrical forceps if only one or two puppies or kittens remain and are within easy reach in the uterine corpus, or by Caesarean operation.

The long-acting ergotamines should never be used in connection with parturition.

***Treatment regimen:***

- Running with the dam, and feathering of the vaginal vault
- A 10% solution of calcium gluconate is given slowly i.v. while carefully checking the heart rate
- The dam is given 30 minutes to respond to treatment. If straining begins, the treatment can be repeated if necessary or continued with oxytocin
- If the calcium infusion has no effect within 30 minutes, oxytocin is given i.v. or i.m.
- The dam again is given 30 minutes to respond to treatment. If straining begins, the treatment can be repeated if necessary, although each additional administration will elicit a weaker response
- If nothing happens within 30 minutes, it is not likely that further treatment will be successful. The dam should be delivered, either by forceps if only one or two fetuses remain and are within easy reach, or by Caesarean operation.

**Obstruction of the birth canal**

Obstruction of the birth canal may be of maternal or fetal origin. Some maternal causes for obstruction are:

- *Uterine torsion*, and *uterine rupture.* These are acute, life-threatening conditions that can occur during late pregnancy or at the time of parturition. Sometimes a few fetuses are born before parturition stops and the condition of the dam may quickly deteriorate. Surgery is always required and a quick diagnosis essential for the survival of the dam
- *Uterine malposition* resulting from inguinal herniation usually is detected around week 4 of pregnancy, as the fetal uterine enlargements are growing and the contour of the abdomen is markedly disturbed. Sometimes the early stages may be mistaken for mastitis of the rear mammary glands. The condition is corrected by surgery, whereby the uterine horn(s) are repositioned and the herniation sutured. In cases with circulatory disturbance and advanced tissue damage the uterus will have to be removed
- *Congenital malformations of the uterus,* e.g. partial or complete aplasia or hypoplasia of one or both uterine horns, or of the corpus uteri or the cervix, are rare causes of maternal obstructive dystocia. Symptoms depend on the character and degree of the malformation. In cases of unilateral aplasia of an entire uterine horn, small litter size may be the only presenting sign. Retained fetuses behind partial occlusions require surgery, and the final diagnosis is usually made during the operation
- *Soft tissue abnormalities* such as neoplasms, vaginal septa or fibrosis of the birth canal may cause obstructive dystocia. The prepartum relaxation of the vagina will often allow the passage of fetuses in cases of neoplasia, especially if the tumour is pedunculated. Vaginal septa may consist of remnants from the fetal Müllerian duct system, but can also occur secondary to vaginal trauma or infection, and if extensive may prevent the passage of the fetuses. Often, however, they are not so extensive, and vaginal relaxation is sufficient to allow the fetuses to pass. Cervical or vaginal fibrosis are usually secondary to trauma or inflammatory processes and in severe cases will cause dystocia. Surgical intervention may save the puppies/kittens in these cases, and tumours and septa formations may be removed. In cases of fibrosis, however, surgery is seldom successful because of new scar formation during the healing process
- *Narrow pelvic canal* causing obstructive dystocia. This may result from previous pelvic fractures, immaturity or congenital malformation of the pelvis. The normal pelvis has a vertical diameter greater than the horizontal (Figure 12.7). Congenitally narrow birth canals exist in some brachycephalic and terrier breeds; in addition, their fetuses have comparatively large heads and wide shoulders. When achondroplasia exists, as in the Scottish Terrier, dorso-ventral flattening of the pelvis modifies the normal pelvic inlet, which creates an obstruction to the engagement of the fetuses. In the Bulldog, the very large, deep chest and pronounced waist results in the gravid uterus dropping down, and the fetuses being presented at a relatively acute angle to the pelvic inlet. Bulldogs sometimes also have slack abdominal musculature, leading to insufficient uterine contractions and abdominal straining to lift the fetus up into the pelvic cavity.

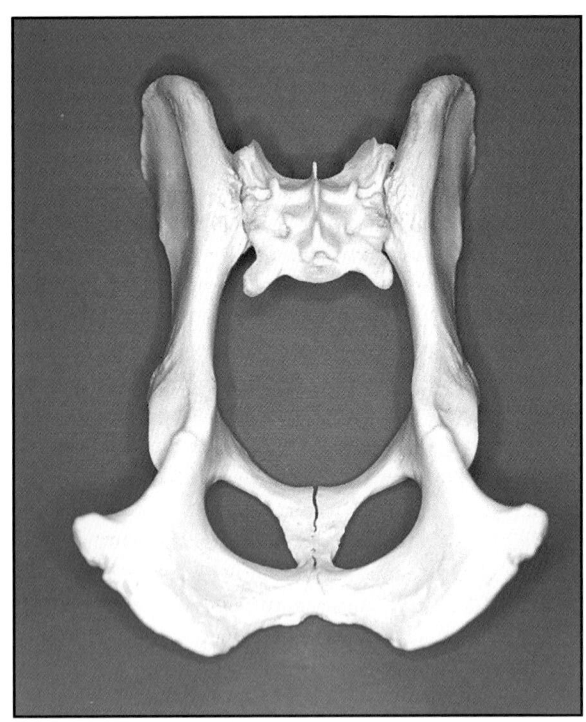

*Figure 12.7: The normal canine pelvis. It is evident that the inner pelvic area is slightly oval. The diagonal is longer than the transverse conjugata; twisting a puppy that is stuck in a diagonal position may therefore help.*

## Fetal causes of dystocia

Fetal obstructions may be caused by oversized fetuses, malpresentations/malorientations or monstrosities, such as hydrocephalus, oedema or various duplications. Oversize in itself is a cause for malpresentation. Fetal death may result in dystocia due to malpositioning or inadequate stimulation for parturition to begin. A healthy fetus is very active during expulsion, extending its head and limbs, twisting and rotating to get through. In most breeds, the greatest bulk of the fetus lies in its abdominal cavity, whereas the bony parts, the head and the hips, are comparatively small. The limbs are short and flexible and rarely cause serious obstruction to delivery in the normal-sized fetus.

### Oversized fetuses

A puppy weight of 4–5% of the weight of the bitch is considered the upper limit for uncomplicated birth. In the absence of monstrosities, oversized fetuses are often associated with small litter size. In breeds in which miniaturizing exists, great disparity of size of fetuses within litters may occur, with some greatly oversized individuals. In brachycephalic breeds like the Boston Terrier, dystocia occurs from the combination of a flattened pelvic inlet and puppies having a large head. The cobby-type cat breeds, such as Persians and Himalayans, with relatively low fertility and therefore few and large fetuses and also being selected for a large, flat head, may be predisposed to obstructive dystocia.

In dystocia due to an oversized fetus, sometimes a portion of the fetus may protrude from the vulva. In anterior presentation the head may be born, and the shoulders and chest cause obstruction, whilst in posterior presentation the hind limbs and hips may protrude.

### Posterior presentation

This is considered normal in dogs and cats, occurring in 40% of fetal deliveries (Figure 12.4). Posterior presentations have, however, been related to a predisposition for dystocia, as mechanical dilation of the cervix may be inadequate, particularly where this involves the first fetus to be delivered. In addition the expulsion is rendered more difficult because the fetus is being delivered against the direction of its hair coat and because the fetal chest instead of being compressed becomes distended by the pressure from the abdominal organs through the diaphragm. Occasionally the fetus may have the elbows hooked around the pelvic brim, preventing further expulsion. Should the fetus become lodged in the pelvic canal, pressure on the umbilical vessels trapped between the fetal chest wall and maternal pelvic floor may cause hypoxia and reflex inhalation of fetal fluids.

### Breech presentation

Breech presentation, i.e. posterior presentation with hind legs flexed forward, can be a serious complication, especially in medium sized and small breeds. Vaginal exploration will reveal a tail tip and perhaps the anus and the bony structure of the pelvis of the fetus.

### Lateral or downward deviation of the head

These are two of the most common malpositionings in the dog and the cat. Lateral deviation is considered to be especially associated with long-necked breeds such as Rough Collies, whereas downward deviation is seen in brachycephalic breeds and long-headed breeds such as Sealyham and Scottish Terriers. In lateral deviation, vaginal exploration will reveal just one front leg, the one contralateral to the direction of the deviation of the head, i.e. when the head is deviated to the left the right front paw will be found and *vice versa*. When there is a downward deviation of the head, either both front legs and sometimes the nape of the neck of the fetus can be palpated, or both front legs may be flexed backwards and only the scull of the fetus be reached.

### Backward flexion of front legs

This condition is especially common when the fetus is weak, or dead, and is sometimes seen in combination with deviation of the head, especially downward. For bitches of the larger or even medium sized breeds, it may be possible to deliver a puppy with one or both front legs flexed.

### Transverse or bi-cornual presentation

Sometimes a fetus, instead of progressing from the uterine horn through the cervix to the vagina, will

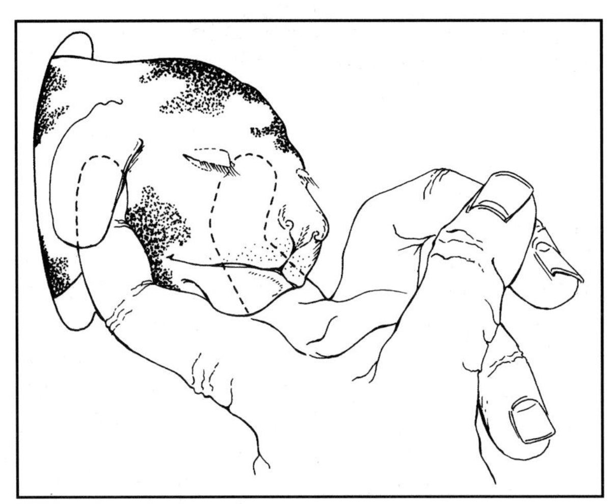

*Figure 12.8: When the puppy's head is within reach, the index and middle fingers are used to grasp it from below, or above, and caudo-ventral traction is applied. Redrawn after Shille (1983).*

proceed into the contralateral uterine horn, possibly due to some obstruction. Another reason may be that the fetus was implanted close to the corpus uteri. If at all possible to reach, the only thing that can be felt is the back, the ribcage or the abdomen of the fetus. These cases always require surgery, as there is no possibility for manual correction.

**Two fetuses presenting simultaneously**

Sometimes one fetus from each horn is presented at the same time, jamming the birth canal. When possible, if one is coming backwards, this one should be removed first, as it occupies more space.

**Management of fetal malpresentations**

If a fetus is present in the birth canal, manipulation by hand or by obstetrical forceps may be attempted, more easily in the bitch than in the queen. In bitches of the larger breeds it may even be possible to insert one hand through the vagina into the uterus and thus extract the puppy.

During natural birth, the puppy will almost make a full somersault, emerging from the loop of the uterine horn, progressing upwards to pass through the pelvic canal and then down again through the long vestibulum of the bitch, the vulva being situated some 5-15 cm below the level of the pelvic floor. In the queen, the vagina and vestibulum are placed more in alignment with the pelvic floor. Thus, after seizing the fetus, a gentle traction in posterio-ventral (bitch) or posterior (cat) direction is applied.

The position of the fetus must be assessed. If it has advanced into and partly through the pelvic canal, it will in the bitch create a characteristic bulge of the perineal region, below the tail. Easing the vulvar lips upwards may reveal the amniotic sac and the position of the fetus. Vaginal exploration and maybe also a radiographic examination will aid in making a diagnosis in the cases when the fetus has not advanced as far.

The narrowest part of the birth canal is within the rigid pelvic girdle. If manipulation is to be attempted, the fetus that cannot easily be pulled out may have to be pushed into the uterus, to lie in front of the pelvic girdle where corrections of position or posture are easier to perform. This should be done between periods of straining of the dam, never working against the uterine contractions. It should also be remembered that the widest part of the pelvic girdle is usually on the diagonal, thus rotating the fetus 45° may create sufficient room for passage. Generous application of obstetrical lubricant, for instance liquid paraffin or petroleum jelly or a sterile water-soluble lubricant, is helpful, especially if the dam has been in second stage labour for some time.

Depending on the position and posture of the fetus, a grip should be applied around its head and neck, from above or below (Figure 12.8), whichever is most convenient, around its pelvis, or around the legs. Care should be taken as the neck and limbs of the fetus are easily torn when pulled. Correction of posture may be obtained by manipulation of the fetus through the abdominal wall with one hand, and concurrent trans-vaginal manipulation with the other. A finger may be introduced into the mouth of the fetus to help correct a downward deviation of the head. Should it be necessary to change the postures of the limbs, a finger is inserted past the elbow or knee and the limb moved medially in under the fetus and corrected.

A gently applied alternating right-to-left traction of the puppy or kitten (Figure 12.9), gently rocking it

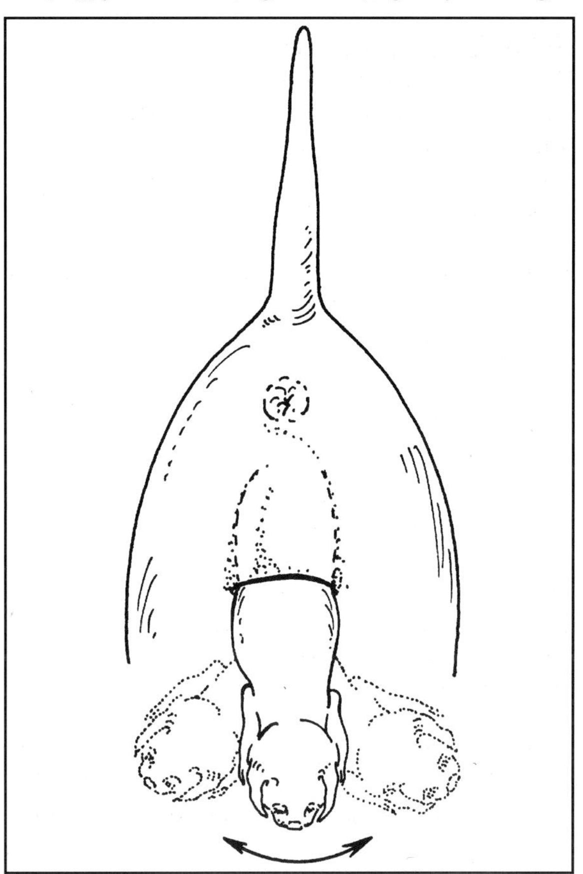

*Figure 12.9: The puppy should be turned alternately to each side to ease the shoulders and the rest of the body free. It should also be twisted into a diagonal position to create more space for delivery.*

back and forth or from side to side and possibly twisting it to a diagonal position within the pelvis will help free the shoulders or the hips one at a time. A slight pressure applied over the perineal bulge may prevent the fetus from sliding back into the uterus again between strainings.

Obstetrical forceps (Figure 12.10) should only be used for assisted traction of a relatively oversized fetus, when it is likely the rest of the puppies in the litter are smaller or when just one or two fetuses remain. Guide the forceps with a finger and never introduce them further than to the uterine body because of the risk of getting part of the uterine wall within the grip, thereby causing serious damage. If the head of the fetus can be reached, the grip should be applied around the neck (Pålsson's forceps) or across the cheeks. In posterior presentation, the grip should be around the fetal pelvis. If the legs can be reached the grip should be around those, not around the feet.

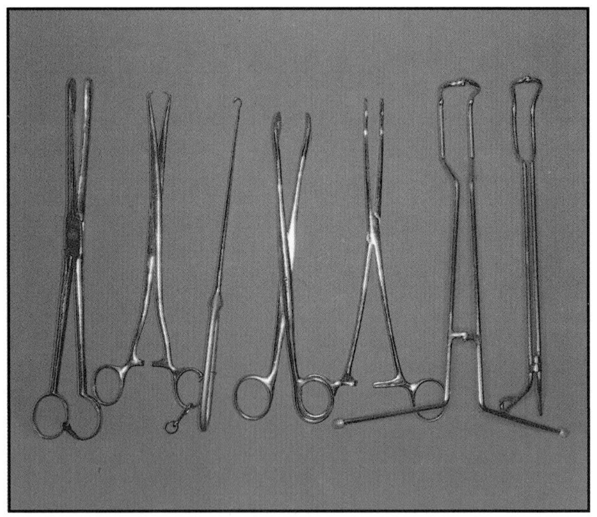

***Figure 12.10:*** *Various obstetrical forceps. From left: Berliner forceps, hook forceps, Albrecht's forceps, another Berliner forceps, Robertson's forceps, and two types of Pålsson's forceps.*

***Outcome of obstetrical treatment:*** Studies reporting on treatment outcome show that digital manipulation including forceps delivery and/or medical treatment for dystocia is successful in only 27.6% of cases in bitches and 29.9% in cats. Around 65% of bitches and almost 80% of queens brought to the clinic because of dystocia thus end up having a Caesarean operation.

In our study, fetal death increased from 5.8% in bitches brought in within 1-4.5 hours after the beginning of second stage labour to 13.7% in the period between 5 and 24 hours. Overall fetal death was 22.3%. Early diagnosis and prompt treatment is therefore crucial in reducing puppy or kitten death rate in cases of dystocia.

## CAESAREAN OPERATION

### Criteria

The indications for Caesarean operation include:

- Complete primary uterine inertia that does not respond to medical treatment
- Partial primary uterine inertia that is refractory to medical management
- Secondary uterine inertia with inadequate resumption of labour
- Abnormalities of the maternal pelvis or soft tissues of the birth canal
- Relative fetal oversize, if considered likely to be repeated in several fetuses
- Absolute oversize, single puppy syndrome or fetal monstrosity
- Excess or deficiency of fetal fluids
- Fetal malposition not amendable to manipulation
- Fetal death with putrification
- Toxaemia of pregnancy and illness of the dam
- Neglected dystocia
- Prophylaxis (history of previous deliveries).

Prophylactic Caesarean operation should be questioned on ethical grounds if it is performed to assist the propagation of a breed line that can not reproduce successfully without intervention.

Once a decision has been made to deliver the litter by Caesarean operation should be carried out without delay. The dam has often endured hours of more or less intensive labour, and may be suffering from physical exhaustion, dehydration, acid-base disorders, hypotension, hypocalcaemia and/or hypoglycaemia. The prognosis for both dam and offspring is good if surgery is performed within 12 hours after the onset of second stage labour; it continues to be fairly good for the dam after 12 hours but guarded for the fetuses. If more than 24 hours have passed after the onset of second stage labour the entire litter is usually dead and further delay compromises the life of the dam.

### Anaesthesia

#### Special considerations

Maternal physiological alterations during pregnancy increase the risk of anaesthetic-related complications for the dam. The functional residual capacity (FRC) is decreased by anterior displacement of the diaphragm by the gravid uterus, and oxygen consumption during pregnancy increases by 20%, which makes pregnant animals susceptible to hypoxia. Consequently supplemental oxygen is advisable during induction and anaesthesia.

Increased sensitivity of the respiratory centre to arterial carbon dioxide tension during pregnancy results in increased minute volume of ventilation and respiratory alkalosis. The maternal pH is not affected due to renal compensation. During first stage labour, a further increase in minute volume due to an increase in respiratory rate, caused by pain and anxiety, may result in an initial rise in pH which, as labour progresses, may change into a metabolic acidosis. The acid–base status of animals presented for emergency Caesarean section varies widely from respiratory alkalosis to metabolic acidosis.

In the normal dam, increased blood volume provides an adequate reserve to compensate for the large quantities of blood and fluid lost at parturition. Normally the bitch and queen are relatively resistant to supine hypotension due to partial vena caval occlusion and aortic compression when placed in dorsal recumbency. However, dehydration, bleeding and inhibition of maternal compensatory mechanisms by anaesthetic drugs increase the risk of supine hypotension. Changes in position (tilting laterally to left or to right), reduction of the time spent in dorsal recumbency and additional cardiovascular support (intravenous fluids and inotropic agents) may be necessary. Regardless of physical status, all animals undergoing Caesarean operation should preferably have crystalloid solutions (e.g. lactated Ringer solution) given i.v. at 10–20 ml/kg/h. Catecholamines with predominantly β-adrenergic activity, such as dobutamine or ephedrine, are preferred as inotropic agents. Dobutamine is administered as a continuous i.v. infusion at 1–10 μg/kg per minute, and ephedrine can be administered at a dosage of 0.03–0.06 mg/kg i.v.

Animals presented for Caesarean operation should be assumed to have a full stomach, as the time of the last feeding is unknown. Pregnant animals also have delayed gastric emptying, resulting in an increased risk of vomiting and regurgitation during anaesthesia.

Anaesthetic techniques for Caesarean operation should provide adequate surgical anaesthesia for the dam while minimizing maternal and fetal depression. Selection of the anaesthetic technique depends on the status of the dam, viability of the fetuses, expected advantages and disadvantages of the techniques available and experience of the veterinary surgeon. Speed is an important consideration; increased anaesthesia time may lead to increased neonatal depression and asphyxia. The surgical preparations of the dam should preferably be performed prior to the induction of anaesthesia. Drugs affect the fetus either directly by crossing the placental barrier or indirectly by altering maternal cardiopulmonary function. Drug-induced cardiopulmonary depression is dose-dependent, necessitating the use of low doses of anaesthetic.

For details of different anaesthetic drugs and techniques a textbook in veterinary anaesthesia is recommended.

**Premedication**

***Phenothiazine tranquillizers:*** Acepromazine or promazine for premedication should only be used at low dosages (0.03–0.05 mg/kg) and when absolutely necessary, since they cause maternal hypotension. They cross the placenta rapidly but do not affect the fetus when given in clinically useful doses.

***Anticholinergics:*** Atropine or glycopyrrolate can be administered as premedication to patients in which excessive salivary secretion or bradycardia is likely, or to prevent vagal reflexes when the gravid uterus is lifted out of the abdomen. The cat is prone to vagal reflexes which makes the administration of anticholinergics beneficial. Doses of atropine are 0.02–0.1 mg/kg in the dog and 0.045–0.1 mg/kg in the cat (s.c.). Glycopyrrolate does not cross the placental barrier and therefore does not alter fetal heart rate which makes it preferable. The recommended dose of glycopyrrolate is 0.01 mg/kg i.m. in the dog and cat.

***Benzodiazepines:*** Diazepam and midazolam produce minimal cardiopulmonary depression in both dam and fetus and are therefore useful premedications. Unfortunately they may some times cause paradoxical excitement in the patient. Diazepam can be given to the cat in doses up to 0.5 mg/kg i.m. and to the dog 0.1–0.5 mg/kg i.v. or from 0.3–1 mg/kg i.m. Midazolam is twice as potent as diazepam and can be administered to the dog in dosages of 0.2–0.5 mg/kg. Flumazenil, a specific benzodiazepine antagonist, may be administered to reverse maternal or fetal lethargy and muscle relaxation in a ratio of 1:13 (flumenazil: benzodiazepine).

**Regional anaesthesia**

Regional anaesthesia by lumbo-sacral *epidural anaesthesia* causes minimal fetal depression. Requirements for sedation of the nervous dam may necessitate the administration of depressant sedative drugs. The spinal cord terminates more caudally in the cat than in the dog making the technique more risky, and in addition the temperament of the cat almost exclusively necessitates sedation. Disadvantages of the technique include prolonged (over one hour) paralysis of the hindlegs, and the possible induction of maternal hypotension from sympathetic blockade. The cranial infiltration of local anaesthetic into the epidural space may lead to paralysis of the respiratory muscles in the dam and thereby respiratory arrest. Elevation of the dam's head, administration of intravenous fluids and avoidance of excessive doses of local anaesthetic minimize maternal and fetal depression. The addition of adrenaline decreases the risk of cranial infiltration, but also prolongs the effect of the local anaesthetic. Recommended doses of lidocaine 2% are 0.2 ml/kg in the cat and 0.1–0.3 ml/kg in the dog, injected into the epidural space in small amounts until the desired effect is achieved. Pregnancy decreases the dose required in the dog, and 1 ml/6 kg is generally satisfactory in bitches presented for Caesarean operation.

Regional anaesthesia by *local infiltration* in the abdominal wall requires sedation of the dam which can be achieved by administration of the α2-adrenoreceptor agonist medetomidine. The depressant effect on the fetuses may be reversed by the application of one drop of the specific antagonist atipamezole on the tongue of the fetus.

**General anaesthesia**

The main disadvantage of general anaesthesia compared to regional anaesthesia is that it induces greater neonatal depression.

Advantages of general inhalation anaesthesia, a technique well known to most veterinary surgeons, include optimum operating conditions and tracheal intubation, which ensures control of the maternal airway and provision of a route for maternal oxygen administration. All inhalant anaesthetics rapidly cross the placental barrier, and the degree of neonatal depression is dependent on the agent, the duration and depth of anaesthesia. Due to pregnancy-induced increases in respiratory rate and decreases in lung volume, alveolar concentrations of inhaled anaesthetics rise more rapidly in pregnant than in non-pregnant animals. The dose required of the inhalant anaesthetic can be decreased by addition of 50% nitrous oxide in a non-rebreathing or semi-closed circuit. Isoflurane is useful for Caesarean operation because it is mostly eliminated through breathing and less than 1% is metabolized in the body.

Induction of anaesthesia by face-mask is an alternative in the quiet bitch but requires adequate gas scavenging. Anxiety and reluctance to breathe through the mask causes uneven supply of the drug to the bitch and may induce hypotension. An option in the queen is to induce anaesthesia in a gas-chamber.

***Barbiturates:*** These are often used as a single bolus for induction of anaesthesia. Barbiturates cross the placental barrier and cause great respiratory depression in the fetuses but short-acting barbiturates such as thiopentone and the ultra-short-acting barbiturate methohexitone affect the fetus minimally if administered in low doses (<8 mg/kg). Pentobarbitone should not be used because of its association with prolonged anaesthetic recovery, neonatal respiratory depression, and high incidence of neonatal mortality.

***Propofol:*** An alkyl phenol anaesthetic, propofol has a short duration of action making it useful for Caesarean operation. In the dog an i.v. bolus (4–6 mg/kg) can be used for induction of anaesthesia. The technique employed is to wait until propofol is eliminated (about 18 minutes) and then maintain anaesthesia by inhalation of gas before removal of the fetuses. This has given good results with live puppies and a rapid recovery in the bitch. For Caesarean operation in the queen (bolus 6–7 mg/kg) propofol seems less useful, giving slower recovery and less vigorous offspring. Propofol can also be given as intermittent injections or as a continuous infusion (0.8 mg/kg/min in the dog, 0.5 mg/kg/min in the cat) but because of placental diffusion this regimen is not advisable for Caesarean operation.

***Saffan*** *(Mallinckrodt Veterinary Ltd)****:*** This is a steroidal preparation of alphaxalone and alphadolone acetate which may be used for induction of anaesthesia in the cat. Recommended doses for the unsedated cat are from 3 mg/kg up to 18 mg/kg i.v. or i.m. depending on the required duration of action (9 mg/kg gives 10–15 minutes). The use of Saffan in the dog is contraindicated. as it causes a massive release of histamine.

***Ketamine:*** Ketamine can be used for general anaesthesia or for induction prior to inhalation anaesthesia. Ketamine should not be used as the sole anaesthetic agent as it causes convulsions and hallucinations, especially in the dog, however the prior administration of medetomidine prevents these adverse reactions. The use of xylazine as a sedative in the dam is not recommended because it induces severe respiratory depression in both the mother and the fetuses, and impairs uterine and placental blood flow by increasing peripheral vascular resistance. Ketamine rapidly crosses the placental barrier, causing dose-dependent fetal depression. It is often used in cats for general dissociative anaesthesia and causes minimal neonatal depression administered in low doses (1 mg/kg i.v., 5 mg/kg i.m.) after sedation with medetomidine (80–100 mg/kg). The dose of medetomidine in the dog varies from 30 to 80 mg/kg, depending on the size of the dog (higher dose to smaller dogs) in the combination of a subsequent i.m. injection of 5 mg/kg ketamine. Atipamezole, an antagonist to medetomidine, given to the fetuses (one drop on the tongue) reverses the depressant effect of medetomidine. The dosage of atipamezole in the cat is 2.5 times the dosage of medetomidine administered and in the dog 5 times (measured in volumes this gives the same volume of atipamezole as of medetomidine in the dog and half the volume of atipamezole as of medetomidine in the cat). The antagonist should not be administered to the dam until the effects of ketamine have disappeared, if the combination is used.

***Narcotic agonists:*** These can be used in combination with sedatives and tranquillizers in neuroleptanalgesic mixtures to produce analgesia and sedation in the dog. The commercially available mixture of fentanyl and fluanison causes less neonatal depression than barbiturates but is said by owners to alter maternal behaviour in the bitch. Recommended dose is 0.3–0.5 ml/kg. The use of fentanyl in the cat is contraindicated, as it causes violent excitement. Opioids readily cross the placenta, causing central nervous system depression and respiratory depression in neonates, which can be reversed by administration of naloxone, an opioid antagonist. Neonates should, however, be closely observed for any signs of relapse of depression as it takes several days for them to eliminate the opioids.

#### Neuromuscular blockers

Neuromuscular blocking agents such as succinylcholine, pancuronium, vercuronium, and atracurium may be useful to improve skeletal muscle relaxation during light anaesthesia. Maternal ventilation and analgesia must, however, be provided. These drugs do not cross the placenta in a quantity sufficient to induce fetal depression. When using succinylcholine, atropine should be given to prevent bradycardia in the dam.

Suggested anaesthetic protocols for Caesarean operation are listed in Figure 12.11.

**Examples of premedication:**

- Acepromazine, 0.03–0.05 mg/kg: tranquillizing effect
- Diazepam, ≤ 1 mg/kg (dog), ≤ 0.5 mg/kg (cat): tranquillizing and muscle relaxing effects
- Atropine, 0.02–0.1 mg/kg (dog), 0.045–0.1 mg/kg (cat) or glycopyrrolate 0.01 mg/kg: anticholinergic effects

**1. Induction of general anaesthesia, without premedication:**

- Gas mask (dog), gas chamber (cat)
- Ultra-short-acting barbiturate, <8 mg/kg
- Propofol bolus, 4–6 mg/kg (dog), 6–7 mg/kg (cat)

Maintenance of general anaesthesia by inhalation of volatile agent; isoflurane (or halothane); and remove the fetuses as soon as possible

**2. Induction of general anaesthesia, with premedication:**

- Saffan in the cat, 3–18 mg/kg
- Short-acting barbiturate, <8 mg/kg
- Ketamine 5 mg/kg (dog), 1 mg/kg i.v. (cat).

Maintenance by inhalation of volatile agent; isoflurane (or halothane); and delay removal of fetuses to allow redistribution/ metabolism of drugs

**3. Lumbo-sacral epidural anaesthesia:**

- Lidocaine 2%, 0.1–0.3 ml/kg (dog), 0.2 ml/kg (cat), injected into the epidural space; often requires premedication

Causes minimal fetal depression, but may cause hypotension in the dam

**4. Reversible, dissociative anaesthesia by the combination of:**

- Medetomidine, 30–80 mg/kg (dog), 70–100 mg/kg (cat)
  AND
- Ketamine, 5 mg/kg i.m. (dog, cat)

Reverse depressant effects of medetomidine with the antagonist atipamezole:

dog: same volume as medetomidine i.m.
cat: half the volume as medetomidine i.m.
neonates: one drop on the tongue

***Figure 12.11:** Suggested anaesthetic protocols for Caesarean operation.*

## POST-PARTURIENT CONDITIONS

It is normal for the bitch and queen to have a slightly elevated rectal temperature, up to 39.2 °C, for a couple of days after parturition. It should, however, not exceed 39.5 °C. Fever during this period usually emanates from conditions of the uterus or the mammary glands.

### Uterine disorders

#### Haemorrhage

Some haemorrhage from the female genital tract after parturition is normal, but maternal blood loss should never exceed a scant drip from the vulva. True haemorrhage should be distinguished from normal vaginal post-parturient discharge. Excessive haemorrhage after parturition may indicate uterine or vaginal tearing, vessel rupture, or may be evidence of an underlying coagulation defect. The haematocrit should be checked, remembering that 30% is normal for the bitch at term. Inspection of the vulva and vagina should be performed in an attempt to locate the source of the bleeding. Oxytocin can be administered to promote uterine involution and contraction of the uterine walls; in more severe cases of uterine haemorrhage an exploratory laparotomy may be necessary. The dam should be monitored closely for signs of impending shock, and blood transfusion may be required while attempting to determine the cause of the haemorrhage.

#### Retained placentas/fetuses

Retained placentas in the bitch or queen may cause severe problems, especially if accompanied by retained fetuses or infection. Clinical signs of retained placenta include a thick dark vaginal discharge. Retained fetuses can be identified by palpation or by ultrasound or radiographic examination. A retained placenta is often palpable, depending on the size of the dam and degree of involution of the uterus, and extraction by careful 'milking' of the uterine horn or by using forceps is sometimes possible. Treatment with oxytocin 1-5 IU (bitch) or 0.5 IU (queen) s.c. or i.m. 2-4 times daily for up to 3 days can help expulsion of retained placentas. The long-acting ergot alkaloids should not be used, as they may cause closure of the cervix. Antibiotic treatment is advisable if the dam is showing signs of illness.

### Acute metritis

Acute metritis is an ascending bacterial infection of the uterus in the immediate post-partum period. Dystocia, obstetrical manipulation, retained fetuses or placental membranes, or parturition in an unsanitary environment predispose to metritis. Metritis may rarely occur after normal parturition, natural mating or artificial insemination or following an abortion. The infection is usually ascending through an open cervix, and is often caused by Gram-negative bacteria. Clinical signs include: fever, dehydration, depression, anorexia, poor lactation and mothering, and a purulent or sanguino-purulent vaginal discharge. A doughy enlarged uterus may be palpated. Abdominal radiographic and/or ultrasonographic examination is indicated to evaluate the uterine size and uterine contents. A vaginal culture is recommended. Vaginal cytology will show large numbers of degenerate neutrophils, red blood cells, bacteria and debris. The complete blood cell count often shows leucocytosis with a left shift. Therapy consists of immediate administration of intravenous fluids and antibiotics, and evacuation of uterine contents. The latter may be accomplished by administering oxytocin or prostaglandin $F_{2\alpha}$. There is one high-dose regimen for administration of prostaglandin $F_{2\alpha}$ (0.1–0.25 mg/kg given s.c. once or twice daily for 3–8 days) and one low-dose regimen (0.025–0.05 mg/kg s.c., 6–8 times daily for 2–3 days). The high-dose alternative may cause adverse reactions such as abdominal pain, increased pulse and respiratory rate, salivary secretion and sweating. These reactions appear within 10 minutes of administration and normally disappear again after 30 minutes to 1 hour. It should be remembered, however, that the prostaglandins are not licensed for use in dogs and cats. In more severe cases ovariohysterectomy is the recommended treatment.

### Subinvolution of placental sites

In the post-parturient period it is normal for the bitch to have a serosanguineous vaginal discharge for up to 3–6 weeks. Normally the uterine involution is completed within 12 weeks after whelping. Subinvolution of placental sites is suspected if a sanguineous vaginal discharge persists for longer than 6 weeks. The aetiology of this condition is unknown, and the bitch often shows no symptoms of illness. Vaginal cytology shows predominantly red blood cells, with syncytial trophoblast-like cells being a useful confirmatory finding. Subinvolution of placental sites almost exclusively affects the young primiparous animal, and in the majority of cases resolves spontaneously, with the prognosis for future pregnancy being good. Because of increased risk of anaemia, secondary bacterial infection or rupture of the affected placental sites with subsequent peritonitis, the bitch should be supervised until the disorder resolves. Ovariohysterectomy is indicated in rare cases of profound permanent bleeding or uterine infection.

### Uterine rupture

Uterine rupture should be considered a possible but uncommon cause of illness in the post-parturient period, less frequently seen in the queen than in the bitch. Uterine rupture can occur when prostaglandin or oxytocin has been administered for treatment of pyometra, metritis or dystocia, or as a complication during ovariohysterectomy. The condition may occur spontaneously as a result of dystocia or during an apparently normal parturition, or due to injury occurring in late pregnancy. The clinical signs of uterine rupture include abdominal pain and distension, and a rapid deterioration of the condition of the dam. The diagnosis is confirmed by exploratory laparotomy, and ovariohysterectomy is the usual treatment combined with intravenous fluids and antibiotic therapy.

### Uterine prolapse

Uterine prolapse is an uncommon complication during parturition. It occurs in primiparous as well as multiparous bitches and queens, but is less frequently manifested in the bitch than in the queen. The prolapse usually occurs immediately or within a few hours after delivery of the last neonate. The prolapse can be complete with both uterine horns protruding from the vulva or limited to the uterine body and one horn. Treatments include manual reposition, reposition by means of laparotomy, and amputation. Ovariohysterectomy is usually performed.

### Toxic milk syndrome

The toxic milk syndrome is poorly documented. Pathological conditions in the uterus of the dam may cause toxins to be excreted in the milk. Suckling offspring that are affected by toxic milk syndrome become vocal and uncomfortable. Other signs are diarrhoea, salivation, bloating and a reddened anus. Treatment consists of removing the offspring from the dam, and the administration of fluid therapy and oral glucose until bloating resolves. If the dam is successfully treated for her uterine condition the litter can be returned to her after 24–48 hours, otherwise it requires hand-rearing.

## Mammary gland disorders

### Agalactia

Agalactia, or absence of milk after parturition, may be due to a failure of milk let-down or milk production. True agalactia, failure of milk production, is uncommon but may be observed after premature parturitions or Caesarean operations. Failure of milk let-down may occur as a consequence of excessive secretion of adrenaline, resulting from fright or pain, which has a blocking effect upon the release of oxytocin. Oxytocin can be administered repeatedly for a few days until milk flow has been established. Primiparous, nervous or confused queens or bitches may experience temporary agalactia. Reassurance by the owner and administration of low doses of azepromazine orally may help the female to settle down, and subsequent suckling by the puppies or kittens will enhance milk flow. Other causes of agalactia are physical exhaustion, undernourishment, shock, mastitis, metritis, systemic infection, and endocrine imbalances.

### Galactostasis

Milk stasis, galactostasis, causes enlarged and oedematous mammary glands which are firm and warm to the touch. The dam shows signs of discomfort and pain, and fails to let down milk. The condition should be differentiated from mastitis and agalactia. The aetiology is unknown in dogs and cats but it mostly occurs in the most productive glands (the two most caudal pairs) in dams with a high milk production and/or few puppies, or in glands with malformed teats that the offspring avoid when suckling. Another cause can be that nursing comes to an abrupt end due to death of the litter or sudden weaning. To relieve mammary congestion the owner can apply gentle massage and warm-water compresses to the mammary glands and careful milking relieves some of the pressure. Sometimes it helps to put older and more aggressively nursing neonates to the dam. Treatment includes reducing food intake and the administration of a mild diuretic. Neglected galactostasis may lead to mastitis or involution of the mammary gland. Cabergoline, a dopamine agonist, at 2.5–5 μg/kg per day orally for 4–6 days reduces prolactin secretion and thereby lactation. The use of dopamine agonists should be restricted to dams who have either lost their litter or have a litter old enough to be weaned.

### Acute mastitis

Acute mastitis in the bitch and the queen occurs from haematogenously spread bacterial infections or from bacteria ascending through the teat orifices. Predisposing factors include mammary gland congestion, trauma and poor sanitary conditions. The mammary glands become hot, painful and enlarged; the milk shows increased viscosity and changes in colour from yellow to brown depending on the amount of blood and purulent exudate present. Clinical signs include fever, anorexia, and depression in the dam. The offspring may be restless and crying. Milk cytology reveals the presence of degenerate neutrophils, red blood cells and bacteria. Culture of the milk often shows growth of *Staphylococcus* spp., *Streptococcus* spp. or *Escherichia coli*. Treatment consists of adequate antibiotics, application of warm-water compresses and massage of mammary glands. If abscessation of the glands occur, surgical debridement and drainage is essential. Untreated acute mastitis may result in gangrenous mastitis and septic shock. Depending on the severity of the condition the litter may stay with the dam or has to be separated and hand-reared.

## Miscellaneous disorders

### Puerperal tetany

An acute decrease in extracellular calcium concentration is the cause of puerperal tetany (eclampsia). Eclampsia occurs most commonly in dogs of the small breeds, usually within the first 21 days after whelping, but occasionally during late pregnancy or at parturition. It has also been reported in cats, but is rare. Early signs are restlessness, panting, pacing, whining, salivation, tremors and stiffness. The symptoms aggravate to clonic-tonic muscle spasms, fever, tachycardia, miosis, seizures and death. Treatment must be instigated immediately and consists of slow intravenous infusion of a 10% calcium borogluconate solution. The dosage required varies from 2 to 20 ml in the dog and from 2 to 5 ml in the cat, depending on the degree of hypocalcaemia and the size of the dam. Careful cardiac monitoring for bradycardia and arrhythmias is important. If arrhythmia or vomiting occurs, the infusion must be temporarily halted, then if necessary resumed at a slower rate. As hypoglycaemia may follow hypocalcaemia, the intravenous administration of a 10% dextrose solution has been recommended. The puppies or kittens should be removed from the dam and hand-fed a canine/feline milk supplement for 24 hours. If the litter is 4 weeks old or more, it is advisable to wean them. Oral supplementation of the lactating dam (having experienced eclampsia) with calcium carbonate at 100 mg/kg body weight per day, divided with meals, and vitamin D is recommended. Prophylactic calcium treatment during the course of pregnancy of bitches expected to develop eclampsia probably is contra-indicated, as it may cause a disturbance of the calcium homeostasis.

### Disturbances in maternal behaviour

Good maternal behaviour includes nest building, nursing, protecting and for the dam to spend most of the time with the litter for at least the first 2 weeks. Most bitches and queens have strong maternal instincts but their behaviour depends strongly upon their hormonal balance, general health and the environment. In some breeds, there is a higher occurrence of bad mothering, suggesting a heritable factor. Sublimation, close emotional attachment to a human being, may cause problems at parturition when the dam may manifest panic and 'regard her offspring with horror and disgust'. In contrast, a dam may resent human intervention, not accepting assisted-birthing and Caesarean-delivered offspring and may sometimes even kill them. Major disturbing factors during and after parturition, mental instability or pain may cause the mother to kill her neonates. Good health, quiet and familiar surroundings and, most important of all, presence of her young will promote normal maternal behaviour.

## REFERENCES AND FURTHER READING


Concannon PW, McCann JP and Temple M (1989) Biology and endocrinology of ovulation, pregnancy and parturition in the dog. *Journal of Reproduction and Fertility*, **39**(Suppl.), 3–25

Darvelid AW and Linde-Forsberg C (1994) Dystocia in the bitch: a retrospective study of 182 cases. *Journal of Small Animal Practice* **35**, 402–407

Ekstrand C and Linde-Forsberg C (1994) Dystocia in the cat: a retrospective study of 155 cases. *Journal of Small Animal Practice* **35**, 459–464

Freak MJ (1948) The whelping bitch. *The Veterinary Record* **60**, 295–301

Gunn-Moore DA and Thrusfield MV (1995) Feline dystocia: prevalence, and association with cranial conformation and breed. *The Veterinary Record* **136**, 350–353

Hall LW and Clarke KW (1991) *Veterinary Anaesthesia, 9th edn.* Ballière Tindall, London

Hellyer PW (1993) Anaesthesia for cesarean section. In: *Textbook of Small Animal Surgery,* ed. D Slatter, pp. 2300–2303. WB Saunders, Philadelphia

Jackson PGG (ed.) (1995) *Handbook of Veterinary Obstetrics.* WB Saunders, London

Jones DE and Joshua JO (1988) In: *Reproductive Clinical Problems in the Dog, 2nd edn,* ed. N King. Butterworth, Oxford

Johnston SD, Root MV and Olson PNS (1996) Canine pregnancy length from serum progesterone concentrations of 3–32 nmol/l (1 to 10 ng/ml). Abstract. *Proceedings, Symposium on Canine and Feline Reproduction, Sydney*

Laliberté L (1986) Pregnancy, obstetrics and postpartum management of the queen. In: *Current Therapy in Theriogenology II,* ed. Morrow, pp. 812–821. WB Saunders, Philadelphia

Linde-Forsberg C and Forsberg M (1989) Fertility in dogs in relation to semen quality and the time and site of insemination with fresh and frozen semen. *Journal of Reproduction and Fertility*, **39** (Suppl.), 299–310

Linde-Forsberg C and Forsberg M (1993) Results of 527 controlled artificial inseminations in dogs. *Journal of Reproduction and Fertility*, **47** (Suppl.), 313–323

Long S (1996) Abnormal development of the conceptus and its consequences. In: *Veterinary Reproduction and Obstetrics, 7th edn,* ed. GH Arthur *et al.,* p 110–133. WB Saunders, London

Mosier JE (1978) Introduction to canine pediatrics. *Veterinary Clinics of North America: Small Animal Practice* **8**, 3–5

Okkens AC, Hekerman TWM, de Vogel JWA and van Haaften B (1993) Influence of litter size and breed on variation in length of gestation in the dog. *The Veterinary Quarterly* **15**, 160–161

Root MV, Johnston SD and Olson, PN (1995) Estrous length, pregnancy rate, gestation and parturition lengths, litter size, and juvenile mortality in the domestic cat. *Journal of American Animal Hospital Association*, **31**, 429–433

Shille VM (1983) In: *Current Techniques in Small Animal Surgery,* ed. MJ Bojrab, pp. 338–346. Lea and Febiger, Philadelphia

Steinetz BG, Goldsmith LT, Hasan SH and Lust G (1990) Diurnal variation of serum progesterone, but not relaxin, prolactin or oestradiol-17beta in the pregnant bitch. *Endocrinology* **127**, 1057–1063

van der Weyden GC, Taverne MAM, Dieleman SJ and Fontijne P (1981) The intrauterine position of canine fetuses and their sequence of expulsion at birth. *Journal of Small Animal Practice* **22**, 503–510

van der Weyden GC, Taverne MAM, Dieleman SJ, Wurth Y, Bevers MM and van der Oord HA (1989) Physiological aspects of pregnancy and parturition in the bitch. *Journal of Reproduction and Fertility* **39**, 211–224

Wallace MS (1994) Management of parturition and problems of the periparturient period of dogs and cats. *Seminars in Veterinary Medicine and Surgery (Small Animal)* **9**, 28–37

Verhage HG, Baemer NB and Brenner RM (1976) Plasma levels of estradiol and progesterone in the cat during polyestrus, pregnancy and pseudopregnanacy. *Biology of Reproduction* **14**, 579–585.

Willis MB (1989) The inheritance of reproductive traits. In: *Genetics of the Dog*, ed. MB Willis, pp. 33–62. HF&G Witherby Ltd, London

CHAPTER THIRTEEN

# The Neonate: Congenital Defects and Fading Puppies

*Tony S. Blunden*

## CONGENITAL DEFECTS

### Introduction

Congenital defects are defined as abnormalities of structure or function present at birth and have been identified in most major breeds of dog and cat. They arise from disruptive events at one or more stages from formation of the blastocyst through embryonic and fetal development. Although these defects are frequently thought to have a genetic origin, they may be the result of maldevelopment through a variety of causes. The term 'congenital' does not mean 'inherited', although a defect may be both congenital and inherited. Many defects cannot be identified without clinical or pathological investigation. It is estimated that congenital defects that are severe enough to threaten the viability of the neonate occur at a rate of about 1-2% of pedigree puppies born, but only a few studies are available which record their frequency.

### Inherited diseases

A sudden increase in the occurrence of congenital defects in a breeding colony or breeding line would warrant further investigation as to the cause. A complete family history and pedigree analysis is required. It is of great importance to recognize which defects are of genetic origin so that controls may be introduced to limit the breeding from affected animals. Although breeding from animals with proven or suspected genetic traits is not recommended, confirmation may be obtained by test matings and the mode of inheritance determined so that possible carrier status may be assessed and testing schemes put in place to detect carriers. Eventually genetic screening tests should become available for identifying major defects.

### Inheritance patterns

The different forms of a gene that can exist at a particular site in a chromosome are called alleles. The particular site of a gene in a chromosome is called the locus. The word 'gene' is commonly used in the sense of either allele or locus. Although any one animal can have a maximum of only two different alleles at a locus, the number of different alleles in a population can be greater than two; the locus is then said to have multiple alleles. The passage of genes from one generation to the next is called inheritance.

Genetic diseases may be caused by a pair of mutant genes, by a single mutant gene or by polygenic inheritance. The phenotypic expression of a genetic defect may be altered by environmental influences or by other genes.

#### Recessive traits

The simple autosomal recessive mode of inheritance is the most common inheritance pattern. The trait is hard to control as the defect only becomes obvious in the homozygous state *(aa)* after two heterozygotes *(Aa)*, apparently normal animals, have been mated. The affected offspring receives the abnormal gene from both parents. If there are continued attempts to breed carrier parents, 25% of the offspring will manifest the trait but 50% will be normal carriers. Table 13.1 shows the predicted outcome of all possible matings in relation to a single autosomal recessive trait.

| Type of mating | Segregation ratio | | |
|---|---|---|---|
| | *AA* | *Aa* | *aa* |
| *AA* x *AA* | 1 | 0 | 0 |
| *AA* x *Aa* | 1 | 1 | 0 |
| *AA* x *aa* | 0 | 1 | 0 |
| *Aa* x *Aa* | 1 | 2 | 1 |
| *Aa* x *aa* | 0 | 1 | 1 |
| *aa* x *aa* | 0 | 0 | 1 |

***Table 13.1:*** *Predicted outcome of all possible matings in relation to a single autosomal recessive trait.*

Eliminating defective animals usually means that recessive type anomalies are kept at a low level.

#### Dominant traits

In contrast, with dominant inheritance the particular trait is expressed in the heterozygous state. Normal animals breed true but abnormal parents may produce both normal and abnormal offspring and therefore dominant diseases can be easily eliminated by removing defective animals from the breeding programme.

These traits, e.g. Factor X and Factor XI coagulopathies, are uncommon.

### No dominance or incomplete dominance

Sometimes genes have several alleles but do not function in the usual dominant/recessive manner, so that the heterozygote exhibits the effect of both alleles. In this case, the trait is controlled by eliminating all abnormal animals from the breeding programme. Normal and severely abnormal animals breed true. Slightly abnormal animals mated together produce litters of 25% normal, 50% slightly abnormal and 25% severely abnormal offspring.

### Incomplete penetrance

Sometimes a gene does not express itself as it should. If a gene is dominant, then a combination such as *AA* should be identical to *Aa* because *A* is dominant to *a*. If the *A* allele is 100% penetrant, three distinct genotypes (*AA*, *Aa* and *aa*) and two phenotypes (*AA* and *aa*) result, and *A* always expresses its dominance. However, if the *Aa* combination sometimes results in a phenotypic appearance of *aa*, then penetrance is not complete. If *Aa* showed the dominant trait 75% of the time, this would be termed 75% penetrance. The reasons for incomplete penetrance are unclear.

### Polygenic inheritance

Polygenic inheritance is where a trait is controlled by a number of genes, each with a relatively small effect, while in addition there are environmental factors that influence the trait to a greater or lesser degree, e.g. conformational traits.

### Threshold traits

These are traits that are controlled by many genes, although exhibiting a narrow range of expression; depending on the number of genes involved there is a threshold at which the feature changes from one form to the next, e.g. patent ductus arteriosus where the ductus may be partially closed or completely open. Other examples include cryptorchidism and umbilical and inguinal hernias.

### Sex-linked traits

Any trait carried on either of the sex chromosomes will, by definition, be sex-linked. However it does seem that the Y chromosome is largely inert. The X chromosome carries the gene for haemophilia A and a few other defects. The dam can transmit the gene to offspring of either sex but a male carrying a certain gene on his single X chromosome can only pass this on to female offspring since male offspring will inherit his Y chromosome.

### Sex-limited traits

A sex-limited trait is one seen only in one sex, e.g. lactational yield is only expressed in females but inheritance is determined by genes transmitted in both sexes; cryptorchidism can be transmitted by females but is only manifest in males. The traits are not carried on the X and Y chromosomes, but the expression is limited to the specific gender.

### Chromosomal aberrations

Chromosomal abnormalities are not commonly reported in the dog and cat. They may arise spontaneously, be inherited or induced by environmental factors. Sometimes a chromosome may be duplicated or lost and an odd number result. The duplication can apply to a single chromosome or to a whole set. Probably most chromosomal abnormalities result in embryonic loss

| Defect | Comment |
|---|---|
| **Head:** | |
| Cleft palate/lip | Various breeds; noted in brachyocephalic types and Siamese. Inherited as simple autosomal recessive trait in English Bulldog; can be caused by hypervitaminosis A or drug therapy such as griseofulvin, corticosteroids |
| Overshot jaw (retrognathia) or undershot jaw (prognathia) | Inherited as autosomal recessive in Longhaired Dachshund and Cocker Spaniel. Prognathism affects Burmese and Persian (accepted breed standard) cats |
| Cranioschisis | Opening of the cranial vault in Cocker Spaniel, may be inherited as a lethal recessive trait |
| Otocephalic syndrome | Reported in the Beagle as an autosomal recessive trait with a low grade form characterized by partial agnathia, hydrocephalus and parietal fontanelle defects and a high grade form with agenesis of all cranial structures anterior to medulla |
| **Axial skeleton:** | |
| Vertebral anomalies: Atlantoaxial instability | Congenital hypoplasia of odontoid process (dens) and/or its non-union with C2, seen in several small breeds (e.g. Pomeranian, Chihuahua) |
| Spina bifida | Uncommon; absence of dorsal portions of vertebrae. Reported in Manx, Maltese and crossbred Siamese cats |
| Dew claws on hind feet | Autosomal dominant trait in most breeds |
| **Muscle:** | |
| Myopathy | In Golden Retriever (onset from 6-8 weeks), there is a severe muscle atrophy of tongue/diaphragm, inherited as sex-linked trait. A myopathy of the Labrador Retriever (onset from 3 months), is inherited as autosomal recessive trait. Also the male Irish Terrier has a sex-linked inherited myopathy manifest from 8 weeks |
| Myotonia | Inherited defect of intracellular calcium pump system in Chow and Staffordshire Bull Terrier |
| Defects in body wall: | |
| Umbilical hernia | Inherited as threshold trait in Basenji, Airedale, Pekingese and Weimaraner |
| Inguinal hernia | Inherited as threshold trait in West Highland White Terrier, Basenji, Cairn, Basset Hound and Pekingese |

***Table 13.2:*** *Inherited defects of the musculoskeletal system.*

| Defect | Comment |
|---|---|
| Gonadal agenesis or hypoplasia | Unilateral or bilateral. Probably not inherited |
| Intersexuality | True or pseudohermaphrodites. Disorders of sexual differentiation involving anomalies of X/Y chromosomes, gonads and phenotype. Intersexuality can be hereditary or induced by non-hereditary factors such as progestogen given in pregnancy |
| Hypospadias | Abnormal location of urinary meatus due to incomplete closure of urethral folds |
| Persistent Mullerian duct syndrome | Usually seen in cryptorchid dogs, but otherwise normal males. Internally both testes attached to cranial ends of bicornuate uterus |
| Cryptorchidism | Unilateral or bilateral, considered to be inherited disorder, common in dwarf and brachycephalic breeds. Persian and other cat breeds |

***Table 13.3:*** *Inherited defects of the reproductive system.*

rather than in congenital defects. Anomalies of chromosome number can also apply to sex chromosomes, leading to XXX, XXY and other combinations such as reported in the Weimaraner and Cocker Spaniel resulting in intersex conditions.

## Non-genetic causes of congenital defects

These include the use of pharmaceutical drugs in pregnancy, e.g. griseofulvin treatment has been linked with microphthalmos in kittens and cleft palate in puppies; progestational agents given to the bitch in early pregnancy have been reported to cause masculinization of the external genitalia of female puppies, with hypertrophy of the clitoris. Corticosteroids are thought to increase the numbers of anasarcous puppies in brachycephalic breeds and have also been linked with skeletal deformities. Anticonvulsant drugs have been associated with cardiac defects, cleft palate, microcephaly and other skeletal anomalies. Therefore careful consideration needs to be given to the use of any drugs during pregnancy and to possible teratogenic effects.

Nutritional components have sometimes been incriminated; hypervitaminosis A (125,000 mg/kg) between days 17 and 22 has been reported to cause cleft palates, kinked tails and deformed auricles in kittens. Excess vitamin D has been linked with tissue calcinosis, premature closure of fontanelles, enamel hypoplasia and supravalvular stenosis.

Environmental chemicals are sometimes suspected teratogens but are difficult to prove as such. Infective agents are occasionally linked with congenital defects such as feline parvovirus which is known to be a cause of cerebellar hypoplasia in kittens. Observations suggest that the effect of a teratogen depends on the stage of fetal development. If exposure occurs during organogenesis in the first trimester the predominant lesions involve cephalic, ocular, otic and/or cardiac structures. If exposure occurs during the transitional period, after day 26, the predominant lesions generally involve the palate, cerebellum, cardiovascular system and/or the urogenital system.

It has to be said that, in many cases, where there is no clearly inherited pattern, the causes of birth defects are not identified or are speculative and the anomalies occur as single, isolated events.

## Types of congenital defect

Congenital defects may affect only a single structure or function, but may also occur as a syndrome of multiple defects and can often be breed-related. The true frequency of congenital defects is unknown as not all defects will be manifest at birth or lead to death in the neonatal period. Some abnormalities, such as cardiac defects, only become apparent as the puppy grows heavier and exercises more, and several eye defects, such as progressive retinal atrophy, have a late onset of progression. Some defects will only be recognized on detailed necropsy or by biochemical/haematological tests in later life.

Inborn errors of metabolism are enzyme deficiencies caused by gene abnormalities. Any metabolic pathway can be affected. The defects are usually autosomal recessive or sex-linked recessive. Two main types of enzyme deficiency exist: one type results in abnormal storage of metabolic intermediates and the second type involves lysosomal enzyme deficiencies which impair degradation of complex carbohydrates. Many abnormalities are likely to go unrecognized or are unreported for economic reasons. The most frequently reported defects involve the central nervous, ocular, musculoskeletal and cardiovascular systems. In the author's own studies on neonatal mortality, skeletal abnormalities, such as cleft palate and hare lip, were the most common defects encountered. Pure bred dogs are more frequently affected by genetically caused congenital defects than are cross-breeds.

The more common congenital defects recognized in the puppy and kitten within approximately the first 3 months are summarized in Tables 13.2–13.12 (references taken from Leipold, 1978; Willis, 1992; Jubb *et al.*, 1993; Casal, 1995; Hoskins, 1995a,b). Where inheritability is known, this is indicated in the Comment column. The prevalence of a condition in a breed is not given in the tables, and the order in which the breeds are given does not reflect prevalence rates. Since accurate incidence rates are not known, an estimate of prevalence is considered unwise. The data in different texts may differ because breed incidences may differ between countries. With some traits definitive breeding studies have elucidated inheritance but in many cases observation of the disease in specific breedlines strongly suggests but does not prove inheritance.

| Defect | Comment |
|---|---|
| **Malformations:** | |
| Cerebellar hypoplasia | One of the most common congenital CNS defects. Evidence of inheritance in Chow. Possibly viral causes such as parvovirus infection in the cat, or toxic agent also suggested in some cases. Cerebellar dysfunction from birth, but not progressive |
| Cerebellar abioatrophy | Premature or accelerated degeneration of formed elements. Ataxia and hypermetria observed from around 12 weeks; recorded in several breeds including Airedale Terrier, Gordon Setter and Border Collie. Inherited striatonigral and cerebello-olivary degeneration occurs in Kerry Blue Terrier, autosomal recessive trait probably |
| Internal hydrocephalus | Noted particularly in Chihuahua, Cocker Spaniel, Bulldog, but may be caused by other processes, e.g. inflammatory. Reported in Siamese and other breeds |
| Spinal dysraphism | Duplication, absence and malformation of central canal, onset at 4-6 weeks; noted in Weimaraner; possibly autosomal recessive trait |
| **Myelinopathies:** | |
| Hereditary myelopathy | Progressive ataxia seen in Afghan Hound (3-12 months); may be inherited as a simple autosomal trait. Myelinolysis and cavitation focused in thoracic spinal cord. |
| Hereditary ataxia | Recorded in Fox and Jack Russell Terriers; inheritance is simple autosomal recessive trait, onset at 2-4 months, rapidly progressive. Demyelination of certain spinal cord tracts |
| Central hypomyelination and dysmyelination | Generalized tremor by 2-3 weeks of age. Reported in several breeds including Chow, English Springer Spaniel, Samoyed, Weimaraner and Bernese Mountain Dog. X-linked inheritance in English Springer Spaniel |
| Hypertrophic neuropathy | Recorded in Tibetan Mastiff. A dysmyelinogenesis, only peripheral nerves affected. Considered primary metabolic defect of Schwann cells. Onset 7-10 weeks, considered to be recessive inheritance |
| **Axonopathies and neuronopathies:** | |
| Progressive axonopathy | Reported in the Boxer, in which it is inherited as an autosomal recessive trait. Lesions most apparent in posterior brain stem and spinal cord. Onset from 8 weeks |
| **Lysosomal storage diseases:** | |
| Globoid cell leucodystrophy galactocerebroside β-galactosidase deficiency | Onset at 3-6 months; inherited in Cairn and West Highland Terriers and Miniature Poodle, as simple autosomal recessive trait, leading to a dysmyelinogenesis. Also seen in Domestic cat with onset from 2 weeks |
| Gangliosidoses | German Shorthaired Pointer, Portuguese Water Dog, Japanese Spaniel and mixed breeds, Siamese, Korat and Domestic cat |
| Glucocerebrosidosis | Sydney Silky Terrier |
| Sphingomyelinosis | Miniature Poodle, Domestic cat, Siamese, Balinese |
| α-L-fucosidosis | Springer Spaniel, inherited as autosomal recessive trait, onset about 6 months |
| α-L-iduronidosis | Plott House, Domestic cat |
| Amylo-1,6 glucosidosis | German Shepherd Dog, Domestic cat |
| Phosphofructokinosis | English Springer Spaniel |
| Ceroid-lipofuscinosis | Many breeds including English Setter, Chihuahua, Dachshund, Saluki, Border Collie, Tibetan Terrier and mixed breeds; Siamese |

***Table 13.4:** Inherited defects of the central nervous system.*

| Defect | Comment |
|---|---|
| **Kidneys:** | |
| Renal agenesis | Bilateral/unilateral. Familial predisposition has been noted in Beagle, Shetland Sheepdog and Dobermann Pinscher. May occur with anomalies of reproductive tract |
| Renal hypoplasia | Sporadic occurrence; appear as miniature replicas of normal kidneys, with reduced numbers of histologically normal nephrons |
| Renal dysplasia and aplasia | Dysplasia refers to segmental abnormality whereas aplasia involves maldevelopment of whole kidney. Renal dysplasia is a familial disorder in soft-coated Wheaten Terrier, Lhasa Apso, Shih-Tzu and Standard Poodle. An hereditary basis is suspected in the Keeshond, Chow and Miniature Schnauzer. Renal dysplasia has also been associated with canine herpes virus infection |
| Polycystic renal disease | Variable number of fluid-filled cysts in renal parenchyma. Individuals may be asymptomatic or develop progressive renal failure. Reported in Persian cat and Cairn Terrier |
| Renal ectopia and fusion | Congenital malposition of one or both kidneys; renal fusion represents union of normally lateralized kidneys. Aetiopathogenesis is unknown |
| Duplex and supernumerary | The presence of one or more accessory kidneys (supernumerary) or an organ comprising two kidneys, pelves and ureters (duplex). Rare |
| Fanconi syndrome | Generalized renal tubular dysfunction described in Basenji, Norwegian Elkhound, Schnauzer and Shetland Sheepdog. Inheritability not confirmed |
| Primary renal glucosuria | Defect of proximal tubular reabsorption of glucose, reported in Scottish Terrier, Norwegian Elkhound and mixed breed dogs. Mode of inheritance unknown |
| Cystinuria | Defect of renal tubular transport of cystine and other dibasic amino acids, reported in many breeds, especially male dogs. Possibly recessive mode of inheritance in Irish and Scottish Terriers |
| Hyperuricuria | Hyperuricuria, production of uric acid due to deficiency of urease; in the Dalmatian is transmitted by a recessive non-sex linked mode of inheritance |
| Primary hyperoxaluria | Acute renal failure due to renal tubular deposition of oxalate crystals. Reported in Domestic Shorthair cats |
| Nephrogenic insipidus diabetes | Severe polyuria and polyuria; nocturia and poor growth in puppies |
| **Ureters:** | |
| Agenesis | Unilateral or bilateral. Unilateral most common, and usually accompanied by ipsilateral renal aplasia |
| Duplication | Associated with duplex and supernumerary kidneys |
| Ureteral valves | Persistent transverse folds of vestigial mucosa and smooth muscle |
| Ectopic ureters | Unilateral or bilateral; may be intramural or extramural and can be associated with other anomalies of the urinogenital tract. Many cases occur in mixed breed dogs but a familial predisposition has been noted in Siberian Husky, Labrador Retriever, Newfoundland, English Bulldog, West Highland White, Skye and Fox Terriers, Welsh Corgi, Golden Retriever and Miniature and Toy Poodle. |
| Ureterocoeles | Congenital cystic dilation of terminal submucosal segment of the intravesicular ureter |

***Table 13.5:** Inherited defects of the urinary tract.*

| Defect | Comment |
|---|---|
| Patent ductus arteriosus | Most common congenital heart defect, and is inherited as a polygenetic threshold trait with a high degree of heritability in Poodles. It frequently occurs in Pomeranian, Collie, Shetland Sheepdog, Maltese and English Springer Spaniels but any breed may be affected. Also Siamese, Persian and other cat breeds |
| Subaortic stenosis | This is the second most common defect, and is inherited as a polygenetic threshold trait in the Newfoundland. It is also common in the Golden Retriever, Rottweiler and Boxer. Most commonly the obstruction is subvalvular |
| Pulmonic stenosis | The third most common defect which occurs most frequently in Beagle, Chihuahua, English Bulldog, Fox Terrier, Samoyed and Miniature Schnauzer. Pulmonic valve dysplasia is the most common cause and is inherited in the Beagle as a polygenetic trait |
| Vascular ring anomalies | Fourth most common defect. Inherited in the German Shepherd Dog and the Great Dane also has a breed predisposition. They are a group of anomalies which result from abnormal maturation of the embryonic aortic arches; persistent right aortic arch (with a left ductus arteriosus) being the most common form |
| Ventricular septal defect | Usually a single defect located high in the septum just below the tricuspid and aortic valves. No breed predilection established. May occur in conjunction with other anomalies |
| Tetralogy of Fallot | This includes a ventricular septal defect, right ventricular outflow obstruction, hypertrophy of the right ventricle and a dextrapositioned aorta that accepts blood from both ventricles. A polygenetic threshold trait has been established in Keeshonds |
| Tricuspid valve dysplasia | Reported in large breed dogs |
| Mitral valve malformation | Reported in Great Dane and German Shepherd Dog |
| Atrial anomalies | Atrial septal defects are usually found in association with other cardiac anomalies |
| Endocardial fibroelastosis | Congenital disorder characterized by proliferation of elastic and collagenous fibres within the endocardium. Seen mainly in young cats, particularly Burmese and Siamese, but does occur in puppies often in association with other heart anomalies |
| Congenital rhythm disorders | Ventricular pre-excitation occurs as isolated abnormality and in conjunction with anatomical congenital lesions. Fatal ventricular arrhythmias are heritable in some bloodlines of German Shepherd Dog. Persistent atrial standstill has been reported in young English Springer Spaniel and Siamese, Burmese and Domestic Shorthair cats, and hereditary stenosis of the AV bundle, associated with sinus arrest, has been described in the Pug |
| Inborn errors of metabolism | Alpha-glucosidase deficiency affecting myocardial function in Lapland Dog, and mucopolysaccharidosis I (lysosomal enzyme $\alpha$-L-iduronidase deficiency) in young Plott Hound |
| Extracardiac arteriovenous fistulas | These may be congenital or acquired and may affect any part of the vascular system, but usually the great vessels, internal organs or distal extremities |

***Table 13.6:*** *Inherited defects of the cardiovascular system.*

| Defect | Comment |
|---|---|
| Juvenile onset diabetes mellitus | Insulin dependent diabetes which occurs before 12 months. In the Keeshond, affected dogs inherit B-cell atrophy as an autosomal recessive trait with incomplete penetrance. It is also believed to be inherited as a recessive trait in Golden Retriever |
| Pituitary gland hypoplasia | Pituitary dwarfism is transmitted as an autosomal recessive trait in German Shepherd Dog and in Carelian Bear Dog |
| Congenital idiopathic central diabetes insipidus | This has been seen in the male Toy Poodle |
| Congenital hypothyroidism | Caused by dysgenesis, dyshormonogenesis, serum transport abnormalities, congenital TSH deficiency, goitrogens and severe iodine deficiency. Mode of inheritance is unclear in dogs; autosomal recessive trait in Abyssinian |

***Table 13.7:*** *Inherited defects of the endocrine system.*

| Defect | Comment |
|---|---|
| Epitheliogenesis imperfecta | Rare congenital discontinuity of squamous epithelium |
| Ichthyosis | Generalized, extreme hyperkeratosis and scaling. Described in Dobermann Pinscher, West Highland White Terrier, Irish Setter, Collie, Bull Terrier, Boston Terrier, Labrador Retriever, Jack Russell Terrier |
| Congenital seborrhoea | Reported in English Springer Spaniel. Puppies develop patches of hyperkeratosis and scale |
| Nevus | Circumscribed developmental defect in the skin arising from epithelial or dermal structures or a combination of both |
| Dermoid sinus or cyst | Occurrence is along the dorsal midline and the sinus runs from skin to supraspinous ligament. It is a neural tube defect resulting from incomplete separation of skin and neural tube during embryogenesis, particularly in Rhodesian Ridgeback in which it is a heritable trait (possibly an autosomal recessive) |
| Hereditary alopecia and hypotrichosis (ectodermal defects and dysplasia) | A rare disease characterized by varying degrees of alopecia from birth. Affected animals have diminished number of adnexal structures in affected areas. The condition may occur alone or in association with other ectodermal defects e.g. abnormal dentition. Male sex-linked inheritance is suspected in Poodle, Basset Hound, Beagle, Labrador Retriever and Bichon Frise. Also reported in a female Labrador Retriever and Rottweiler; Sphinx, Cornish, Devon Rex, Mexican Hairless, Siamese and Birman |
| Ehlers-Danlos syndrome (cutaneous asthemia) | Structural defect of collagen leading to hyperextensibility of skin, seen in English Springer Spaniel, Beagle, Boxer, German Shepherd Dog, Greyhound, Dachshund, St Bernard and mixed breeds; Himalayan and other cat breeds. It may arise as spontaneous mutation or inherited as a dominant autosomal trait |
| Acrodermatitis | American Bull Terrier. Puppies show growth retardation and lighter skin pigmentation than normal, and cannot swallow well. By 6 weeks lesions appear on footpads, ears, muzzle and around orifices |
| Junctional epidermolysis bullosa | Reported in neonatal Toy Poodle |
| Pigmentary defects: | |
| Linked with deafness | Partial or complete albinism may be linked to deafness in dogs, especially the white Bull Terrier, Sealyham, white and merle Collie and Dalmatian and white cats, and may be associated with ocular defects. Autosomal dominant with incomplete penetrance |
| Familial vitiligo and poliosis | Seen in the Rottweiler, usually adults but sometimes young puppies |
| Leucotrichia (poliosis) | Reported in a litter of Labrador Retriever puppies |

***Table 13.8:*** *Inherited defects of the skin.*

| Defect | Comment |
|---|---|
| Hypoplasia of larynx | Inherited as a simple autosomal recessive trait in Skye Terrier |
| Tracheal hypoplasia | Seen in first 2 months of life, especially English Bulldog |
| Congenital diaphragmatic hernia | May be inherited as simple autosomal recessive trait |
| Primary ciliary dyskinesia | Abnormal ciliary function in respiratory epithelium leading to reduced mucociliary clearance. English Pointer, English Springer Spaniel, Border Collie, English Setter, Dalmatian, Dobermann Pinscher, Chihuahua, Golden Retriever, Old English Sheepdog, Chow, Bichon Frise |

***Table 13.9:*** *Inherited defects of the respiratory system.*

| Defect | Comment |
|---|---|
| **Digestive Tract:** | |
| Congenital mega-oesophagus | It is considered to be a neuromuscular developmental disorder or immaturity and is reported in Great Dane, German Shepherd Dog and Irish Setter, in particular, but also several other breeds. The condition appears to be heritable, with a pattern in the Miniature Schnauzer compatible with a simple autosomal dominant or a 60% penetrance autosomal recessive mode of inheritance |
| Segmental aplasia of the alimentary tract | Death in early neonatal period. The author has noted segmental aplasia of the small intestine in particular |
| Imperforate anus | |
| Congenital pyloric stenosis | Noted in Boxer and Boston Terrier; Siamese |
| **Liver:** | |
| Portosystemic venous shunt and intrahepatic arterioportal fistula | Most common congenital abnormalities of hepatobiliary system are aberrations of the portal circulation |
| Congenital metabolic abnormalities | Mucopolysaccharide storage disorders. Copper storage disorder in Bedlington Terrier (autosomal recessive trait) |

***Table 13.10:*** *Inherited defects of the alimentary system.*

| Defect | Comment |
|---|---|
| **Congenital Coagulopathies:** | |
| Factor VIII deficiency (Haemophilia A) | One of the most common inherited haemostatic defects; X-linked recessive mode of inheritance in Irish Setter, St Bernard, Shetland Sheepdog, Beagle, Collie, German Shepherd Dog, English Setter, Greyhound, Vizsla, Weimaraner, Chihuahua, Cairn Terrier, Samoyed and Husky. Defect is mild in cats |
| Factor IX (Haemophilia B) | X-linked recessive disorder, not as common as Factor VIII. It has been described in the Cairn Terrier, St Bernard, as well as a mixed breed dog; British Shorthair, Siamese, Domestic Shorthair |
| Factor VII | Miniature Schnauzer, Alaskan Malamute, Boxer, Bulldog and Beagle; a mild coagulation disorder |
| Factor X | Inherited as an autosomal dominant trait in Cocker Spaniel, with a severe bleeding diathesis in newborn and young adult dogs, but mild in mature adults. Rare disorder |
| Factor XI (Plasma thromboplastin antecedent) | Inherited as autosomal dominant trait with incomplete expression or as incompletely recessive trait in English Springer Spaniel. It is characterized by mild bleeding episodes but severe post-surgical bleeding. It has also been reported in Pyrrenean Mountain Dog, Weimaraner and Kerry Blue Terrier and is relatively rare |
| **Congenital Extrinsic Platelet Function Disorder:** | |
| Von Willebrand's Disease (VWD) | Due to defective or deficient Von Willebrand factor (Factor VIIIR). This is the most common inherited bleeding disorder in dogs; also seen in Himalayan and other cat breeds. The most common mode of inheritance is autosomal with incompletely dominant expression (most dog breeds); also autosomal recessive mode of inheritance. Factor VIIIR is important in mediating the adhesion of platelets to subendothelial surfaces |
| **Congenital Intrinsic Platelet Function Disorders:** | |
| Canine thrombopathy | Described in Basset Hound |
| Thrombasthenic thrombopathy | Inherited autosomal recessive trait in Otter Hound. Bizarre giant platelets seen |
| Spitz thrombopathy | Described in two Spitz females |
| Cyclic haematopoiesis | Autosomal recessive disorder in grey Collie characterized by cyclic fluctuations of circulating neutrophils, reticulocytes and platelets due to bone marrow stem cell defect. Most puppies die before 6 months |
| **Anaemia:** | |
| Pyruvate kinase deficiency | Inherited as simple autosomal recessive trait in Basenji. Non-spherocytic haemolytic anaemia |
| Haemolytic anaemia | Red blood cell life spans are decreased in association with an autosomal recessive-transmitted chondrodysplasia. Specific to Alaskan Malamute |
| Non-spherocytic haemolytic anaemia | More severe in Poodle but also the Beagle, in which it is fatal by 3 years. Cause unknown |
| Phosphofructokinase deficiency | Chronic haemolysis with haemolytic crises and mild myopathy. Specific to English Springer Spaniel |

***Table 13.11:*** *Inherited defects of the blood and lymphatic system.*

| Defect | Comment |
|---|---|
| **Eyelid:** | |
| Ectropion | Inherited in St Bernard, Bloodhound, Bulldog, Chow, Irish Setter and Cocker Spaniel but may be secondary to trauma/conjunctivitis |
| Entropion | Inherited in Chow, Bloodhound, Great Dane, Labrador, Bulldog, Bull Mastiff, Spaniel (Springer and Cocker), Papillon, St Bernard, Golden Retriever and Pomeranian; may be acquired. Also Persian cat |
| Diamond eyelid malformation | Predisposition in St Bernard, Clumber Spaniel |
| Districhiasis | Double row of eyelashes; inherited in Pekingese, Poodle, Cocker Spaniel, Shetland Sheepdog, Miniature Longhaired Dachshund |
| Trichiasis | Misdirected eyelashes, although arise from normal position on lid; noted in Pug, Pekingese, but may be acquired after mild entropion |
| Nictitating membrane scrolling | Predisposition in German Shepherd Dog, Great Dane |
| Agenesis | Absence of varying segments of eyelid margin; may be accompanied by other ocular defects such as iris coloboma and dermoids. Various dog breeds, also Domestic Shorthair and Persian cats |
| Micropunctum or imperforate lacrimal punctum | Bedlington Terrier, Cocker Spaniel, Sealyham Terrier, Golden Retriever |
| **The Globe:** | |
| Microphthalmos (complete absence is rare) | May be linked with multiple ocular defects; inherited in several breeds of dogs such as Miniature Schnauzer, Old English Sheepdog, Akita, Cavalier King Charles Spaniel and in merle to merle matings. In the Australian Shepherd Dog microphthalmos is associated with multiple colobomas and inherited as an autosomal recessive trait linked with coat colour. Microphthalmos has been reported in offspring of pregnant cats treated with griseofulvin |
| Strabismus - divergent | Recognized in brachyocephalic dogs, e.g. Boston Terrier |
| Strabismus - convergent | Siamese |
| Nystagmus, spontaneous | Siamese, results from an anomalous visual pathway development |
| **The Cornea:** | |
| Deep corneal opacity | Due to remnants of embryonic pupillary membrane that adhere to inner cornea; inherited in Basenji. Sporadic in other breeds |
| Epibulbar dermoids | Any breed, but St Bernard, German Shepherd Dog, Dachshund, Dalmatian appear to have predisposition. Also Birman, Burmese and Domestic Shorthair cats |
| Corneal dystrophy | Familial bilaterally symmetric corneal opacities unrelated to previous ocular disease. Most canine stromal corneal dystrophies (with deposition of lipids) are not manifest until after 1 year of age. A progessive corneal dystrophy has been described in Manx cats, with stromal oedema and ulceration, that is manifest as early as 4 months. Possibly inherited as autosomal recessive trait |
| **Anterior Uvea:** | |
| Persistent pupillary membranes | Inherited in Basenji |
| Iris cysts | Floating fluid-filled vesicles arising from posterior iris epithelium; usually found in anterior chamber |
| Pupillary anomalies | Corectopia (eccentric pupil) may accompany multiple ocular defects; inherited in Australian Shepherd Dog |
| Heterochromia of the iris | Variation in iris colour: occurs commonly in subalbinotic animals, reported in Persian and Angora cats. Multiple ocular defects reported in dog in association with partial albinism and deafness |
| Iridocorneal abnormalities | Congenital mesodermal remnants present in iridocorneal angle. Basset Hound |
| **Lens and Vitreous:** | |
| Alteration of size & shape of lens | As embryonic lens influences development of optic cup, multiple ocular defects are often associated with lens abnormality |
| Congenital cataracts | These may be inherited or secondary to intra-uterine influences. Congenital cataracts have been reported in the Beagle, Cocker Spaniel, Cavalier King Charles Spaniel, Old English Sheepdog, Australian Shepherd Dog, Bedlington Terrier, Sealyham Terrier, Labrador Retriever. Juvenile cataracts develop from the neonatal period until 6 years of age. Hereditary is a major factor, but inflammatory, metabolic, nutritional, toxic and traumatic causes must also be considered. Also reported in Domestic Shorthair, Persian, Birman and Himalayan cats |
| Hyaloid remnants | Most common vitreous anomaly |
| Persistent, hyperplastic primary vitreous secondary glaucoma | Fibrovascular membrane present on posterior lens surface; various breeds. Inheritance demonstrated in Dobermann Pinscher, Staffordshire Bull Terrier and Bouvier des Flandres. Autosomal dominant Border Collie and Terrier breeds, especially Wirehaired Fox, Jack Russell, Sealyham, Tibetan and Wirehaired and Smoothhaired Miniature Bull Terriers. Autosomal recessive trait |
| Pectinate ligament goniodysgenesis with glaucoma | Inherited in Basset Hound, Siberian Husky, American Cocker Spaniel, Cocker Spaniel, Dandie Dinmont Terrier, Elkhound, Great Dane, Welsh Springer Spaniel, Welsh Terrier |
| **Retina and Optic Nerve:** | |
| Collie eye anomaly | Inherited in Rough and Smooth Collie, Shetland Sheepdog, Border Collie, Australian Shepherd Dog, as an autosomal recessive trait with chorioretinal hypoplasia, optic disc colobomas and retinal detachment |
| Multifocal retinal dysplasia | Inherited in English Springer Spaniel, Labrador Retriever as autosomal recessive trait, with multifocal retinal folds and retinal detachment. Also reported in several other breeds, with or without other ocular defects: American Cocker Spaniel, Beagle, Akita, Australian Shepherd Dog, Dobermann Pinscher, Old English Sheepdog, Rottweiler, Yorkshire Terrier, German Shepherd Dog, Cavalier King Charles Spaniel, Hungarian Puli, Elkhound and Field Spaniel. Congenital retinal dysplasia may also be non-inherited |
| Total retinal dysplasia | Inherited in Bedlington Terrier, Labrador Retriever and Sealyham Terrier |
| Hemeralopia | Day blindness (cone absence) in Alaskan Malamute, normal fundus. Reported from 8-20 weeks; also Miniature Poodle by 12 weeks. Autosomal recessive trait |
| Congenital stationary night blindness | Evident by 6 weeks in the Briard and Tibetan Terrier. Possibly an autosomal recessive trait |
| Optic nerve hypoplasia | Any breed may be affected. Small optic disc. Bilateral/unilateral |
| Generalized progressive retinal atrophy (PRA) | Inherited condition; generalized PRA commences from a few months to few years depending on breed; present in Rough Collie, Miniature Schnauzer, Gordon Setter, Irish Setter, Miniature and Toy Poodles, American Cocker Spaniel, Norwegian Elkhound, Longhaired Miniature Dachshund, Chesapeake Bay Retriever, Golden Retriever, Tibetan Spaniel, Cardigan Corgi, Irish Wolfhound and Japanese Akita. The mode of inheritance is autosomal recessive. Inherited retinopathy in the cat has only been adequately studied in the Abyssinian breed, in which there are two types - an early onset rod-cone dysplasia inherited as an autosomal dominant trait where affected cats are blind by a few months of age, and a late onset (5-10 years) degeneration affecting rods earlier than cones, inherited as an autosomal recessive trait |

***Table 13.12**: Inherited defects of the ocular system.*

## FADING PUPPIES

### Definition

The fading puppy syndrome has been a recurring problem for breeders and veterinary surgeons for decades. Various theories have been formulated to account for this syndrome and some have doubted that it is a genuine disease entity. One source of confusion has been to incorporate all conditions leading to poor weight gain and ill-thrift in the first months of life under this one syndrome. In fact, a study of neonatal mortality by the author shows that there is a narrow age band when most of these puppy mortalities occur, between 3.5 and 5 days after birth. The puppies are of adequate birth weight for breed and would be considered to have good survival prospects but subsequently fail to thrive and then die, usually between days 3 and 5, without obvious cause. The dams are in good health with unremarkable pregnancies of normal gestational length. There is no association with dystocia, poor mothering or obvious lactational deficit.

### The vulnerability of the neonate

In order to appreciate the normal appearance and responses of the neonate at clinical examination it must be remembered that the newborn puppy is an immature animal, dependent on its dam for survival in the first 3 weeks and is particularly vulnerable because of four major factors, listed below:

- *Thermoregulatory mechanisms are poorly developed.* At 24 hours the body temperature is 35.5 °C, rising to 38 °C by the seventh day and 38.5 °C by the fourth week. The zone of thermal neutrality falls between 27.5 and 36 °C. Initially, brown fat metabolism, under the control of the sympathetic nervous system, is important in heat production (non-shivering thermogenesis). Shivering mechanisms begin around days 6–8 and by the fourth week the puppy is a good homeotherm. Probably as long as the dam is an attentive mother and the puppies can maintain close contact with the mammary area, there will be sufficient heat in their microenvironment to maintain thermal balance. Where this natural system is disrupted, such as occurs with poor mothering or hand-rearing, there is a danger of chilling and hypothermia.

- *Risk of dehydration.* As 82% of body weight is water and kidney function is immature, the neonate is particularly susceptible to dehydration. Glomerular filtration increases from 21% at birth to 53% at 8 weeks and tubular secretion matures at 8 weeks. The water requirement is 60–90 g/450 g body weight per day and the water turnover is about twice that of an adult. Glucosuria is common in neonates until 2 weeks of age. It is important that the neonate should feed regularly from birth to maintain hydration.

- *Risk of hypoglycaemia.* The puppy is born with a relatively small reserve of glycogen, mainly in the liver. Failure to suck results in rapid depletion of this reserve and development of hypoglycaemia by the second day. The puppy should make regular daily weight gains from day 1, with the birthweight normally doubled by day 10.

- *Immunological immaturity.* It is important that colostrum is ingested during the first 12–24 hours of life, as only 5% of the maternal antibody is acquired across the placenta. The immunological system is considered to be immature and although capable of stimulation, it probably does not become fully competent until 3–4 months.

### Normal behaviour and appearance of the neonate

The neonate spends long periods in deep sleep, interrupted by short periods of body twitching; this behaviour disappears after 4 weeks. Loss of activated sleep patterns indicates ill health. Considerable maturation of the central and peripheral nervous systems occurs during the first 3 weeks of life and therefore neurological responses are different from the adult. Flexor dominance involving body and neck muscles at birth is replaced by extensor dominance by the third day of life. Puppies can stand at 3 weeks with normal tone and postural reflexes. The eyes do not open until 10–15 days and sight is poor until 4–5 weeks. The external and auditory canal does not open until 12–14 days, after which time there is a definite startle response. The healthy neonate only cries when roused or hungry; increased vocalization indicates some abnormality. The puppy exhibits 'rooting' behaviour as it crawls purposefully and searches for a teat; the sucking reflex is dependent on the contact of the teat with the oral cavity. The normal puppy has a rounded abdomen with a stomach full of milk but is not tympanic, a sleek hair coat, warm body and elastic skin.

The immaturity, small size and vulnerability of the neonate renders inappropriate diagnostic techniques used in the adult. Therefore observation of behavioural changes, evidence of dehydration, chilling and loss of body weight or failure to gain weight are particularly helpful signs when assessing neonatal health. Examination should also involve checks for congenital defects, e.g. cleft palate, imperforate anus, evidence of traumatic injuries (areas of swelling, rib/limb fractures, haemorrhage) and evidence of infection of the umbilicus, docking sites and eyelids (ophthalmia neonatorum). The neonate is dependent on the dam for licking stimulation around the urogenital region, for regular urination/defecation and for cleaning and grooming. A dirty, unkempt coat indicates poor mothering.

## The fading puppy – clinical and post-mortem findings

Although there is no obvious abnormality at birth, the puppies begin to lose weight (often in the first 24 hours), show poor sucking responses and display either lassitude or unusual restlessness, with plaintive and persistent crying; there is a gradual progression towards generalized weakness and death. Sudden and unexpected death is not typical. Post-mortem examination reveals that body weight is well below birth weight, the stomach and intestines are largely devoid of contents and there is a general absence of gross lesions or defects. The liver to body weight ratio changes from 1:10 to 1:20. Histopathological examination of all main organ systems does not reveal any evidence of an infective aetiology or other specific lesions.

## Effects of management systems

In the author's own study, a variety of breeding management systems were involved and no common management-linked factors were associated with the mortalities, although there is a strong tendency for certain dams within particular kennels to have successive fading litters, whilst other bitches experience no problems.

## Approach to investigation

As clinical signs are not specific, post-mortem investigation is ideally required to rule out other causes of neonatal mortality, e.g. septicaemia, congenital defects, maternal trauma. It is necessary to examine puppies from as many affected litters as possible over a long enough period in order to obtain a representative view of the causes of perinatal mortality. Puppies even within one litter may die of different causes. The puppies should be kept at +4 °C until post-mortem investigation can be undertaken (*not* frozen because of freezing/thawing artefacts). The most common causes of death in the author's own study, which accounted for about 50% of mortalities in liveborn puppies, were: 1) infections (mostly bacterial); 2) maternal and management-related factors, e.g. poor mothering qualities in the dam; 3) low birthweight; and 4) congenital abnormalities.

A conclusion as to the cause of death needs to be made in the context of clinical signs and evaluation of kennel management practice. Some time should be spent reviewing the quality of management, especially at time of whelping and through the neonatal period. Important factors to consider are:

- Kennel construction and whelping facilities
- Type of heating, risk of draughts and heat loss
- Whelping routine and supervision, especially over the first 2–3 days
- Hygiene practice
- Quality of staff
- Presence of disease vectors, e.g. birds, mice
- Risk of infection through introduction of outside animals, e.g. stud dogs
- Nutritional and health status of breeding stock
- Worming programme
- Vaccination policy.

The first five factors probably have most significance for the neonate as they have bearing on the most critical early days of life and interrelate with maternal influences.

It is essential that the breeder keeps accurate records of all litters for inspection, including details of birthweights and daily weight gain for the first 3 weeks, so that a complete picture of the problems can be obtained. The examination of one moribund puppy after months or years of high mortality rates is unlikely to yield meaningful results.

However, in spite of the author's own detailed investigations, about 50% of cases showed no specific cause of death. Although signs and post-mortem findings were not diagnostic, it was established that there was a common pattern to the clinical presentation and that this group of mortalities constituted a large group within a narrow time frame, without evidence of an infective aetiology and probably death results from a process initiated on day 1 (or even just before birth). Thus until demonstrated otherwise, it would seem reasonable to term these 'fading puppies'.

## Possible causes of fading puppy syndrome

A whole range of aetiological possibilities have been considered to be a cause of fading, including maternally related factors such as poor mothering, inadequate lactation, trauma and inadequate nutrition of the dam in pregnancy, and also congenital anomalies, neonatal isoerythrolysis, low birthweight, thymic atrophy and infectious agents including bacteria, viruses, protozoa and parasites. However, approximately 50% of neonatal deaths in the author's investigation could not be ascribed to these causes. It has also never been established whether the aetiology is related to one predominating factor or if it is multifactorial.

In a study of lung surfactant composition, there was found to be a significant reduction in the phosphatidylcholine (lecithin) component, which presented similarities to findings in babies dying of sudden infant death syndrome. Lung surfactant is required for normal respiratory adaptation and maintenance following birth. Abnormality of lung surfactant composition could predispose to hypoxia and breathing/suckling difficulties. However, it was not determined if abnormal surfactant was fundamental to the death of the puppies or a byproduct of some other pathological process or defect, such as a central respiratory drive deficit.

As the initial signs of fading can often be detected in the first 24 hours of life, it is possible that this group of puppies is not completely viable from birth and that there may be prenatal factors involved which are not immediately apparent. As mentioned, some dams appear to be more at risk from fading litters than others. A genetic predisposition is possible although the occurrence of large numbers of puppies (often whole litters) affected in many breeds, without evidence of linked close breeding, does not support an inherited defect alone.

### Approach to treatment

There are no consistent reports of successful treatment for fading puppies, and antibiotics appear to be of little value. As these puppies probably enter a fatal cycle of dehydration, hypoglycaemia and weight loss soon after birth, it is possible that there is some fundamental problem of bonding with the dam and the suckling response. Research into the physiology of birth adaptation and feeding responses in the first 24–48 hours might provide more answers to this very old problem. Until more information is available, *early* supplementary feeding of suspect fading litters might improve survival. In practice, supportive therapy is often given too late to alter survival outcome, and there is often insufficient investigation of neonatal mortalities and management to assess causes of breeding loss accurately.

## REFERENCES


Blunden AS (1983) *Neonatal and Perinatal Mortality in the Dog: Clinical, Pathological and Managemental Studies.* PhD Thesis, London

Blunden AS (1988) Diagnosis and treatment of common disorders of newborn puppies. *In Practice* **10**, 175–184

Blunden AS, Hill CM, Brown BD and Morley CJ (1987) Lung surfactant composition in puppies dying of fading puppy complex. *Research in Veterinary Science* **42**, 113–118

Casal, ML (1995) Feline paediatrics. *Veterinary Annual* **35**, 210-228

Detweiler DK, Hubben K and Patterson D (1960) Survey of cardiovascular disease of dogs. *American Journal of Veterinary Research* **21**, 329-359

Evans JM (1978) Neonatal mortality in puppies. In: *Refresher Course in Canine Medicine.* Proceedings No 37. University of Sydney, Sydney, pp 127–139

Fox MW (1970) Inherited structural and functional abnormalities in the dog. *Canadian Veterinary Journal* **11**, 5

Hodgman SFJ (1963) Abnormalities and defects in pedigree dogs. I. An investigation into the existence of abnormalities in pedigree dogs in the British Isles. *Journal of Small Animal Practice* **4**, 447

Hoskins, JD (1995a) Congenital defects of the cat. In: *Textbook of Veterinary Internal Medicine: Disease of the Dog and Cat, 4th edn*, ed. SJ Ettinger and EC Feldman, pp. 2106-2114. WB Saunders, Philadelphia

Hoskins, JD (1995b) Congenital defects of the dog. In: *Textbook of Veterinary Internal Medicine: Disease of the Dog and Cat, 4th edn*, ed. SJ Ettinger and EC Feldman, pp. 2115-2129. WB Saunders, Philadelphia

Hoskins JD (1995c) Puppy and kitten losses. In: *Veterinary Pediatrics, 2nd edn*, ed. JD Hoskins, pp. 51-55. WB Saunders, Philadelphia

Jubb KVF, Kennedy PC and Palmer N (1993) *Pathology of Domestic Animals, 4th edn.* Academic Press, London

Leipold HW (1978) Nature and causes of congenital defects of dogs. *Veterinary Clinics of North America* **8**, 47–77.

Mulvihill JJ and Priester WA (1971) The frequency of congenital heart defects (CHD) in dogs. *Tetratology* **4**, 236

Nicholas FW (1996) *Introduction to Veterinary Genetics.* Oxford University Press, Oxford, pp. 97–104

Priester WA, Glass AG and Waggoner NS (1970) Congenital defects in domestic animals: general considerations. *American Journal of Veterinary Research* **31**, 1871

Roth JA (1987) Possible association of thymic dysfunction with fading syndromes in puppies and kittens. *Veterinary Clinics of North America: Small Animal Practice* **17**, 603-616

Willis, MB (1992) *Practical Genetics for Dog Breeders.* HF & GF Witherby Ltd, London, pp 131–162.

CHAPTER FOURTEEN

# Care of Neonates and Young Animals

*Paula Hotston Moore and Kit Sturgess*

## INTRODUCTION

By definition, the neonatal period covers the first 7–10 days of life and is characterized by poor neurological function, the progressive development of spinal reflexes and a total dependency on the dam. This period is followed by a transitional period (10–21 days of age) characterized by the development of a competent audio-visual system, further neurological development and an increasing independence from the dam. Puppies and kittens enter a period of socialization from 3 weeks of age lasting until about 3 months of age, during which time feeding and sleeping occupy progressively less of the day, being replaced by social activity. There is maturation of the nervous system and hepatic and renal functions develop. For the purposes of this discussion, disease problems of young puppies and kittens will be considered up to the point at which maternally derived antibodies are thought to be no longer protective, at around 5-6 weeks of age.

CHAPTER FOURTEEN (i)

# Care and Management of the Neonate

*Paula Hotston Moore*

The neonate is defined, in this instance, as a newly born puppy or kitten, up to the age of 10 days. The newborn is totally dependent on the dam not only for nutrition but for warmth, comfort, encouragement to urinate and defecate and general well being. It is important to remember the nursing considerations not only of the neonate which is born and reared naturally, but also those of the neonate delivered by Caesarean operation and of the orphaned neonate.

## NATURAL BIRTH

Following a delivery *per vaginum*, the dam usually attends to the neonate herself. The following steps should be performed by the dam:

- Licking the fetal membranes away from the mouth and nose
- Licking and nuzzling to maintain body temperature
- Biting off the umbilical cord
- Encouraging the neonate to suck.

Should any of these steps not be performed by the dam, assistance is required to ensure neonatal survival.

The removal of fetal membranes is performed by briskly wiping the neonate with a towel, with particular attention to the mouth and nose.

The umbilical cord should be clamped with a pair of sterile forceps 4 cm from the umbilicus. The forceps should be removed after 10 minutes and if haemorrhage occurs a ligature of absorbable suture should be placed around the umbilical stump.

A healthy neonate at an ambient temperature will be active and will find the dam's teat and begin to suck. If one of the newborn has difficulty finding the milk supply, its mouth should be placed over the teat. By pressing on either side of the teat with a finger and massaging the mammary gland, milk is released and the neonate will be able to feed.

Should the dam reject one of her offspring, rubbing placental fluids on the neonate may help her to recognize it as her own. The neonate should be monitored to ensure acceptance by the dam and also that feeding is occurring every 2–3 hours.

Occasionally the dam will attack the neonate. In this situation, it is necessary either to tranquillize or muzzle her for a short period of time until the problem of rejection is overcome.

As the newborn feeds, its metabolic rate increases, which raises the body temperature. The neonate has come from within the uterus, where the temperature remains constant, to a variable outside environment. It has little subcutaneous fat, so its insulation is poor, and as heat loss is greatest in smaller individuals with a larger surface area per unit of body weight, it is therefore important to help the neonate maintain its body temperature.

The environmental temperature for the initial 24 hours should be maintained at 30–33°C, which can be reduced to 26–30°C over the following 3–4 days. Bedding in the dam's pen should be suitable for nesting; this allows the neonate to have good insulation. When the newborn huddles with the dam, this reduces exposed surface area and thus reduces heat loss.

The healthy neonate should increase in body weight by 5–10% per day. Daily recording of body weight ensures that progress is closely monitored. Temperature generally falls from approximately 36°C at birth to 30°C within a few hours and then increases to 38°C over the first 7 days. Respiratory rate should be 15–40 per minute, depending on size, with a regular rhythm and without respiratory noise.

No discharge should be present from the eyes, ears or nose. A healthy neonate will crawl around and be very mobile. Ten days after birth it will stand and at 3 weeks of age begin to walk. Ten to 14 days after birth, the eyes will open; although the cornea may be cloudy, this will clear during the next 2 weeks. Strabismus is common in the neonatal kitten, but declines gradually after 8 weeks of age.

The dam should not be allowed to become overheated. Ideally, the enclosed housing area for neonate and dam should have a part which is not directly heated, enabling her to seek a cooler environment, if necessary. Shredded newspaper is preferred as a bedding material, as it is both cheap and easily disposed of. Soiled material should be removed and replaced frequently to provide clean and dry bedding for both neonate and dam.

Initially, the dam will lick the perineal area of her offspring to stimulate urination and defecation. At 2–3 weeks of age, the neonate begins to urinate and defecate voluntarily. As soon as possible, it should be encouraged to urinate and defecate away from the nesting area, making it easier to keep the bedding clean and dry; this will also hasten toilet training.

All neonates, whatever the method of delivery, should be examined for any gross clinical abnormalities such as cleft palate, atresia ani or limb deformity.

## SPECIAL CARE OF THE NEONATE DELIVERED BY CAESAREAN OPERATION

Following delivery, the airways of the neonate should be cleared immediately of any fetal membranes. A soft towel can be used gently to wipe away mucus. Fluid can be removed from the nose and mouth by suction using a bulb syringe or alternatively, a length of soft narrow tubing attached to a syringe. The neonate should be cupped in the nurse's hands with the head tilted downwards and the hindquarters elevated; this position allows fluid to drain away from the respiratory tract. It should have its head and neck supported and then be rocked forwards and backwards in an arc, releasing fluid and mucus from the respiratory tract. Gentle, brisk rubbing of the newborn animal with a towel, especially over the thorax, stimulates breathing. By removing the amniotic membranes and fetal fluids from the hair of the neonate, heat loss due to evaporation is avoided.

Survival of the neonate is dependent upon the rapid onset of normal, spontaneous respiration. If the neonate is not breathing, respiration can be stimulated pharmacologically, using doxopram hydrochloride (20 mg/ml) 1–2 drops sublingually, having ensured the airways are clear. Oxygen can be administered by mask or tracheal catheter intubation.

A regular, strong heartbeat should be detectable over the thoracic wall. If a heartbeat cannot be found, external cardiac massage should be gently attempted. Artificial respiration, when required, is performed by gently blowing into the mouth and nose of the animal. Revival techniques should be performed for 5 minutes post-delivery, when required.

If the dam was given opioids prior to delivery, naloxone hydrochloride (0.4 mg/ml) should be administered to the neonate. Two to 5 drops can be given sublingually to reverse respiratory depression and the neonate should then be monitored for several hours to ensure respiratory depression does not recur.

At delivery, a clamp is placed around the umbilicus. The clamp is removed 10 minutes post-delivery and the umbilical stump dipped in a solution of chlorhexidine. If haemorrhage occurs, the umbilicus should be ligated with an absorbable suture.

Hypothermia will occur quickly in neonates exposed to a cool temperature. The newborn should be placed in a pre-warmed environment of 30–33°C. A paediatric incubator purchased second hand is an ideal environment, since it can be thermostatically controlled. Alternatively, the neonate can be placed in a box on a heat pad covered with towels. When a heat pad or other direct source of heat is used to supply warmth, it must be covered with a towel or blanket to prevent contact with the skin. A thermometer is placed at the level of the neonate to enable constant monitoring of the temperature.

When both the dam and her offspring have recovered from the anaesthetic, the neonate is placed with the dam, who should be in an enclosed environment such as a whelping pen, make-shift kennel or a cardboard box.

The dam is observed to ensure a smooth recovery from both surgery and anaesthetic.

## CARE OF THE ORPHANED NEONATE

An orphaned neonate needs to have all its requirements supplied by a source other than its mother.

The best possible nutrition for the neonate comes from the dam's milk. When this is unavailable, a well formulated milk replacer should be used. It is important for the milk replacer to be of a similar gross composition to the milk of the dam. The *quality* of cow's milk is the same as milk from a bitch or queen; however, the *quantity* of protein, fat and carbohydrate differs in each. The milk replacer should be a 'perfect' substitute for the mother's milk. Commercially available species-specific milk replacers should be used in preference to home-made preparations. In an emergency, when a commercial milk replacement product is unavailable, a home-made recipe can be fed to the orphan.

An example of a home-made diet for a puppy is: 1 litre whole cow's milk, 4 egg yolks and 1 tablespoon corn oil.

An example of a home-made diet for a kitten is: 3 oz condensed milk, 3 oz water, 4 oz plain full fat yoghurt and 3 large egg yolks.

Diets using cow's milk have a higher lactose content than maternal milk and often result in diarrhoea. A home-made diet should only be fed when no alternative is available and only for a few hours, until a commercial milk replacer is purchased.

Correctly formulated milk replacer will support the neonate during the critical first few weeks of life. The milk replacer has been manufactured to provide a highly digestible diet which closely resembles the composition of dam's milk. There are a number of such products available (Table 14.1).

The optimum caloric intake has been studied for each of the commercially available milk replacers and therefore the suggested feeding guide should be fol-

| Product | Species | Manufacturer |
|---|---|---|
| Welpi | Puppies | Pet Life |
| Cimicat | Kittens | Pet Life |
| Esbilak | Puppies | PetAg Distributed in UK by Kruuse |
| K.M.R. (Kitten Milk Replacer) | Kittens | PerAg Distributed in UK by Kruuse |
| Whiskas® Instant Milk Substitute | Kittens | Pedigree Petfoods |
| Pedigree® Instant Milk Substitute | Puppies | Pedigree Petfoods |
| Lactol Milk Powder | Puppies/ kittens | Shirley's |

***Table 14.1:*** *Recommended milk products for orphaned puppies and kittens.*

lowed. The milk replacer should be made up and fed to the manufacturer's recommended quantity and frequency. It is important to increase the amount of food given as the neonate grows.

A nursing bottle with an appropriately sized teat should be used to feed the neonate. If such a nursing bottle is not available a 1 ml syringe or dropper can be used until a bottle can be purchased. The neonate is held in one hand with the head held up and slightly stretched out. The teat is placed in the animal's mouth and is pulled away slightly: this elevates the head and encourages sucking. Care should be taken not to tilt the head too far back or to pour in the milk too fast, as excess milk will come out of the nose. This is to be avoided, as aspiration can occur. The neonate will reject the bottle when it has received enough feed.

A common problem in orphaned neonates is diarrhoea, usually an indication of overfeeding.

When the neonate is approximately 14 days old, it can be encouraged to lap from a shallow bowl. The muzzle is placed in the bowl and milk splashed onto the lips to encourage lapping. To begin with, the feed may have to be finished by bottle but it will soon lap each meal voluntarily.

The growth rate of a hand-reared neonate will be lower than that of an animal reared naturally. Once the hand-reared neonate is a few months old, however, it should be a normal, healthy animal and indistinguishable from a naturally reared puppy or kitten of the same age.

The animal orphaned at birth is of particular concern because it is deprived of colostrum and is more susceptible to disease. In this situation, it must be closely monitored and at the first sign of ill health or reluctance to feed should be brought for veterinary attention.

An important nursing procedure for the orphaned neonate is to encourage urination and defecation manually. The dam's licking and stimulating the genital area of the neonate is copied by rubbing the area with a piece of damp cotton wool. This is carried out after each feed, until the neonate is able to urinate and defecate voluntarily.

In conclusion, it is of great importance to provide a controlled environment in which the neonate can thrive. Nursing the neonate is very time consuming and has its difficulties but is, however, extremely rewarding.

CHAPTER FOURTEEN (ii)

# Infectious Diseases of Young Puppies and Kittens

*Kit Sturgess*

Puppy and kitten mortality is around 15-40% in the first 12 weeks of life, with the majority of deaths occurring in the first week. Infectious disease accounts for a relatively small percentage of deaths, but it is a major cause of morbidity in paediatric patients. There is nothing unique about the spectrum of infections which can occur in neonates, however the prevalence, clinical signs and prognosis of various diseases in young animals may vary considerably from those in adults.

## REVIEW OF PHYSIOLOGY

During the first few weeks of life, significant physiological changes occur which will directly affect the clinical signs shown and the ability of the neonate to respond to disease. A basic understanding of neonatal physiology is vital to enable clinical findings to be evaluated accurately. With separation of the placenta, the peripheral resistance increases and hypoxia develops rapidly inducing gasping respiration. Constriction of the umbilical vein squeezes significant quantities of blood from the placenta into the neonate and therefore, when possible, should be left intact. In response to the increasing oxygen tension the ductus arteriosus narrows (complete closure in 1-2 days) and the pulmonary vessels dilate. Increased left sided pressure results in the closure of the foramen ovale between the atria. Fetal $PO_2$ rises from 20 mm Hg to around 50-60 mm Hg. With time, the $PO_2$ rises further, correcting the acidosis that develops in the newborn.

### Rectal temperature and thermoregulation

The rectal temperature of dry, healthy day-old puppies and kittens is around 35.5°C (±0.8) rising gradually over the first week of life to around 37.5°C. Adult temperature is achieved by approximately 4 weeks of age (Table 14.2). Thermoregulation in the newborn is poor as their ability to shiver and vasoconstrict in response to falling body temperature is limited. The ability to shiver develops at around 6-8 days of age, prior to which, brown adipose tissue is the main source of thermogenesis.

### Haematology and biochemistry

Normal ranges for haematological and biochemical parameters are different from adults, therefore results need to be interpreted with reference to the age of the puppy or kitten (Tables 14.3 and 14.4). It must be remembered that the circulating blood volume of puppies and kittens can be small (25-40 ml in a 4-week-old kitten). Repeated blood sampling can cause severe anaemia and should, therefore, be kept to a minimum.

### Glucoregulation

Newborn puppies and kittens have limited reserves of glycogen and poor hepatic gluconeogenic responses to low blood glucose. A fasted puppy, however, is able to maintain blood glucose concentrations for as long as 24 hours provided it remains healthy.

### Hepatic and renal function

Hepatic microsomal enzymes which are involved in many metabolic functions including drug metabolism may not be fully functional until 4-5 months post partum, though near normal liver function is probably present from around 8 weeks of age. Albumin levels in neonates are significantly lower than in adults which

| Age (days) | Rectal temperature (°C) | Heart rate (beats/min) | Respiratory rate (/min) | Environmental temperature (°C) |
|---|---|---|---|---|
| 0-7 | 36 ± 1 | 200-250 | 15-35 | 29-32 |
| 8-14 | 38 | 70-220 | 15-35 | 27 |
| 15-28 | - | 70-220 | 15-35 | 27 |
| 29-35 | Adult | 70-220 | 15-35 | 21-24 |
| >35 | Adult | 70-220 | Adult | 21 |

***Table 14.2:*** *Physiological values in young puppies and kittens.*

| Parameter | Puppies (mean ± SD or range) | | | | Kittens (mean or range) | | | |
|---|---|---|---|---|---|---|---|---|
| | 0–3 days | 0–2 weeks | 2–4 weeks | 6 weeks | 0–3 days | 2 weeks | 4 weeks | 6 weeks |
| PCV (%) | 46.3 ± 8.5 | 33-52.5 | 27-37 | 34 | 41.7 | 33.6-37.0 | 25.7-27.3 | 26.2-27.9 |
| Hb (g/dl) | 15.8 ± 2.9 | 14-17.5 | 8.5-11.6 | 9.59 | 11.3 | 11.5-12.7 | 8.5-8.9 | 8.3-8.9 |
| RBC (x$10^6$/l) | 4.8 ± 0.8 | 3.6-5.9 | 3.4-4.9 | 4.91 | 5.11 | 5.05-5.53 | 4.57-4.77 | 5.66-6.12 |
| MCV (fl) | 94.2 ± 5.9 | 89-93 | 78-83 | ND | 81.6 | 65.5-69.3 | 52.7-55.1 | 44.3-46.9 |
| MCH (pg) | 32.7 ± 1.8 | 28-30 | 23-25.5 | ND | 24.6 | 22.4-23.6 | 18.0-19.6 | 14.2-15.4 |
| MCHC (g/dl) | 34.6 ± 1.4 | 32 | 32 | ND | 27.3 | 33.7-35.3 | 32.5-33.5 | 31.3-32.5 |
| WBC (x$10^3$/l) | 16.8 ± 5.7 | 6.8-23 | 23-25.5 | 15.00 | 7.55 | 9.1-10.2 | 14.1-16.5 | 16.1-18.8 |

***Table 14.3:*** *Haematological values in young kittens and puppies.*

*ND = No data*

| Parameter | Puppies (range or median) | | Kittens (range) | |
|---|---|---|---|---|
| | 0–2 weeks | 6 weeks | 2 weeks | 4 weeks |
| Total protein (g/l) | 34-52 | 44.5 | 40-52 | 46-52 |
| Albumin (g/l) | 15-28 | 26 | 20-24 | 22-24 |
| Sodium (mmol/l) | ND | 148 | ND | 149-153 |
| Potassium (mmol/l) | ND | 5.3 | ND | 4.0-4.8 |
| Chloride (mmol/l) | ND | 105 | ND | 120-124 |
| Inorganic phosphate (mmol/l) | ND | 2.96 | ND | 2.03-2.41 |
| Calcium (mmol/l) | ND | 3.53 | ND | 2.35-3.24 |
| Urea (mmol/l) | ND | 1.2 | <5 | <5 |
| Creatinine (μmol/l) | ND | 36 | ND | 36-54 |
| Cholesterol (mmol/l) | 2.93-9.01 | 4.11 | 4.29-11.59 | 0.58-11.36 |
| ALP (IU/l) | 176-8760 | 131.5 | 68-269 | 90-135 |
| ALT(IU/l) | 10-337 | 16.5 | 11-24 | 14-26 |
| Creatinine kinase (IU/l) | ND | 210 | ND | ND |
| Glucose (mmol/l) | 4.16-11.68 | 10.08 | 6.08-10.32 | 7.92-8.96 |
| Bilirubin (μmol/l) | 1.7-16.9 | 5.1 | 1.7-16.9 | 1.7-3.4 |
| Bile acids (μmol/l) | <10 | ND | <10 | <10 |

***Table 14.4:*** *Serum biochemistry values in young puppies and kittens.*

*ND = No data*

can result in increased circulating drug levels. Glomerular filtration rate is approximately one-fifth of adult levels and tubular secretion mechanisms do not mature until approximately 8 weeks of age. This means that glucosuria is common and urine specific gravity is low (1.006–1.007). Puppies and kittens have a limited ability to conserve fluid and fluid requirements are therefore high at around 120–180 ml/kg per day.

## Immune function

Neonates possess a degree of immune competence, but do not have a fully matured spectrum of immune responses. IgM (rather than IgG or IgA) antibodies predominate with reduced T cell activity in response to stimulation, this is probably due to the thymus not being fully mature until 12 weeks of age. Reduced activity of cells involved in non-specific immune responses, such as neutrophils, are also reported in some species though relatively little work has been done in puppies and kittens. With a poorly functioning immune system in terms of speed, magnitude and breadth of response, the importance of passively acquired immunity in providing resistance to infectious disease cannot be over-emphasized.

## Passive immunity

More than 90% of passive immunity in the dog and cat is provided from colostral intake; however, the protection afforded depends on the previous exposure of the dam and her immune status. Gut permeability to immunoglobulins begins to decline within 8 hours of birth and no further absorption is possible after 48-72 hours. Passive immune protection of the intestinal tract continues during the whole period of suckling as IgA antibodies resist gastric degradation and can bind potentially harmful pathogens in the gut lumen, preventing them from attaching to or penetrating the intestinal mucosa. Colostrum also contains cellular components though their precise role is unclear.

As the majority of cases of neonatal infections are not caused by agents for which vaccines are available, it is vital that the puppies or kittens are born into the same environment as the one in which the dam has been housed. If queens or bitches are to be moved into a new environment before parturition, sufficient time must be allowed for them to develop an immune response to new environmental pathogens prior to delivery. Giving colostrum-deprived puppies or kittens plasma from adults or commercial immunoglobulin preparations has not been shown to be beneficial.

## Cardiovascular function

Unlike adults, heart rates in newborn puppies may respond to hypoxia by falling rather than rising. This is likely to be a protective mechanism to reduce oxygen demand in the face of hypoxia. However, a falling heart rate has to be interpreted with care as it may signify hypoxaemia and distress (though not necessarily deteriorating condition). Normal blood baroreceptor function is present by 4 days of age. Responses to cardioactive drugs are less profound than those of adults until puppies are 9-10 weeks of age.

## Neurological development

Over the first 11-12 weeks of life, puppies and kittens develop normal adult reflexes and responses. Until that time, they display primitive reflexes which gradually disappear, while the adult reflexes gradually become dominant. The time sequence of these events has been studied in puppies and is presented elsewhere. Not only are reflexes immature but behaviour patterns tend to be much simpler, being driven by hunger and the search for warmth. Newborn puppies and kittens will sleep for more than 80% of the time and will tend to lie quietly when replete and warm. When stressed (for whatever reason) they will cry and crawl around, making side-to-side head movements.

# EVALUATION OF THE NEONATE

## Case history

Basic information is essential to evaluate clinical and laboratory findings. The history should include a breeding history of the household, kennel/cattery management (hygiene, worming, vaccination), the health of the dam during the pregnancy, the health of the remainder of the litter, age of the puppy/kitten and pattern of the illness to date.

## Clinical examination

Examination of the neonate can be difficult, becoming more rewarding as the puppy/kitten gets older and begins to develop adult responses. Neonates tend to show limited responses to disease, initially becoming agitated and crying, progressing to inactivity, hypothermia and loss of the sucking reflex. Such changes can be very rapid in the face of severe infection and it is important that the owner is made aware of the potential significance of these signs to allow aggressive, early treatment. Lack of weight gain can be a sensitive indicator of developing problems and can be easily measured by the owner. Failure to gain weight over any 24-hour period is worthy of further investigation (Figure 14.1).

As with any clinical examination, a systematic approach is essential in order to make sure all parts of all systems are examined.

- *External features*: body weight; hair coat (amount, condition, parasites); state of hydration; signs of injury; appearance of umbilicus; dew claws; (docking site); discharge from nose; urine staining (patent urachus); diarrhoea/rectal patency; congenital malformation
- *Eyes:* swelling under lids indicates pus formation (often *Staphylococcus* spp., very rarely *Chlamydia psittaci*); eyes open between 5 and 14 days in kittens and between 10 and 14 days in puppies; pupillary light response is usually present within 24 hours of the eyes opening, mild corneal cloudiness is usually present as eyes open
- *Ears*: external auditory meatus is closed at birth and opens between 6 and 14 days; check for mites; middle ear infection indicated by a

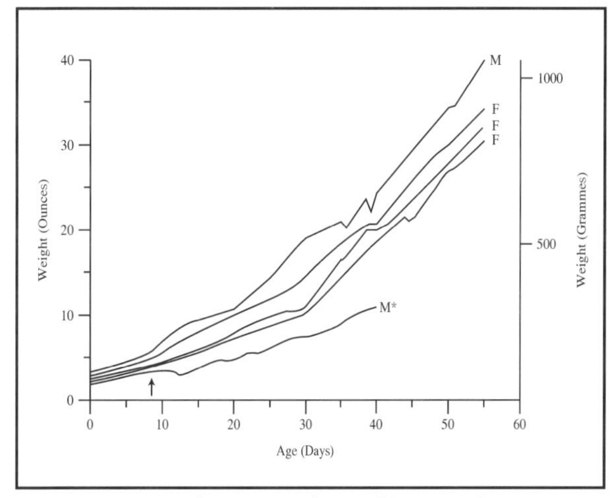


*Figure 14.1: Weight gain in a litter of Burmese kittens.*

*M, male; F, female. *This male kitten began to lag behind the other kittens from 8 days of age and despite treatment eventually died. On post-mortem, extensive lung abscessation was found.*

bulging tympanum
- *Mouth*: mucous membrane colour; evidence of cleft palate
- *Thorax*: heart rate around 200–220 beats per minute; respiration 15–35 per minute; regular rhythm; heart murmurs may be functional (usually soft); lung sounds difficult to distinguish but should be present; check for symmetry or malformation of the thoracic cavity
- *Abdomen:* should feel full but not swollen or tight; liver and spleen not palpable; intestines soft, mobile and non-painful; urinary bladder freely movable
- *Neurological assessment:* alertness, response to stimulation, sucking reflex; other reflexes appropriate to age (see above); gait (walking from around 4 weeks old); posture. Flexor and extensor dominance appears more variable in kittens than in puppies.

## INVESTIGATION OF SUSPECTED CASES OF NEONATAL INFECTION

Routine biochemistry and haematology can be performed from a very early age on blood obtained by jugular puncture (Figure 14.2) but must be interpreted using reference ranges for the same age group (Tables 14.3 and 14.4). Many infectious diseases develop too rapidly to obtain results quickly enough to be of value to that individual, especially bacterial culture and sensitivity. However, as infections are frequently a litter problem, laboratory data can improve management and treatment of any subsequent cases. Radiographs can be difficult to evaluate in young puppies and kittens as mineralization of the skeleton is poor and they are easily over exposed but they may provide useful information. The kV should be reduced to half that used for an adult of similar body thickness. Faecal examinations can be easily performed and are of particular value where protozoan parasites are suspected. Many young animals, however, will die before an infection is diagnosed. Maximum information can be obtained from post mortem if the carcass is fresh and if not immediately available for post mortem, the body should be stored in the refrigerator and not the freezer. A systematic approach to the examination is vital and details should be recorded.

## CONSIDERATIONS AFFECTING THE TREATMENT OF NEONATAL DISEASE

There are special considerations when giving drugs or fluids to paediatric patients and the following should be borne in mind:

- Absorption, distribution, metabolism and excretion of drugs can be significantly different from in adults

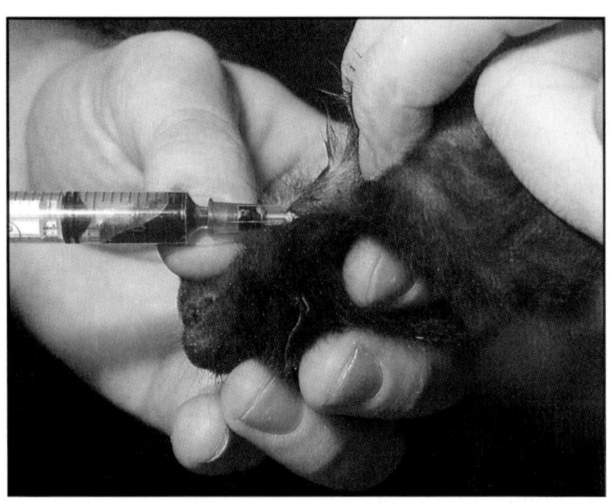

***Figure 14.2:*** *Blood sampling a 7-day-old kitten.*

- Few drugs have had dose rates calculated for use in the neonate. Therefore an increase in initial dose may be required with a lengthening of the interval between doses
- Great care should be taken when administering some types of antibiotics orally because of the potentially adverse effects on the developing gut microflora
- Subcutaneous and intramuscular absorption of drugs is slower and less reliable than in adults
- Antibiotics administered to the dam do not reach therapeutic concentrations in the milk
- Fluid requirements are higher in neonates than adults BUT total volumes can be low
- Nutritional requirements are high in the face of sepsis (about 1.5 times maintenance energy requirements) and every effort should be made to provide nutritional support either by naso-oesophageal or gastric intubation.

For many of the bacterial infections encountered in puppies and kittens, clavulanic acid potentiated amoxycillin represents a logical first choice drug in the absence of culture and sensitivity results. When administering fluids or drugs to critically ill neonates the intravenous, intraosseus or intraperitoneal routes should be considered.

### Fluid therapy

In many cases, the neonate will be acidotic but reduced hepatic function can mean that it is less able to metabolize lactate into bicarbonate. In the majority of cases lactated Ringers solution with added maintenance potassium (at 20 mmol/l) is suitable, if not ideal. If acidosis is likely to be severe, then bicarbonate can be given. In the absence of a known acid–base status then approximately 2 mmol/kg can be given over 10–15 minutes. Glucose can be replaced using a 5% dextrose solution mixed 50:50 with lactated Ringers or by giving 1–2 ml of 10–25% glucose intravenously to a profoundly depressed puppy or kitten.

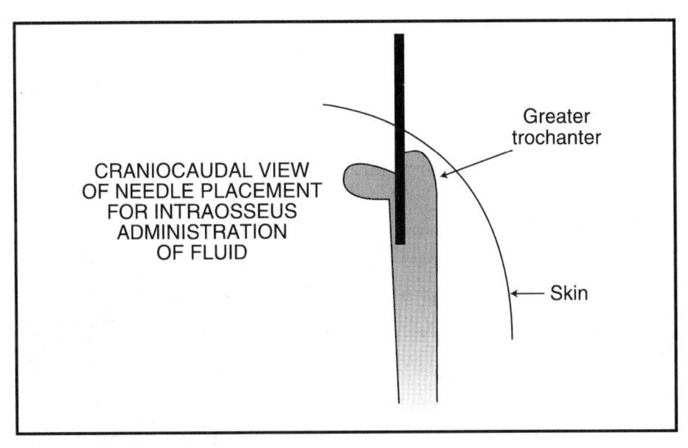


***Figure 14.3:*** *Intraosseus fluid administration into the proximal femur.*

Fluid requirements of a neonate are high, but total volumes are low. A syringe pump can be of great value and is significantly cheaper than fluid pumps; otherwise a burette with a paediatric giving set (60 drops per ml *versus* 10-15 drops per ml for a conventional adult giving set) will ensure that the puppy or kitten is not over-hydrated.

### Example

Two-week-old kitten weighing 250 g and estimated to be 6% dehydrated

Maintenance 6 ml/kg/h =1.5 ml/h
Deficit 0.06 x 250 = 15 ml
Total requirement for first 6 hours = maintenance (9 ml) + 50% of deficit (7.5 ml) = 16.5 ml
Hourly requirement = 2.75 ml/h or 2.75 drops/min (paediatric giving set)

## Methods of drug and fluid administration

- *Intravenous* - a 23 or 25 gauge catheter can be placed in the cephalic vein of many small puppies or kittens; however, their short legs can make the catheter placement and flow difficult to maintain
- *Intraperitoneal* - not ideal, as absorption can be relatively slow especially in the face of hypovolaemia and this route is poorly suited to long-term fluid therapy. Risks of puncturing viscera are low but aseptic technique is mandatory. Fluid requirements should be calculated and the volume divided, to be given 2-3 times daily
- *Intraosseus* - in the absence of venous access, intraosseus therapy represents a logical alternative. In most young puppies and kittens, the cortical bone is sufficiently soft that a hypodermic needle (18-19 gauge) can be placed. The area should be surgically prepared and the needle placed in either the proximal tibia or proximal femur (Figure 14.3). Only one attempt should be made at each site as if the bone is already punctured it will result in fluid leaking out. Fluids, drugs or whole blood can be given at the same rates as for i.v. therapy.

# SPECIFIC INFECTIONS AFFECTING THE NEONATE

## Ectoparasites

Ectoparasites commonly affecting young puppies and kittens are listed in Table 14.5.

A variety of products is available for treatment of ectoparasites (Table 14.6) but not all are licensed for use in puppies and kittens less than 12 weeks old. The 'cascade system' of drug usage should be followed whenever possible. Flea control can often be achieved using a flea comb and treatment of the dam and environment. Fipronil provides a safe treatment in puppies over 2 days and kittens over 7 days old, though alcohol toxicity has been reported in neonates sprayed in baskets. Ivermectin, which has been successful in the treatment of ear mites, *Demodex* and *Sarcoptes* in adults, is unlikely to be contraindicated in puppies and kittens, although idiosyncratic reactions can occur. Kittens are relatively less susceptible to ivermectin toxicity, though this has been reported. Ivermectin can be particularly useful if large numbers of cats need to be treated, e.g. for ear mites, with the dose being repeated in 2-3 weeks. Other products that can be used include amitraz and selenium sulphide. When treating a *Sarcoptes* infestation, all members of the litter plus the dam should be treated

| Parasite | Comments |
|---|---|
| Common flea | Commonly encountered, more rarely clinically significant. Can result in anaemia if burden heavy |
| Ear mites | Very common in neonates, passed on from dam, rarely cause severe disease |
| *Demodex* | Rare in kittens, more common in puppies. Primary immunodeficiency has not been demonstrated. 30-50% of puppies less than 1 year old recover spontaneously. Generalized or localized. Typically scaling, seborrhoeic and pustular lesions. Puppies usually infected within first few days of life. Diagnosis on skin scraping. Secondary pyoderma common |
| Sarcoptes | Lesions especially on ventrum, legs and ears, very pruritic papulocrustous, erythematous. Diagnosis on skin scraping |
| Lice | Generally indicates poor husbandry. Pruritic |

***Table 14.5:*** *Ectoparasites commonly affecting young puppies and kittens.*

and environmental decontamination considered. The use of anti-seborrhoeic preparations prior to acaricidal shampoos can be helpful. The treatment of *Demodex* is outwith the scope of this discussion; concomitant antibiotic use is recommended.

In general, organophosphorus-containing compounds are best avoided in the treatment of ectoparasites in neonates. Pyrethrins and pyrethroids are safer but overdosage is still common.

The incidence of protozoan parasites (Table 14.7) will be minimized by good husbandry, hygiene, daily disposal of faeces and avoiding both overcrowding and the use of outside runs which cannot be effectively disinfected. Potentiated sulphonamides need to be used with care in neonates due to their reduced hepatic metabolism and they are not recommended as a routine antibiotic.

## Endoparasites

*Toxocara* spp. and *Toxascaris leonina* are almost ubiquitous parasites of puppies and kittens, being transmitted in the dam's milk. Heavy infestations are associated with unthriftiness, diarrhoea, poor coat condition and a 'pot-bellied' appearance. Rarely, complete bowel obstruction is caused. Regular worming should be performed in all puppies and kittens. Piperazine, though widely used, is potentially toxic, particularly in cats, and overdosage is common due to the size of the tablets. The benzimidazole group or pyrantel embonate are more efficacious, safer and easily administered.

## Bacterial infections

A variety of bacterial infections have been associated with neonatal septicaemias in puppies and kittens. Bacteria involved are usually common, such as *Staphylococcus, Escherichia, Klebsiella, Enterobacter, Streptococcus, Enterococcus, Pseudomonas, Clostridium, Bacteroides, Fusobacterium* and *Salmonella* spp. Of these, Gram-negative bacilli are found most frequently. Death can occur suddenly with few clinical signs but more commonly frequent crying, restlessness, hypothermia, diarrhoea, dyspnoea, haematuria and cyanosis are seen. In more chronic disease, animals fail to gain weight as expected. Diagnosis is based on history and clinical examination. Ideally, blood cultures should be obtained but their results are only likely to be of value to other litter mates, as those first infected are likely to die before a diagnosis can be established and appropriate treatment given. Treatment needs to be aggressive, with antibiotics, fluids, glucose and oxygen.

| Drug | Dose rate in puppies | Dose rate in kittens | Use | Licensing |
|---|---|---|---|---|
| Ivermectin | 200 μg/kg | 200–400 μg/kg | Sarcoptic mange, demodectic mange, ear mites | NLC/D |
| Amitraz | As adult | **LETHAL** | Demodectic mange | NL <12<br>NLchi |
| Selenium sulphide | As adult | As dog | Licensed for seborrhoeic dermatitis but effective in treatment of *Cheyletiella* | NLC |
| Clindamycin | 11 mg/kg bid | 11 mg/kg bid | Toxoplasmosis | LP |
| Trimethoprim/ sulphadiazine | 30 mg/kg sid; 15 mg/kg bid | 30 mg/kg sid (kittens >1 kg)* | Coccidiosis | LP |
| Metronidazole | 25 mg/kg bid for 5 days then 10 mg/kg bid | As puppy | *Giardia* | NLC/D |
| Fenbendazole | 50 mg/kg sid for 3 days | As puppy | *Giardia* | LP |
| Doxycycline† | N/A | 10 mg/kg sid | *Haemobartonella felis* | LP |
| Clavulanic acid potentiated amoxycillin | 12.5–25 mg/kg bid | As puppy | Broad-spectrum antibiotic | LP |
| Augmentin | As above | As puppy | Intravenous use | NLC/D |

***Table 14.6:*** *Suggested dose rates for drugs in puppies and kittens.*

** Tablets should not be divided. Smaller kittens can be dosed orally with the 24% (not 7.5%) injectable solution of trimethoprim–sulpha though this is a non-licensed use.*
*† Potential for tooth discoloration in very young kittens.*
*NLC/D = not licensed for cat or dog; NL<12 = not licensed for animals less than 12 weeks old; NLchi = not licensed for Chihuahuas; NLC = not licensed for cats; LP = licensed product.*

| Parasite | Comments | Treatment |
|---|---|---|
| Coccidia | Usually seen in large breeding establishments. Acute diarrhoea often haemorrhagic coinciding with bouts of oocyst production. Acquired from dam | Trimethoprim/ sulphonamide |
| *Neospora* | Hindlimb paresis developing in puppies at 5–8 weeks of age. Antibodies detected using immunofluorescent antibody testing. Polyradiculoneuritis and granulomatous polymyositis at post mortem | ? Clindamycin or potentiated sulphonamides |
| *Toxoplasma* | Puppies born to infected bitches show diarrhoea, respiratory distress and ataxia with extensive inflammatory change in the CNS at post mortem. Kittens from infected queens most commonly show anorexia, lethargy and hypothermia. Field cases of neonatal toxoplasmosis are not common as most queens and bitches will have been exposed prior to pregnancy | Unlikely to be effective; ? clindamycin |
| *Giardia* | Acute small intestinal diarrhoea. Diagnosis can be difficult, repeated faecal examination using zinc sulphide flotation | Metronidazole (neurological side-effects not uncommon) or fenbendazole |
| *Haemobartonella felis* | Depression, lethargy, dyspnoea and severe anaemia in young kittens | Blood transfusion Doxycycline and prednisolone 2 mg/kg sid and reducing |

***Table 14.7:*** *Protozoan parasites common in young puppies and kittens.*

Diarrhoea is common in neonates but the role of bacteria is less clear. The majority of cases are self-limiting and can be treated using dietary manipulation (starvation followed by frequent, small, bland meals) and fluid therapy (oral electrolyte solutions). Antibacterial agents should be avoided if possible as they can further disrupt the developing bowel microflora.

There have been a number of reports of *Bordetella bronchiseptica* infection causing fatal respiratory disease in kittens. Clinical signs reported include acute dyspnoea and cyanosis. In older kittens a more chronic mixture of upper and lower respiratory tract signs occurs. Diagnosis is based on culture of oropharyngeal swabs using selective media. *B. bronchiseptica* has, however, been found as a normal commensal in adult cats.

## Viral infections

Viral infections are uncommon in puppies and kittens until maternally derived immunity begins to wane at around 5–6 weeks. Feline immunodeficiency virus (FIV) and feline leukaemia virus (FeLV) can both infect kittens transplacentally as well as perinatally via body fluids and milk. Rarely are clinical signs of FeLV or FIV infection evident by 5–6 weeks of age. Feline infectious peritonitis (FIP) has been reported in America as a possible cause of a kitten mortality complex but, like the other viral diseases, most kittens are protected by passively derived immunity.

Canine herpes virus is a well reported cause of neonatal death, with infection occurring *in utero*, at whelping, from other dogs or via fomites. Affected puppies show abdominal pain and cry continually. Death occurs in 24–48 hours, with treatment rarely being effective. Post-mortem findings include petechial haemorrhages, an enlarged spleen and excessive pleural and abdominal fluid. Occasionally, neonatal infection is associated with the ocular form causing panuveitis and retinitis. Older puppies tend to acquire a much milder mucosal form with upper respiratory tract signs predominating.

Transplacental spread of feline parvovirus (feline panleucopenia, feline infectious enteritis) is well documented, causing cerebellar ataxia. Many kittens appear normal at birth, with their neurological deficits becoming obvious as they develop greater mobility. The condition is non-progressive and many kittens will learn to cope well, making good pets. Frequently, cases of transplacental spread are associated with the use of modified live vaccines in pregnant queens. Canine parvovirus does not appear to cross the placenta; infection in young puppies is now very rare but does still occur, particularly in colostrum-deprived individuals, and is associated with myocarditis. In very young puppies, this causes sudden death due to arrhythmias and mortality can reach 70%. As puppies become older, infection results in more slowly progressive heart failure which may not become evident until 1-2 years of age. Puppies over 5-6 weeks of age usually develop the enteric form of the disease.

Transplacental spread of canine distemper virus (CDV) will occur causing the development of neurological signs in puppies at 4-6 weeks of age. Depending on the stage of gestation, abortion, stillbirths or fading puppies are also associated with CDV infection. Puppies which survive may suffer from permanent immunodeficiencies. CDV infection in young puppies can also result in severe damage to the dental enamel of the permanent teeth.

## Fading puppies and kittens

Fading puppy and kitten syndromes cover a multitude of infectious and non-infectious conditions of the neonate which cause animals born apparently healthy to gradually become inactive, loose their suckle reflex and die in the first 2 weeks of life (see Chapter 13). Localizing signs are usually absent. This condition represents a clinical description rather than a diagnosis and requires investigation as outlined above.

## CHAPTER FOURTEEN (iii)

# Vaccination

*Kit Sturgess*

A wide choice of vaccines is available for use in puppies and kittens (Table 14.8). The precise choice of vaccine will vary between practices and will depend on client/animal factors, disease prevalence within the practice area and practice purchasing considerations. Vaccines (with the possible exception of intranasal products) do not prevent infection but are designed to produce a rapid and strong anamnestic response to the pathogen on field challenge. This enables an individual to develop a protective immune response within the incubation period of the disease, and hence it will not show clinical signs of the disease. Therefore despite being fully vaccinated, infection can still occur but with very low grade or inapparent clinical signs meaning that animals can still become carriers and shedders of the infectious agent. This means that it is not necessarily safe to house a vaccinated animal with a susceptible individual. Due to strain variations between isolates, infectious dose and differences in an individual's immune response, no vaccine can be expected to be 100% effective. Vaccination should form an integral part of a wider disease control strategy.

Two basic types of vaccines are available:

- Attenuated live vaccines
- Dead vaccines: whole killed adjuvanted or subunit adjuvanted.

## MODIFIED LIVE VACCINES

Modified live vaccines generally provide a longer, stronger more complete immune response. These vaccines undergo some multiplication within the host causing a low grade (usually unnoticeable) infection. Most attenuated vaccines are replication deficient; the

| Agent/Disease | Vaccines available |
|---|---|
| Feline herpes virus type-1 | ML; KA; SU; i/n |
| Feline calicivirus | ML; KA; SU; i/n |
| Feline parvovirus (feline infectious enteritis/panleucopenia) | ML; KA |
| *Chlamydia psittaci* | ML; KA |
| Feline leukaemia virus | SU; KA |
| Feline infectious peritonitis virus* | i/n |
| Distemper | ML; measles (ML) |
| Canine parvovirus | ML; KA |
| Canine adenovirus-1 (canine infectious hepatitis) | CAV-2 ML; KA |
| Canine adenovirus-2 (CAV-2) | ML; KA |
| *Leptospira* (*L. canicola; L. icterohaemorragiae*) | KA |
| Parainfluenza virus | ML |
| *Bordetella bronchiseptica* | i/n |
| Rabies† | KA |

***Table 14.8:*** *Vaccines available in UK (December 1996).*

** Not currently available in UK*
*† Only on named animal basis prior to export or in quarantine*
*ML – modified live; KA – killed adjuvanted; SU – subunit; i/n – intranasal*

strain is fully pathogenic but being slower to divide, the host immune response becomes dominant. It is essential, therefore, that the host has a competent immune system; otherwise overt, clinical disease can develop. Live vaccines are contraindicated in immunosuppressed animals and immune incompetent animals, e.g. the very young or during pregnancy or following recent parturition. Contamination of live vaccines with other infectious agents has very occasionally been reported. Modified live vaccines are generally preferred, particularly when the infectious agent is poorly immunogenic, e.g. respiratory viruses or *Chlamydia*.

Currently available modified live vaccines acting at the mucosal surface have attenuated pathogenicity. As they are temperature-sensitive mutants and therefore incapable of invading the body, the immune status of the individual is less critical. Such vaccines may prevent infection by blocking the sites of entry of the pathogen and can therefore provide very rapid protection.

## KILLED VACCINES

There are many types of killed vaccine with very diverse formulations, and therefore very different degrees and types of protection. Originally, killed vaccines were inactivated whole organisms plus an adjuvant. Now, this group contains many subunit vaccines which may be genetically engineered and presented in more immunogenic forms, e.g. immune stimulating complexes (ISCOMS). Most killed vaccines contain adjuvant. Adjuvanted vaccines are associated with many reports of adverse drug reactions, primarily malaise and inappetance for 24–48 hours following vaccination and local inflammation at the injection site. In the 1992 report of suspected adverse reactions, 135 cases of drug reactions were recorded in cats, of which 69 (51%) were associated with vaccination, the majority with inactivated vaccines (47 inactivated; 22 live).

## VACCINATION SCHEDULES

Most manufacturers recommend an initial vaccination at 6–9 weeks of age and a second injection at 12–14 weeks. Essentially, this schedule is designed to begin vaccination as soon as maternally derived immunity has waned in a significant number of individuals and to give a final injection when almost all individuals will respond. Recent improvements in vaccine immunogenicity have led to fewer problems associated with the persistence of maternally derived antibodies in 12-week-old puppies or kittens blocking their response to vaccination. Vaccination can be performed earlier than 6 weeks in high risk groups, but should be as an additional immunization and is outside the manufacturer's recommendations. The use of some of the relatively poorly attenuated live vaccines in puppies and kittens younger than 6 weeks should be avoided, as there is a risk of vaccine-induced disease. Because the efficacy of intranasal vaccines is unaffected by maternally derived immunity or immune competence, these vaccines can be used from birth if required (Table 14.9).

In situations where repeated outbreaks of a specific vaccinable disease occurs, increased frequency of vaccination may not necessarily be the best solution. The likely source of infection should be considered and the susceptible puppies/kittens isolated. In many cases this is likely to be the dam, in which case early weaning at 3–4 weeks may provide a better approach to disease control.

## VACCINE CHOICE

In general, modified live vaccines, where available, are preferred for routine vaccination. However, certain modified live panleucopenia vaccines are not recommended for use in oriental kittens. When vaccinating puppies and kittens younger than 6 weeks, the use of dead vaccines may be more appropriate. The choice of which vaccines to use will depend on geographical area, likely lifestyle of the puppy or kitten and owner finances and is reviewed elsewhere.

### FeLV vaccines

The field efficacy of currently available FeLV vaccines have not been established, nor have they been directly compared using a natural challenge system. Like all vaccines, it is unlikely that any will provide 100% protection, particularly against transient infection and hence there is a possibility of latent or focal FeLV occurring. The following should be considered:

- Only immunity to subgroup A virus is required for protection
- There is no evidence that anti-FOCMA (feline oncornavirus cell membrane-associated antigen) antibodies are involved in protection against persistent infection

| **Age** | 0–3 days | 1–3 weeks | 3–6 weeks | > 6 weeks |
|---|---|---|---|---|
| **Dose** | 1 drop | 3 drops | 1/4 vial | Full dose |
| **Route** | Intraocular and intranasal | Intraocular and intranasal | Intranasal | Intranasal |

***Table 14.9:*** *Suggested vaccination regimen for intranasal respiratory virus vaccine in kittens. Repeat vaccination every 3 weeks from starting age to 12 weeks.*

- Cellular and humoral immunity to viral proteins other than gp70 (surface protein) may be relevant to protection.

Until the field efficacy of vaccination can be established, unnecessary exposure of vaccinated cats should be avoided and periodic testing of colonies should be continued.

## PRE-VACCINATION TESTING FOR FeLV

The possibility that a kitten is persistently viraemic at presentation for vaccination (following intrauterine or postnatal infection) needs to be considered. Ideally all cats should be tested prior to vaccination, as vaccination of a persistently infected cat:

- Wastes the client's money
- Exposes the cat to all the risks associated with vaccination without benefit
- Risks the cat being considered safe to be mixed with other cats
- Risks FeLV not being considered as the cause of disease that subsequently develops
- Does not allow the true efficacy of the vaccine to be established.

Such arguments only hold true if FeLV positive cats can be identified with complete reliability, unfortunately, this is not the case. Whilst the sensitivity and specificity of currently available ELISA tests are high (98-99%), because the prevalence of infection in the healthy cat population is low, the risk of a result being a false positive in a healthy kitten may be as high as 25-50%. Any healthy kitten with a positive ELISA test should have the result confirmed either by virus isolation or by retesting in 1-3 months.

## POST-VACCINATION ANTIBODY TITRES

Measurement of antibody titres following vaccination is generally unnecessary and, in most cases, if the efficacy of vaccination is in question then it is less expensive to re-vaccinate the puppy or kitten than to measure antibody titres. Further, what constitutes a protective titre is not always clear and in many cases, the induction of cell mediated responses is more important than circulating antibody levels.

## REFERENCES AND FURTHER READING


Blunden T (1988) Diagnosis and treatment of common disorders of newborn puppies. *In Practice* **10**, 175-184

Boothe DM and Tannert K (1992) Special considerations for drug and fluid therapy in the pediatric patient. *Compendium on Continuing Education for the Practicing Veterinarian* **14**, 313-328

England GCW (1994) Obstetric and paediatric nursing of the dog and cat. In: *Veterinary Nursing*, ed. DR Lane and B Cooper, pp. 409-430. Butterworth-Heinemann, Oxford

Fisher EW (1982) Neonatal diseases of dogs and cats. *British Veterinary Journal* **138**, 277-284

Hoskins JD (1990) Clinical evaluation of the kitten: From birth to eight weeks of age. *Compendium on Continuing Education for the Practicing Veterinarian* **12**, 1215-1225

Hoskins JD (1993a) Feline neonatal sepsis. *Veterinary Clinics of North America: Small Animal Practice* **23**, 91-100

Hoskins JD (1993b) Fading puppy and kitten syndromes. *Feline Practice* **21**(5), 19-22

Johnson CA, Grace JA (1994) Care of neonatal puppies and kittens. In: *Nursing Care in Veterinary Practice*, pp. 207-211. Veterinary Learning Systems, Trenton, New Jersey

Miller E (1995) Diagnostic studies and sample collection in neonatal dogs and cats. In: *Kirk's Current Veterinary Therapy XII*, ed. JD Bonagura and RW Kirk, pp 27-30. WB Saunders, Philadelphia

Monson WJ (1987) Orphan rearing of puppies and kittens. *Veterinary Clinics of North America: Small Animal Practice.* **3**, 567-576

Monson WJ (1993) Care and management of orphaned puppies and kittens. In: *Nursing Care in Veterinary Practice*, pp. 101-105. Veterinary Learning Systems, Trenton, New Jersey

O'Brien D (1994) Neurological examination and development of the neonatal dog. *Compendium on Continuing Education for the Practicing Veterinarian* **12**, 1601-1609

Poffenbarger EM, Ralston SL, Olson PN and Chandler ML (1991) Canine neonatology. Part II. Disorders of the neonate. *Compendium on Continuing Education for the Practicing Veterinarian* **13**, 25-37

Scott FW, Weiss RC, Post JE, Gilmartin JE and Hoshino Y (1979) Kitten mortality complex (Neonatal FIP?). *Feline Practice* **9**(2), 44-56

Sturgess CP (1996) Feline vaccination. *Veterinary Annual* **36**, 202-216

Tennant B (1994) Nutrition of companion animals. In: *Veterinary Nursing*, ed. DR Lane and B Cooper, pp. 355-398. Butterworth-Heinemann, Oxford

CHAPTER FIFTEEN

# Surgery of the Genital Tract

*Robert N. White*

## INTRODUCTION

It is not the purpose of a manual of this size to describe all the commonly performed genital tract surgical procedures in great detail. For instance, each individual surgeon has their own preferred technique for performing the procedures of castration and ovariohysterectomy; to describe them all would not be possible. Some of these techniques would probably not be considered correct to the surgical technique purist, but in the hands of that particular operator are techniques that work in practice. Rather, this chapter will attempt to review the commonly performed procedures, highlighting the possible complications and their treatment.

## OVARIOHYSTERECTOMY

Ovariohysterectomy is the most common abdominal surgical procedure performed in veterinary surgery. Ovariectomy on its own is an uncommon operation in the bitch or queen, and it is usually assumed that a hysterectomy is an indispensable part of the procedure. In fact, for elective neutering it is only necessary to remove the ovaries because pyometritis, the commonest possible spontaneous uterine disorder, is dependent on cyclical ovarian activity. Ovariectomy is widely practised in Europe, although largely for historical reasons; ovariohysterectomy is performed in the UK.

### Indications

#### Elective

Elective indications for ovariohysterectomy include:

- Preventing unwanted pregnancies
- Preventing the unwanted behavioural signs which can be associated with oestrus
- Protection from subsequent development of certain mammary neoplasms.

#### Therapeutic

Therapeutic indications for ovariohysterectomy include:

- Pyometritis
- Neoplasms of the ovaries and uterus
- Stabilization of bitches suffering from diabetes mellitus
- Uterine torsion
- Uterine prolapse
- Uterine rupture
- Prevention of clinical signs associated with pseudopregnancy (bitch)
- Prevention of the recurrence of vaginal hyperplasia (bitch).

### Procedure in the bitch

Elective ovariohysterectomy should not be performed during oestrus because of the increased vascularity and turgidity of the genital tract at this time. Ideally, neither should the procedure be performed in dioestrus (the luteal phase) since it is possible that this will result in the development of clinical pseudopregnancy (pseudocyesis) (see Chapter 4).

The approach to the peritoneal cavity when undertaking the ovariohysterectomy procedure is a matter of personal preference, although the majority of surgeons employ a ventral midline coeliotomy. The urinary bladder should be expressed prior to surgery since bladder distension results in cranial displacement of the trigone with slackening of the ureters, making the possible inadvertent ligation of a ureter potentially more likely. The distension will also reduce the exposure of the uterine body.

1. The proposed surgical site is clipped giving generous margins and cleaned.
2. The bladder should be manually expressed.
3. The bitch is moved to the operating theatre, positioned in dorsal recumbency and the surgical site prepared for surgery.
4. A skin incision is made in the midline from the level of the umbilicus in a caudal direction for a distance sufficient to allow access and exteriorization of the reproductive viscera.
5. The subcutaneous fascia and fat are cleared from the linea alba.
6. The linea alba is incised to allow access and exteriorization of the reproductive viscera.
7. A uterine horn is located and exteriorized; its associated ovary is grasped and pulled through

the wound with gentle traction.

8. It is usual to break the ovarian suspensory ligament either manually or with the aid of haemostatic forceps.
9. Theoretically, correct surgical technique would suggest that three haemostatic forceps (three-forceps technique) are now placed on the ovarian blood vessels; the most proximal being removed so that a ligature may be tied snugly around the crushed tissues. The lack of space, amount of fat and friability of the blood vessels often allows only one or two forceps to be placed adjacent to the ovary, with the ligature being tied so that it cuts into the fatty pedicle containing the vessels. Alternatively, the ligature may be placed through the perivascular tissues of the pedicle and tied as a transfixing suture. In general, a single ligature of appropriate suture material will be sufficient.
10. Once the ligature is tied the pedicle is transected between the two proximal forceps, or between the ligature and the forceps, allowing the ovary to be lifted clear of the abdomen.
11. The pedicle is grasped with thumb forceps and the haemostatic forceps remaining on the ovarian vessels are removed, allowing an assessment of haemorrhage to be made.
12. The pedicle is replaced into the peritoneal cavity.
13. The broad and round ligaments of the uterine horn are torn manually; any vessels encountered may require ligation or electrocautery (some surgeons prefer to ligate the broad ligament on both sides in its entirety).
14. A similar procedure is carried out on the opposite ovary and uterine horn.
15. Both horns of the uterus should now be outside the peritoneal cavity, and with gentle traction the cervix and cranial vagina should be visible.
16. The uterine vessels situated on both lateral aspects of the cervix should be ligated at the level of the proximal cervix (this procedure can be accomplished in many ways, most including a transfixion suture which also ligates the cervical lumen). The use and manner of application of crushing forceps on the uterus and cervix is a matter of personal preference (remember, it is possible for forceps used on tissue of this thickness to cut rather than to clamp).
17. The uterus is transected either on the cranial pole of the cervix or through the body of the cervix.
18. The cut end of the cervix does not require closure or inversion and once haemorrhage has been assessed, the 'stump' may be replaced in the peritoneal cavity.
19. The peritoneal cavity, ovarian pedicles and cervical stump should be assessed for haemorrhage, a swab count performed, and the peritoneal cavity closed in a routine manner.

## Procedure in the queen

Ideally, the procedure should not be performed in the queen in oestrus because of the increased vascularity and turgidity of the genital tract. Unfortunately, the cat's oestrous cycle is such that it is not uncommon for the procedure to be performed during this period.

Similarly to the bitch, the approach to the peritoneum is a matter of personal preference, with some surgeons employing a flank approach, while others prefer a ventral midline coeliotomy. It must be remembered that should a surgical emergency develop, it is more likely to be successfully resolved if the abdomen has been approached via the midline, as the exposure achieved with this approach is superior to that obtained via the flank approach. In the cat, it is considerably more difficult to reach and exteriorize the cervix via the flank approach. It is often the case, therefore, that uterine tissue remains *in situ* following ovariohysterectomy via this approach, making the occurrence of uterine stump pyometra a significant possibility. In 'colour point' cats, the ventral midline coeliotomy will ensure that any discoloured hair regrowth occurs symmetrically and ventrally. The ovariohysterectomy procedure carried out via the midline approach is similar in procedure to that described for the bitch. The incision is made an equal distance caudally and cranially from a point midway between the umbilicus and brim of the pelvis. Great care should exercised if the uterine body is to be clamped in the queen; in the author's experience, it is not uncommon for the tissue to be friable and the haemostatic forceps to cut into and transect the structure.

The flank approach can be described as follows:

1. The proposed surgical site is clipped, giving generous margins and cleaned.
2. The bladder is manually expressed.
3. The cat is moved to the operating theatre, positioned in lateral recumbency (either side) and the pelvic limbs extended in a caudal direction.
4. The surgical site is prepared for surgery.
5. A vertical skin incision is made 2–4 cm caudal to the last rib (the dorsoventral level of the incision is often estimated by considering the incision site to be the apex of an imaginary isosceles triangle whose base is a line between the crest of the ilium and the greater trochanter of the femur).
6. The subcutaneous fat is incised or a small portion is resected, revealing the fascia of the external abdominal oblique muscle.
7. The external abdominal oblique, internal abdominal oblique and transverse abdominal muscles are separated, in turn, along their natural muscle fibre planes.
8. The peritoneum is commonly incised by blunt penetration with atraumatic thumb forceps.
9. The greater omentum should be retracted in a cranial direction.

10. By gently grasping intra-abdominal tissue in a dorsal direction, the nearest uterine horn, or its broad ligament, can be exteriorized (some surgeons prefer to use a spay hook for this procedure).
11. Gentle traction will reveal the ovary and its suspensory ligament.
12. The ovarian vessels should be ligated in a similar manner to that described for the bitch.
13. The broad ligament may be carefully torn on both sides with little risk of haemorrhage.
14. It is often easier to ligate and transect the uterine body without prior placement of haemostatic clamps, because these will commonly transect the tissue especially in young cats and individuals in oestrus.
15. A single encircling ligature will suffice.
16. The uterus should be excised as close to the cervix as possible.
17. The individual muscle layers may be closed separately, but commonly, they are closed as one; subcutaneous fascia and skin are closed in a routine manner.

## Complications

The more common complications associated with ovariohysterectomy include:

- Haemorrhage
- Iatrogenic injury to the urinary tract
- Uterine/ovarian stump granuloma and sinus tract development
- Uterine stump pyometra
- Ovarian remnant syndrome
- Urinary incontinence
- Weight gain/obesity.

As well as the complications listed above, which are specifically associated with the ovariohysterectomy procedure, there are a number of complications which are associated with surgical wounds of the abdomen. These include haematoma and seroma formation, infection, wound dehiscence and herniation, and tissue reaction to the suture materials. The reader is referred to standard surgical textbooks for further discussion on the management of these general surgical complications.

Haemorrhage is reported as the most common cause of death following ovariohysterectomy. Bleeding is associated with ligature failure of the ovarian or uterine vessels, but may also occur in some individuals as a result of poor haemostasis of the smaller vessels within structures such as the broad ligament. Haemorrhage may be detected visually intraoperatively or haemoperitoneum may develop in the post operative period. Post operative haemorrhage should be suspected in the individual that takes longer than expected to recover from the surgical procedure. Clinical signs associated with severe haemorrhage may be present; these include pale mucous membranes, delayed capillary refill time, weak pulses, tachycardia, tachypnoea, hypothermia, abdominal distension and leakage of blood from the abdominal wound. Abdominocentesis or lavage may be performed to confirm the presence of blood within the peritoneal cavity. During active haemorrhage, the packed cell volume (PCV) of this fluid will be similar to that of a venous blood sample.

Once intraperitoneal haemorrhage has occurred, it is often difficult to assess whether conservative therapy will result in the cessation of the bleeding or not. If there are any doubts as to whether the haemorrhage is ongoing, it is advisable that surgical intervention is undertaken following the stabilization of the patient. Intravenous fluid administration should be instigated in an attempt to stabilize the haemodynamic status of the individual prior to surgical intervention.

Surgically, the approach to the abdomen is the same as that for the original ovariohysterectomy procedure, although the length of the incision should be extended so that the ligature sites can be clearly examined. Some form of suction system is extremely useful for removing the blood which has accumulated within the peritoneal cavity. It is often necessary to exteriorize the spleen and the small intestines to allow for better, unobstructed visualization of the ovarian pedicles. These structures should be placed carefully on sterile saline-soaked abdominal swabs.

The right ovarian pedicle may be approached by locating the duodenum, which lies against the right lateral abdominal wall, and retracting it towards the left abdominal wall. The mesoduodenum acts like a dam, trapping the abdominal contents medial to itself and allowing an unobstructed view of the right ovarian pedicle. A similar procedure can be carried out using the descending colon and the mesocolon to retract the abdominal contents towards the right abdominal wall, thus exposing the left ovarian pedicle.

The uterine stump may be approached by exteriorizing the urinary bladder and retracting it in a caudal direction. The uterine stump may have retracted to an intrapelvic position requiring its manual retraction into the caudal abdomen before the uterine vessels can be assessed for haemorrhage. Haemorrhage at any of these sites will require the placement of further ligatures. Ideally, these ligatures should be placed as transfixing ligatures so that they cannot slip from their placement site. All other bleeding points should be assessed and the haemorrhage stopped either with electrocautery or ligation. Care should be taken to assess the remnants of the torn broad ligament for any evidence of haemorrhage. Should multiple bleeding points be found, especially if they are at different sites, consideration should be given to the animal's blood clotting status.

Inclusion of either the proximal ureter in an ovarian pedicle ligature or the distal ureter in the uterine pedicle ligature is an uncommon, but well recognized possible complication of the surgical procedure. If both ureters are involved, the obstruction will cause anuria and a post-renal azotaemia. More commonly, only one of the ureters is involved and the functional efficiency of the contralateral kidney makes observation of clinical signs associated with renal impairment unlikely. There may be behavioural changes attributable to abdominal discomfort and there may be a reduction in urine production. Hydronephrosis and the back pressure of urine may produce nephromegaly of the associated kidney in the longer term. These changes may be detected radiographically or with abdominal ultrasonography. The affected kidney may develop abscessation and necrosis. Complete blood counts may be useful in assessing such processes. Unfortunately, in most instances, by the time the problem has been diagnosed, the affected kidney and ureter are severely damaged and treatment consists of nephroureterectomy. In cases diagnosed shortly after the surgery, the ligature may be removed and kidney function may return. Stricture formation of the ureter at the site of the injury may subsequently lead to a further obstruction. Inclusion of the distal ureter in the uterine ligature may also result in the formation of a ureterovaginal fistula. These individuals are likely to present with a sudden onset urinary incontinence a few days after the ovariohysterectomy. Contrast radiographic studies may be used to confirm the diagnosis. A ureterovaginal fistula may be successfully treated by ligation of the fistula and performing a ureteroneocystostomy.

There is the possibility for a uterine stump pyometra to develop in a dog or cat in which a portion of the uterus is left *in situ* at the time of ovariohysterectomy. The development of the pyometra requires endogenous or exogenous progesterone priming of the uterine tissue. The source of progesterone is commonly either residual ovarian tissue or the exogenous administration of the progestational drugs. Clinical signs are similar to those seen with a 'standard' pyometra, except that an obvious abdominal mass cannot be palpated. Radiographically, a soft tissue mass may be seen between the urinary bladder and the descending colon. Ultrasonography will often confirm questionable cases. As with other cases of pyometra, the patient should be stabilized prior to surgical exploration. At surgery, the pus-filled residual uterine body should be removed and, if suspected as the underlying cause, residual ovarian tissue should be located and removed. Uterine stump pyometra can be avoided by ensuring that all uterine and ovarian tissue is removed at the time of ovariohysterectomy. It must be remembered that this is technically much easier to achieve via a midline coeliotomy than via a flank approach to the peritoneal cavity.

Uterine and ovarian stump granuloma result from the use of non-absorbable suture material in combination with poor surgical technique. Poor asepsis and the vascular compromise of remaining tissue may result in abscess formation at the ligature sites. This abscess may be contained within the abdominal cavity or a sinus tract may develop which will commonly burst on the flank should an ovarian pedicle ligature site be the source of the problem. Radiography and ultrasonography may be useful in confirming the diagnosis. Contrast radiography of a discharging sinus will often reveal the site of the source. Treatment consists of stabilizing the individual prior to surgical exploration and removal of the offending ligatures. All non-essential infected and devitalized tissue should also be debrided; the tissue may be cultured and bacterial sensitivities assessed.

Ovarian remnant syndrome will most commonly result from improper surgical technique resulting in one or both ovaries being incompletely resected. It is possible that the syndrome may result from the vascularization of a 'dropped' portion of ovarian tissue into the peritoneal cavity during the ovariohysterectomy procedure (the author knows of a number of surgeons who consciously place one of the ovaries within the omentum at the time of ovariohysterectomy in an attempt to mimimize the development of post-ovariohysterectomy conditions associated with the loss of ovarian hormonal activity). A third cause of the syndrome is the presence of ovarian 'rest' tissue in a location other than the immediate ovarian area. This tissue may become functional at almost any time.

Diagnosis of ovarian remnant syndrome should be straightforward so long as the dog is not receiving exogenous oestrogens. Diagnosis may be based on the appearance of more than 80–90% superficial cells on vaginal cytology during a period of oestrus or the demonstration of elevated oestrogen and progesterone concentrations. The subject is more fully covered in Chapter 4.

Treatment consists of the surgical removal of the remaining ovarian tissue which may be easier to locate when the animal is in oestrus or early dioestrus. If no obvious ovarian tissue can be located, it is prudent to resect both ovarian pedicles in the hope that these will include any potential ectopic ovarian tissue. Care should be taken not to include the ureters in this resection.

Urinary incontinence may occur following ovariohysterectomy for various reasons, but the commonest cause of incontinence in spayed dogs is urethral sphincter mechanism incompetence (SMI). The aetiopathogenesis of this condition is multifactorial, complex and poorly understood, but ovariohysterectomy appears to be a contributing factor. Although the mechanism is not known, possible explanations include neurological, vascular and hormonal changes following surgical removal of the reproductive tract in the bitch. The majority of bitches that develop urethral SMI post-spaying will develop signs of urinary incontinence within one year of the procedure.

In bitches with suspected urethral SMI, a complete clinical examination should ideally be undertaken prior to instituting any medical management. This should include routine urinalysis, abdominal radiographs, urethral catheterization and more involved contrast radiographic studies. The main aim when treating SMI is to restore continence by increasing urethral tone. This may be achieved medically using exogenous oestrogens to improve smooth muscle contractility and sensitize the urethra to $\alpha$-adrenergic stimulation or sympathomimetic drugs, such as ephedrine and phenylpropanolamine (1.5 mg/kg twice daily) to act on the $\alpha$-receptors within the urethral smooth muscle resulting in increased urethral tone. Surgical options for the condition include colposuspension, cystourethropexy, sling urethroplasties and the creation of an artificial sphincter. The reader is referred to the Further Reading section for texts covering this subject in more detail.

Studies indicate that ovariohysterectomy may adversely affect the ability of dogs to regulate food intake making them likely to overeat and, thus, predispose them to obesity, but when food intake is controlled and dogs are regularly exercised, no significant increase in weight should occur.

# CASTRATION

## Indications

### Elective

Elective indications for castration include:

- Sterilization
- Preventing the unwanted behavioural signs associated with a sexually mature male.

### Therapeutic

Therapeutic indications for castration include:

- Testicular neoplasia
- Cryptorchidism
- Benign prostatic hyperplasia
- Perineal herniation
- Testicular torsion
- Orchitis.

## Procedure in the dog

### Open technique

1. The proposed surgical site including the scrotum is clipped; care should be taken when clipping and scrubbing the scrotal skin.
2. The dog is moved to the operating room, positioned in dorsal recumbency and the surgical site prepared for surgery.
3. The scrotal wall is grasped over one testis, forcing the gonad to a point in the midline cranial to the base of the scrotum.
4. With the testis in this position, a skin incision is made in the median raphe by cutting down on to the gonad.
5. The incision with the blade is extended through the tunica vaginalis to expose the testis; pressure exerted behind the testis will result in its emergence from the vaginal sac.
6. The mesorchium is divided and the scrotal ligament is separated from the caudal pole of the testis.
7. The ductus deferens and blood vessels in the spermatic cord are ligated and transected (three-forceps technique) separately; commonly, all these structures are ligated together using a single or double ligature. The ligature may be placed as a transfixing ligature by passing it through the perivascular and periductal tissues prior to tying.
8. The cavity of the scrotal vaginal sac may be obliterated with sutures to prevent abdominal organ herniation.
9. The second testis is removed in the same manner through the same skin incision.
10. The subcutaneous tissues and skin are closed in a routine manner.

### Closed technique

The technique is initially the same as that performed for the open procedure:

1. The skin incision in the median raphe is not extended to penetrate the vaginal tunic, rather, the subcutaneous tissues are bluntly dissected to expose the vaginal sac on one side.
2. Firm pressure on the testis will release the sac through the skin incision.
3. The sac may be grasped and pulled in a caudal direction to place the entire scrotal portion exterior to the body cavity; this procedure may be made easier by breaking the fascial attachments bluntly with the aid of a swab.
4. The vaginal sac is exteriorized sufficiently to expose the tunic covered spermatic cord.
5. The spermatic cord portion is cleared of extraneous fascia and ligated by placing an absorbable transection suture through the cremaster muscle prior to ligating the remaining cord.
6. Clamping with haemostatic forceps (two or three forceps technique) may be undertaken, but is often considered unnecessary.
7. The spermatic cord distal to the ligature is clamped with a single pair of forceps prior to transecting the cord between these forceps and the ligature. During this transection procedure, the cord proximal to the ligature may be grasped with atraumatic forceps so that it can be safely observed for haemorrhage following the transection procedure.

There should be minimal tension on the cord during this assessment and should haemorrhage be observed, a further ligature should be placed.
8. The ligated spermatic cord can then be allowed to retract through the incision towards the external inguinal ring.
9. The castration is completed by displacing the remaining testis to the skin incision, incising the overlying fascia and repeating the procedure described above.
10. The subcutaneous tissue and skin are closed in a routine manner.

**Scrotal ablation (scrotectomy)**

Under certain circumstances, the scrotum may be removed as an integral part of the castration procedure. In the aged dog, the scrotum has often become a pendulous structure which may become oedematous and swollen in the days following routine castration. In the longer term, the dog may be left with an unsightly scrotal sac which may be liable to inadvertent trauma. Both these possibilities can be prevented by performing a scrotectomy at the time of castration. If there is any disease process involving the scrotum such as testicular neoplasia or the animal is prone to scrotal dermatitis, a scrotectomy should be considered part of the castration procedure.

1. The dog is positioned in dorsal recumbency, similarly to a routine castration.
2. The whole of the scrotum and surrounding skin is prepared for aseptic surgery.
3. An elliptical skin incision is made around the base of the scrotum.
4. The subcuticular fascia is undermined and incised, freeing the edges of the scrotal skin.
5. Each testis is identified within its tunic and separated from the other.
6. The vaginal tunic is cleared of extraneous fat and the castration proceeds on both sides in a similar manner to that described for closed castration.
7. After both cords have been ligated and transected any remaining fascia is separated allowing the two testes and scrotum to be removed together.
8. The subcutaneous fascia is closed using a simple continuous suture pattern of synthetic absorbable material.
9. The skin wound is closed routinely.

## Procedure in the tom (open and closed)

1. The anaesthetized animal is either placed in lateral recumbency with its legs tied forwards or it is placed in dorsal recumbency.
2. The scrotal fur is either plucked or clipped and the area prepared for surgery.
3. The scrotum is opened from dorsal to ventral with a scalpel blade on both sides; the depth of incision should be sufficient to divide the skin and the underlying fascia.
4. Most surgeons perform an 'open' castration and continue the incision through the vaginal tunic, allowing the testicle to be exteriorized by gentle pressure behind the testicle.
5. The spermatic cord structures may be ligated with an absorbable suture material, but many surgeons prefer to separate the ductus deferens from the testicular vessels and then tie these two structures together into a square knot.
6. A similar procedure is undertaken on the opposite side.
7. The tied spermatic cord will usually retract into the vaginal tunic following the procedure, and no attempt is made to close the scrotal incisions with sutures.

'Closed' castration may be carried out by not incising into the vaginal tunic.

1. The tunic-covered testicle is exteriorized by manually squeezing it out of the scrotum and applying traction to it until the tunic-covered spermatic cord is exposed.
2. The spermatic cord is ligated and transected.
3. Clamping with haemostatic forceps (two or three forceps technique) may be undertaken, but is often not required for structures of this size.
4. A similar procedure is undertaken on the opposite side.
5. The tied spermatic cord will retract into the vaginal tunic following the procedure and no attempt is made to close the scrotal incisions with sutures.

## Complications

Complications associated with castration include:

- Haemorrhage
- Incisional irritation and scrotal swelling
- Indirect scrotal herniation.

Haemorrhage from the testicular vessel following castration may present in differing ways depending on the manner in which the castration was performed. In the dog, the bleeding may either enter the abdominal cavity resulting in haemoperitoneum or it may enter the scrotum resulting in a scrotal haematoma. In the cat, the open scrotal incisions commonly result in haemorrhage from the site of incision. Scrotal haematomas are generally treated conservatively, although in severe cases they may require evacuation and ligation of the offending spermatic artery. An Elizabethan collar may be required to prevent self-trauma to the incision and scrotum. Should severe intra-abdominal haemorrhage occur, the dog will require surgical exploration of the caudal abdomen to locate and ligate the retracted bleeding spermatic artery.

Indirect scrotal herniation is a rare occurrence in the dog, and is unlikely to occur in the cat due to the long length of the inguinal canal in this species. It can be prevented by performing a closed castration, but even when open castration is performed the occurrence of the problem is very rare.

## Cryptorchid syndrome

Testicular descent is normally complete very shortly after birth. Testes not located within the scrotum by 6 months of age are generally considered permanently retained. Unilateral or bilateral cryptorchidism, especially in the dog, is a common finding. It is considered hereditary in the dog, and in affected breeds is transmitted in a sex limited autosomal recessive manner. The right testis is more commonly retained than the left (ratio of 2.3:1). The inguinally or abdominally retained testis is predisposed to malignant changes (seminoma and Sertoli cell tumour) in later life and it is, therefore, advisable that the dog be castrated to remove the retained testis (Figure 15.1) and also to prevent the possibility of the dog fathering further litters with the condition.

The diagnosis is usually made following clinical examination and a palpable absence of one or both testes. Spasm of the cremaster muscle can cause apparent elevation of a normally descended testis in some young puppies. Therefore, several examinations may be required to confirm the diagnosis of cryptorchid syndrome. The cryptorchid testis may be found anywhere from just cranial to the scrotum in the subcutaneous tissue of the groin region, to the position in embryonic organogenesis, just caudal to the kidney. Careful palpation will usually detect most testes which are placed distal to the superficial inguinal ring in the subcutaneous tissue of the groin.

Abdominally placed testes will require exploratory coeliotomy for confirmation of their position and their removal. Surgical approach is via a parapenile skin incision and a ventral midline coeliotomy. It is common to find the testes located in a mid-abdominal position. Location may be aided by initial recognition of the ductus deferens at the level of the prostate. This structure may be followed from the prostate on the relevant side and will lead to the cryptorchid testis.

Examination may reveal the testes to be in a caudal abdominal position at the level of the inguinal ring. Again, the ductus deferens may be traced proximally to locate the testes.

Once the cryptorchid testis is located, testicular vessels and ductus are isolated, triple clamped and ligated in a similar fashion to that described for routine castration. If the retained testis is located subcutaneously through the inguinal ring, it may be removed by a standard pre-scrotal incision with manipulation of the testis into the incision by digital pressure.

Agenesis of the testis has been reported, but is very rare, and a surgeon should be completely confident that

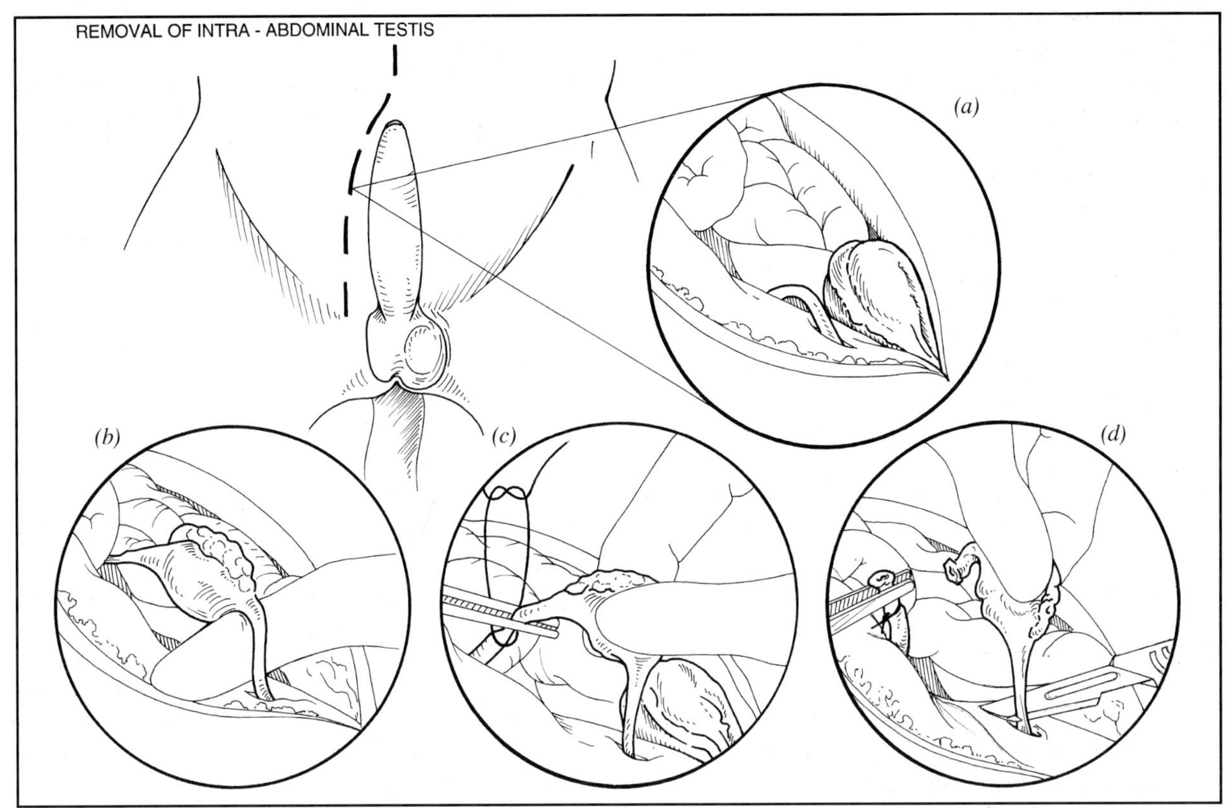


*Figure 15.1: Removal of intra-abdominal testis. (a) A parapenile skin incision reveals the scrotal ligament as it passes through the inguinal canal. (b) Alternatively, the testis may be located by initial recognition of the ductus deferens at the level of the prostate; this may then be followed on the relevant side and will lead to the cryptorchid testis. (c) The spermatic cord is clamped and ligated. (d) The spermatic cord is transected and the testis removed by severing the scrotal ligament.*

a retained testis does not exist before this diagnosis is made. Assessment of endogenous testosterone can most readily be accomplished by performing either a gonadotrophin-releasing hormone (GnRH) or human chorionic gonadotrophin (hCG) stimulation test. The GnRH stimulation test is preferred. For the GnRH stimulation test, 0.5–1.0 µg GnRH per kilogram of body weight is administered i.m., and blood for testosterone determination is obtained immediately prior to and 1 hour after GnRH administration. In a normal dog, pre- and post-GnRH testosterone concentrations are 1.7–17 mmol/l (0.5–5.0 ng/ml) and >17 mmol/l (>5 ng/ml), respectively.

## CAESAREAN OPERATION

### Indications

The following have been cited as guidelines for the use of the Caesarean operation:

- Prolonged gestation; normal parturition is not likely to occur after 70 days gestation
- Complete primary inertia
- Incomplete primary inertia, especially if several puppies remain and ecbolic agents have failed
- Secondary uterine inertia when several fetuses remain
- Abnormalities of the maternal pelvis or soft tissues which may impede the passage of a fetus (pelvic fractures, vaginal tumours or imperforate hymen)
- Relative and absolute fetal oversize
- Fetal monstrosities
- Fetal malpositioning that cannot be corrected to allow delivery through the vagina
- Fetal death with putrefaction.

In both the cat and the dog, it is often not possible to identify the cause of dystocia and therefore the decision to perform a caesarean operation is based largely on subjective assessment of the circumstances of each individual case, rather than on a precise diagnosis of the cause of the dystocia. This assessment is based on criteria such as the duration and progress of the whelping, the number and viability of the fetuses born and still unborn, the nature of vaginal discharges, changes in the pattern of straining and the often uninformative findings on vaginal examination. The assessment process and anaesthesia are covered in Chapter 12.

### Procedure in the bitch and queen

The approach to the peritoneal cavity is by a ventral midline coeliotomy. This surgical approach is similar to that described for ovariohysterectomy in the bitch. Operative speed is important during the procedure, because surgical delay and prolonged time from anaesthetic induction to fetal delivery is associated with increased fetal asphyxia and depression. Compression of the posterior vena cava with the gravid uterus reduces caudal vena cava venous return. In humans, this condition is described as supine hypotension syndrome. In the small animal species, it is generally not considered a problem, although it may possibly occur in dogs heavier than 30 kg and can be prevented by tilting the patient through 10–15° of lateral tilt from dorsal recumbency.

Care should be taken during linea alba incision to ensure that the gravid uterus is not inadvertently incised. The skin and subcutaneous incisions should be performed carefully to ensure that hypertrophied mammary gland tissue is not invaded. The length of abdominal incision depends entirely on the estimated size of the uterus and the fetuses it contains. Ideally, the uterus should be exteriorized and packed off with moistened laparotomy sponges to prevent abdominal contamination with uterine fluids. Both horns may be exteriorized at once, although some surgeons prefer to move only the horn from which they are removing the fetuses.

The uterus is incised with a scalpel in a relatively avascular area on the dorsal or ventral aspect of the uterine body. Care must be taken not to lacerate a fetus with the blade. Scissors may be used to extend the incision to a sufficient length for easy removal of the fetuses. In dystocic individuals, it is best to remove the fetus within the uterine body first.

The fetuses may be brought to the uterine incision by gently 'milking' them along the uterine horn. Once at the uterine incision, the fetus may be gently grasped and the amniotic sac ruptured to allow removal of the fetus. As each amniotic sac is broken the fetal fluids should ideally be removed by suction. The umbilical vessels are clamped some 2–3 cm from the fetal abdominal wall and severed, allowing the neonate to be passed to an attendant for resuscitation. After each neonate is removed, the associated placenta should be removed by gentle traction. Placentae that are well attached may be left *in situ*. This procedure is repeated until all fetuses and placentae have been removed.

It is very important that all fetuses are removed, and this can be ensured by careful assessment of both uterine horns, the uterine body and the vagina for remaining fetuses. Preoperative and/or postoperative radiography may be useful in ensuring accurate counting of the fetuses and also their entire removal.

It is possible to remove the fetus and placenta with the umbilical cord and fetal membranes still intact. The amniotic sac is ruptured and the cords are severed once the neonate has been handed to an attendant. This method of delivery possibly results in more maternal haemorrhage.

The uterus should be assessed after the removal of the fetuses and ovariohysterectomy is recommended if uterine viability is questionable. Some surgeons suggest that the Caesarean operation should not be combined with an elective ovariohysterectomy, avoiding longer anaesthetic time, and reducing fluid and blood losses.

A technique of *en bloc* resection of the ovaries and uterus with the fetuses and the subsequent removal and resuscitation of the neonates by assistants has recently been reported. The ovarian and uterine pedicles are isolated and the pedicles clamped just prior to removal of the uterus. Neonatal survival has been reported to be similar to conventional Caesarean operation, but a considerable number of assistants are necessary to resuscitate the entire litter simultaneously.

Following a conventional Caesarean operation, once all the fetuses have been removed, the uterus should rapidly begin to contract. It is recommended that if the uterus has not begun to contract at the time of closure, oxytocin (5–20 IU i.m.) should be used to promote uterine involution. This uterine contraction is important in arresting any haemorrhage.

The uterus may be closed in a single or double layer closure pattern. Most surgeons prefer a double layer closure and an inverting continuous pattern such as a Cushing or Lembert is used. An absorbable suture material such as polyglactin 910, glycolic acid, poliglecaprone 25 or polydioxanone is preferred. Hopefully, abdominal contamination with fetal fluids will be minimal, but should contamination have occurred, the abdomen should be liberally lavaged with warm sterile saline solution prior to coeliotomy closure. Omentum may be placed over the uterine incision to minimize adhesion formation. The coeliotomy may be closed in a routine manner, although some surgeons prefer to place buried subcuticular sutures rather than skin sutures to minimize neonate interference with the wound.

There are few data on the effect of Caesarean operation on postoperative fertility, but it appears that fertility is not markedly affected by surgery. There is certainly an increased likelihood that subsequent litters will require delivery by Caesarean operation, and owners of breeding bitches should be informed appropriately.

Milk production should normally start within hours of the Caesarean operation. Agalactia may be treated with oxytocin (0.5 IU/kg i.m.). Ovariohysterectomy does not affect milk production, as prolactin and cortisol maintain lactation.

### Neonatal resuscitation

Neonatal mortality is primarily associated with hypoxia, which can be minimized by the rapid removal of the fetuses from the amniotic sac and effective resuscitation.

The assistant resuscitating the neonate should ensure that all fetal membranes are removed. The oral cavity and nostrils should be cleared of fluid using suction, cotton swabs, or by gently swinging the neonate in a downward arc to remove the fluids from the upper airways by centrifugal force. The chest should be palpated for evidence of a heart beat. The neonate should be vigorously dried because skin stimulation reflexly stimulates respiratory drive. In the normal situation, the neonate should be breathing by this stage and should possess a strong pulse. If spontaneous respiration is still not evident, endotracheal intubation with a plastic intravenous catheter (16–20 gauge) or mouth-to-mouth, or mouth-to-nose, resuscitation may be attempted. Respiration can be stimulated with doxopram (1–2 drops orally) and opioid narcotics may be reversed using the antagonist naloxone (0.01 mg/kg). Oxygen supplementation should be provided in any case showing respiratory compromise. The haemostatic clamp is removed from the umbilical cord, which is checked for haemorrhage. Should bleeding occur the cord may be ligated with a suitable suture. The neonates should be placed in a warm environment until they can be returned to the dam. Early discharge of the dam and offspring from the practice is recommended to minimize exposure of neonates to hospital pathogens.

It is important that the dam is returned to her litter as soon as she has recovered from the anaesthetic to ensure her acceptance of the young. All the neonates should be thoroughly inspected for any evidence of congenital abnormalities prior to discharge from the practice. It is important that the neonates receive colostrum from the dam and her milk let-down should be assessed prior to discharge.

## SURGICAL CONDITIONS OF THE FEMALE

The majority of the following conditions are only observed in the bitch and the descriptions should be considered specific to this species unless otherwise stated.

### Vulva, clitoris, vestibule and vagina

#### Rectovaginal and rectovestibular fistulae

These are congenital conditions occurring in both the bitch and queen and are commonly seen in conjunction with an imperforate anus. There is an abnormal communication between the rectum and either the vagina (Figure 15.2) or the vestibule. The vulva serves as the common orifice for both the urogenital and the gastrointestinal tracts. In individuals that cannot evacuate faecal material, a megacolon will develop. This is especially likely in individuals in which atresia ani is also present. Diagnosis can usually be made on clinical examination and conformation of the site of the fistula into the urogenital tract may be confirmed with either a positive contrast enema or a positive contrast retrograde vaginourethrogram.

Surgical treatment consists of locating the fistula between the ventral rectum and the dorsal vagina via a

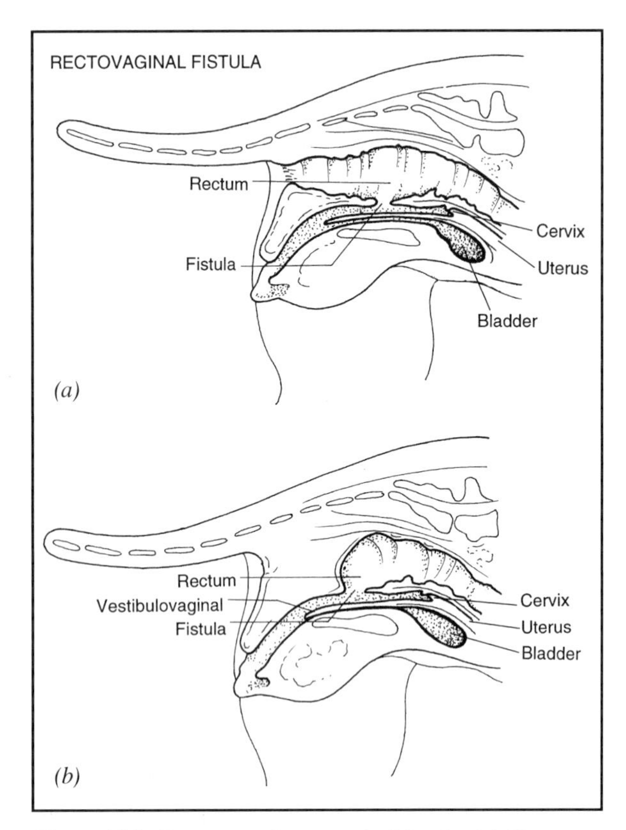


*Figure 15.2: Two types of rectovaginal fistula. (a) The rectum communicates directly with the vagina but terminates in a normal anus. (b) The rectum terminates at its communication with the vagina and there is atresia ani.*

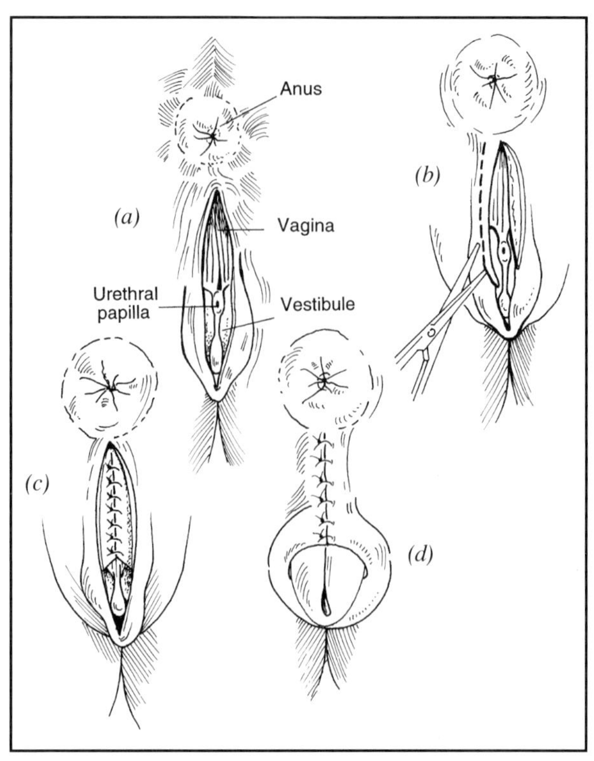


*Figure 15.3: Repair of an anovulvular cleft. (a) There is incomplete closure of skin from the anus to the dorsal vulvar commisure. (b) Incisions are made bilaterally along the mucocutaneous junction between the anus and the vulvar labia, joining at the dorsal vulvar commisure. (c) The vestibular wall is closed using simple interrupted sutures. (d) Simple interrupted suture closure of the skin.*

perineal incision between the anus and the vulva. The fistula should be divided and oversewn, so that it cannot redevelop at a later time. In the larger individual the procedure is relatively straightforward, but in the puppy it is technically demanding. If atresia ani is also present, this condition should be corrected at the same time. The prognosis is guarded especially in individuals with chronic megacolon in which it is common for the colonic function not to return.

### Anovulvar cleft

An anovulvar cleft is a rare congenital abnormality occurring in both the bitch and queen which results from failure of the urogenital folds to fuse. The incomplete closure of the skin between the dorsal vulvar commissure and the anus allows the vestibular floor and clitoris to be directly observed. The defect is sometimes seen in the intersex individual but can also occur in the normal female. Clinical signs are often related to the soiling of the clitoris and the vestibule with faecal material and the exposure of these structures to the air.

Treatment attempts to create a dorsal vestibular wall by making an inverted 'V' perineoplasty incision along the mucocutaneous junction between the anus and the vulva. The two sides of the 'V' are subsequently closed in two layers to create a dorsal vestibular wall (Figure 15.3).

### Clitoral hypertrophy

Clitoral hypertrophy in the dog may be associated with intersexuality and an os clitoris or os penis may be present, although hypertrophy of the clitoris may also occur in the normal female. Animals are often presented for aesthetic reasons or because the animal continually licks at the inflamed tissue. An individual presenting with an enlarged clitoris should be carefully assessed for evidence of intersexuality. The hypertrophy of the clitoris will often fail to regress following the removal of the gonads in either the intersex or the normal female individual.

The clitoris can be dissected from its fossa using a combination of sharp and blunt dissection. The exposure can be greatly enhanced if an episiotomy is performed. This will allow the clitoris to be safely removed from the ventral floor of the vestibule without damaging the more cranial urethral opening. Great care should be taken to ensure that this stoma is undamaged during the dissection procedure. Once the clitoris has been resected, the clitoral fossa may be oversewn with a continuous pattern absorbable suture.

### Persistent hymen

This is a relatively common congenital lesion in the bitch which results from failure of the paramesonephric ducts to unite with each other, the failure of their fusion or cannulation with the urogenital sinus. The incomplete perforation of the hymen results in stenosis of the vaginovestibular junction, and can take two forms; an annular fibrous stricture and a vertical septum.

Often the condition will only be recognized following failed attempts to breed from the individual or, sometimes, at the time of whelping when birthing difficulties are encountered. The condition may be associated with a chronic vaginitis due to inadequate drainage of the cranial portion of the vagina. The diagnosis may be confirmed by digital examination of the vagina. Further confirmation can be achieved following visual examination of the vestibule/vagina using a speculum or endoscope. Contrast radiography may be helpful in detecting individuals with a severe form of the abnormality in which a double vagina is present.

In many individuals, the condition will be detected as an incidental finding requiring no treatment, but in animals in which a medically unresponsive vaginitis is present for which no other underlying cause can be found, and in those individuals showing breeding interference, the surgical correction of the lesion may be indicated. Simple dorsoventral bands may be broken down manually per vaginam. Annular strictures may require exposure via an episiotomy (Figure 15.4). Generally, they can be excised at the vestibulovaginal junction and the vaginal mucosa either left to granulate or closed with an absorbable suture if haemorrhage is present. Care should be taken not to damage the opening of the urethra at the urethral tubercle.

### Vaginal prolapse

Vaginal prolapse is an uncommon condition in the bitch. It can present as either a complete or a partial prolapse. In both instances a doughnut-shaped eversion of the complete vaginal circumference protrudes through the vulvar labia. Careful examination reveals the urethral tubercle to be included in the prolapse. In cases of complete vaginal prolapse, the cervix is exteriorized in the prolapse process. It is important that the condition of vaginal prolapse is differentiated from vaginal oedema (hyperplasia) and vaginal and vestibular neoplasms. Brachycephalic breeds are apparently predisposed to the condition of prolapse, and a hereditary weakness of the perivaginal tissue is considered a predisposing factor. It has been suggested that forced separation of a tied mating pair, abdominal straining

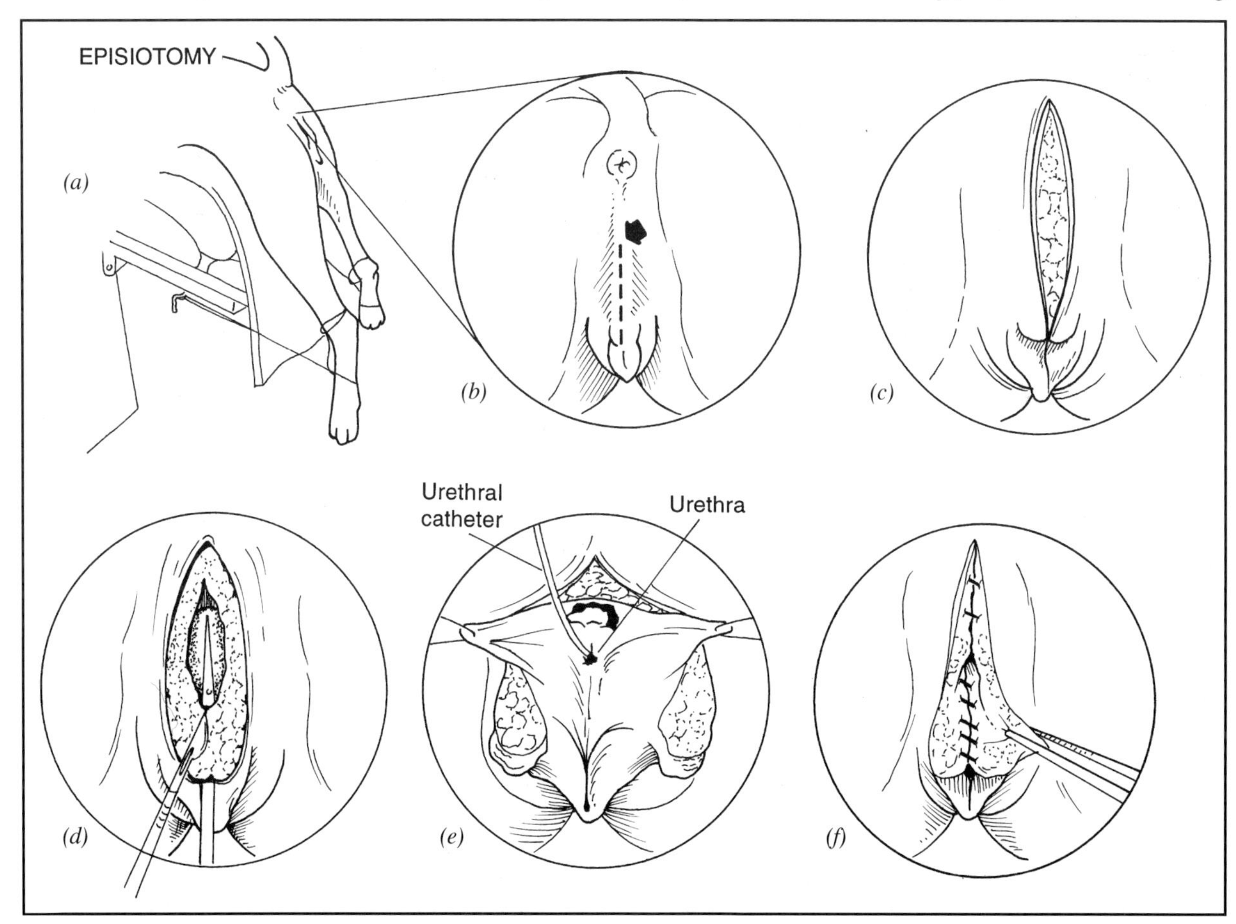


***Figure 15.4:*** *Episiotomy. (a) The animal is positioned in sternal recumbency, with the table tilted head-down and the pelvis elevated with sandbags. (b,c) A purse-string suture is placed around the anus. A skin incision is made from the vulval lips towards the anus, terminating 1–2 cm distal to the anus. (d) Scissors may be placed in the vagina to guide the incision through the vestibular muscle and mucosa. (e) The cut edges may be retracted using stay sutures and the urethral tubercle exposed, allowing placement of a catheter. (f) The vestibular mucosa, vestibular muscle/subcutaneous fat and the skin are closed in separate layers.*

resulting from constipation, and mating with a disproportionately large male may have a role in the aetiology of the condition. Generally, the condition occurs during normal oestrus when oestrogen levels are high.

In cases of mild prolapse, no treatment may be necessary, and the prolapse may regress during the dioestrus period. In the more severe prolapse, the vaginal tissue should be protected until the spontaneous resolution of the condition can occur in dioestrus. Protection will usually require replacement of the vaginal tissue under general anaesthesia or heavy sedation. The cleansed tissue may be manually returned and maintained in position by placing a number of sutures across the vulvar labia. It may be necessary to place a urinary catheter to ensure that the urethra is unobstructed during the resolution period. Longer term prevention of further prolapse may require uteropexy to the abdominal wall via a coeliotomy.

Chronic or very severe vaginal prolapse may require the surgical resection of the traumatized and devitalized vaginal tissue. The entire circumference of the prolapsed vagina is resected and the cut internal and external mucosal edges sutured together (Figure 15.5). The urethral tubercle should be protected by placing a urinary catheter and the exposure can be greatly improved by performing an episiotomy. Breeding is undesirable in animals with a tendency to prolapse and, therefore, ovariohysterectomy should be performed if at all possible.

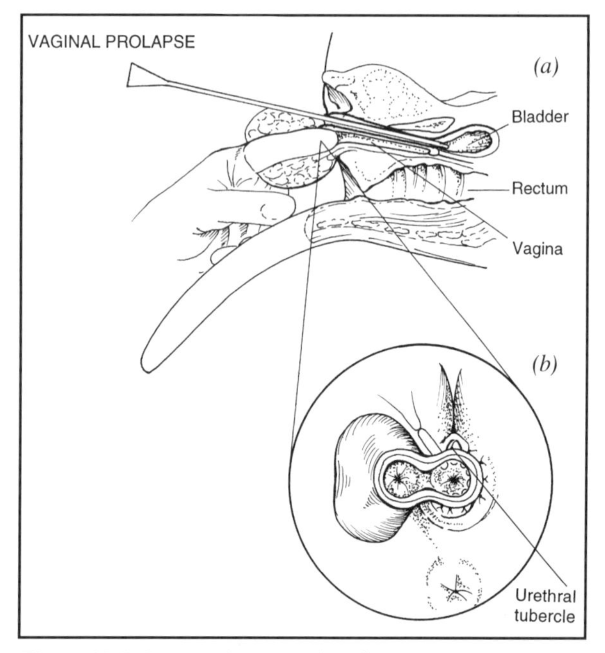


*Figure 15.5: Repair of a vaginal prolapse. (a) A urinary catheter is placed to ensure that the prolapsed vaginal tissue is resected distal to the urethral tubercle. A finger may be inserted into the lumen of the prolapse to give further guidance to the level of resection. (b) The prolapse is resected by making staged full-thickness incisions in the circumference of the vaginal wall. Horizontal mattress sutures are placed to close the incised edges of the vaginal wall. It should be noted that the circumferential incision is distal to the urethral tubercle.*

### Vaginal oedema (hyperplasia)

Vaginal hyperplasia is a condition of the bitch in which there is a thickening and an increased vascularity to the vaginal and vestibular tissue during the follicular phase of the oestrous cycle. In some bitches, this thickening becomes exaggerated and the ventral vagina cranial to the urethral tubercle becomes so oedematous that it protrudes from the vulvar labia. This oedematous process is most commonly seen in a bitch's first oestrus, but regardless at which cycle in the dog's life the problem occurs, it usually spontaneously regresses during the subsequent luteal phase. Once the condition has occurred, it is likely to recur at subsequent oestrous periods. It is important to differentiate this condition from other conditions with similar appearance such as vaginal prolapse and vaginal/vestibular neoplasia. With vaginal oedema, the urethra does not become exterior and can usually be catheterized without difficulty. The condition may be hereditary as it is seen most commonly in brachycephalic breeds such as English Bulldogs and Boxers.

Management of vaginal oedema can be difficult. Virtually all prolapses will shrink and disappear during dioestrus. Therefore, conservative management is aimed at keeping the prolapsed tissue clean and non-traumatized. The ability of the bitch to pass urine should be continually assessed, as it is possible for the oedematous process to involve the urethral papilla. The condition is prevented in the long term by ovariectomy or ovariohysterectomy, although it is unlikely that performing this procedure during an episode of vaginal oedema will hasten the shrinkage of the mass. Prophylactically, in individuals with a predisposition to the condition, the synthetic progestogen megoestrol acetate (2 mg/kg daily orally for 7 days) may be administered in early pro-oestrus to antagonize the effects of oestrogens on the vaginal tissue. This therapy cannot be recommended because of the potential side-effects and complications associated with use of such agents.

The surgical management of the condition consists of performing an episiotomy, placing a urinary catheter into the urethra and resecting the redundant, oedematous vaginal tissue at its base on the vaginal floor cranial to the urethral tubercle. The vaginal mucosal defect is closed with absorbable sutures (Figure 15.6).

### Vulvar hypoplasia (infantile vulva)

The clinical signs seen with this condition are most commonly associated with a localized vulvar and perivulvar dermatitis. The condition may be seen in bitches prior to their first season, but is most frequently observed in the spayed individual. In some individuals, the onset of their first oestrus will resolve the condition. In others, the high circulating oestrogen levels at this time will not improve the situation. When the condition is seen in the immature bitch, it is

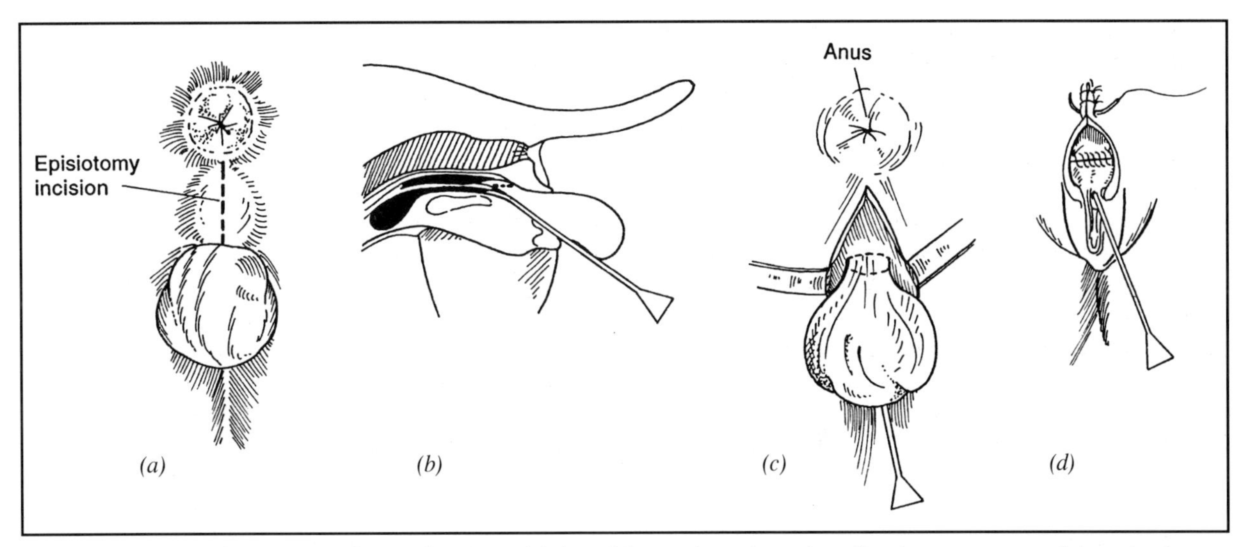


***Figure 15.6:*** *Surgical correction of vaginal oedema. (a) An episiotomy is performed to allow better exposure of the redundant vaginal tissue. (b) The mass is lifted off the vestibular floor, exposing the urethral tubercle, and a urinary catheter is placed into the urethra. (c) The mass is removed by making a transverse elliptical incision (dotted line) at its base, just cranial to the urethral tubercle. (d) The vaginal mucosal defect is closed with absorbable sutures, followed by closure of the episiotomy.*

recommended that she be left to undergo her first oestrus. In intractable cases, episioplasty should be considered with the resection of the perivulvar folds.

### Trauma

Vulvar and vaginal trauma is a rare occurrence in both the bitch and the queen, and most cases are associated with the use of obstetrical instruments during assisted parturition. Superficial mucosal tears may be left untreated; full thickness tears should be more carefully assessed with surgical debridement and closure in many instances.

### Neoplasms

Excluding mammary gland neoplasia, vaginal and vestibular tumours are the most common female reproductive tract neoplasms in the bitch. Vaginal neoplasms in the queen are considered very rare. Many different vaginal neoplasms have been reported, but the commonest is the benign leiomyoma. In the United Kingdom the transmissible venereal tumour is only occasionally seen in imported bitches, although in countries where the disease in enzootic, it is probably the second most common vaginal tumour. It is suggested that the leiomyoma is hormonally influenced by oestrogen in the bitch, but reports are conflicting and since the tumour may occur in both the spayed and intact bitch, some authorities conclude that there is no connection between blood oestrogen concentrations and the incidence of this neoplasm. On the other hand, it is not uncommon for a leiomyoma to develop rapidly following an oestrous cycle.

Clinical signs associated with vaginal neoplasia are variable, but are mostly related to size and position of the mass. Signs may include bulging of the perineum, prolapse of tissue from the vulva, dysuria, stranguria, tenesmus and obstruction to copulation in intact females. In some individuals a secondary vaginitis may also be present.

Surgical excision via an episiotomy should be curative in individuals with a leiomyoma, as long as an ovariohysterectomy is performed at the same time. Leiomyomas are usually well encapsulated, allowing their blunt dissection from the vaginal wall. Malignant neoplasms of the vagina are rarely amenable to surgical resection.

## Uterus

The uterus comprises the cervix, a body, and two horns. The horns lie within the abdominal cavity and extend from the uterine tubes to the uterine body. The uterine body is located partly in the pelvic cavity and lies between the urinary bladder and the descending colon.

### Congenital anomalies

Congenital abnormalities of the uterus in the bitch or queen are uncommon. Reported anomalies include hyperplasia, agenesis, atresia, septate uterine body, double cervix and various degrees of cornual fusion. Of these anomalies, uterus unicornis is the most common. Most anomalies are incidental findings diagnosed during routine ovariohysterectomy.

### Pyometra

The aetiology and pathophysiology of cystic endometrial hyperplasia and pyometra are discussed in Chapter 4, where the treatment of the condition and its complications are well reviewed.

The diagnosis of pyometra in dogs and cats is usually uncomplicated. The diagnosis is based on the clinical history, the clinical findings on examination, radiography and laboratory findings. In cases in which these routine diagnostic methods are not definitive, vaginal cytology and ultrasonography may be necessary to confirm the diagnosis. Ideally, an exploratory laparotomy should not be used as an aid to diagnosis.

There are numerous perioperative problems associated with pyometra in both the dog and cat. Animals with pyometra should be assessed rapidly and thoroughly so that medical therapy and fluid therapy can be initiated without delay. Once the individual is considered stable, surgery to perform an ovariohysterectomy is recommended as the most appropriate treatment. Medical management of the condition is discussed in Chapter 4. The most frequently encountered perioperative problems include:

- Fluid, electrolyte and acid-base imbalances
- Renal dysfunction
- Sepsis
- Hypo- or hyperglycaemia
- Liver damage
- Cardiac dysrhythmias
- Clotting abnormalities.

Intravenous fluid therapy is indicated in any individual with suspected pyometra, to support the cardiovascular system and combat electrolyte and acid-base disturbances. In general, a balanced electrolyte solution such as Hartmann's solution is suitable. The rate of administration of the fluid should be assessed on an individual basis, but it is not uncommon for animals to present with cardiovascular collapse and hypovolaemic shock requiring an initial infusion rate of up to 90 ml/kg/h to restore a circulating volume.

A metabolic acidosis is the most common acid-base abnormality associated with pyometra. This may sometimes result in a compensatory respiratory alkalosis. The lactate bicarbonate precursor in Hartmann's solution will generally be sufficient to maintain the blood pH within acceptable limits. Should the blood pH drop below 7.2, bicarbonate therapy may be indicated.

Ideally, the electrolyte status of the animal's blood should be assessed. If the animal is hypokalaemic, potassium should be added to the maintenance fluids. The amount of potassium added to the fluids and the rate of administration of these fluids after the first hour should be based on the serum potassium concentrations. These data are summarized in Table 15.1

The use of corticosteroids in hypovolaemic and septic shock is controversial. A short-acting, intravenous agent such as methylprednisolone sodium succinate (30 mg/kg) is probably the drug of choice.

Renal function should be assessed preoperatively and supported with intravenous fluid therapy. Plasma creatinine and urea concentrations should be measured along with a full urinalysis (care should be taken not to perforate the distended uterus if the urine sample is to obtained by cystocentesis). Ideally, urine output should be maintained at 1-2 ml/kg/h. Restoring fluid balance and supporting kidney function will allow the kidneys

| **Serum $K^+$ (mEq/l)** | **mEqKCl/ 250 ml fluid** | **Maximum* infusion rate (ml/kg/h)** |
|---|---|---|
| 2.0 | 20 | 6 |
| 2.1-2.5 | 15 | 8 |
| 2.6-3.0 | 10 | 12 |
| 3.1-3.5 | 7 | 16 |
| | | *Do not exceed 0.5 mEq/kg/h |

***Table 15.1**: Composition of maintenance fluids.*

to correct many of the acid-base and electrolyte abnormalities *in vivo*. Assessment of renal function should continue into the recovery period until the measured parameters fall within their normal ranges.

Sepsis and septic shock in animals with pyometra is common and may result in a depletion of glycogen stores, an increased peripheral use of glucose and a decrease in gluconeogenesis leading to hypoglycaemia. Any concurrent acidosis will also impair gluconeogenesis, possibly worsening the hypoglycaemia. It is important, therefore, to measure the blood glucose concentration in individuals with pyometra and treat any hypoglycaemia detected. Treatment should include correction of any fluid, electrolyte or acid-base abnormalities and in addition glucose may be administered intravenously. This can often be achieved by the addition of dextrose (2.5-5.0% solution) to the crystalloid fluid being administered. In severe cases, 50% dextrose may be administered directly as a slow bolus at a rate of 1-2 ml/kg. Blood glucose measurements should be repeated frequently and should be continued into the post operative period.

*Escherichia coli* is the most commonly cultured organism from the uterus of the dog with pyometra, but other organisms such as *Staphylococcus* spp., *Streptococcus* spp., *Pseudomonas* spp., *Proteus* spp. and *Klebsiella* spp. have also been isolated. Intravenous broad-spectrum bactericidal antibiotics with both Gram-negative and Gram-positive activity should be routinely administered. A combination of first generation cephalosporin, cefuroxime at 20 mg/kg and the fluoroquinalone enrofloxacin at 20 mg/kg will often prove effective.

Hyperglycaemia may also occur as a result of the bacterial sepsis, but is usually a transient, self-limiting problem requiring no treatment. Should the problem persist, the animal should be assessed for evidence of diabetes mellitus.

Liver damage can occur with pyometra, and is associated with toxicity secondary to sepsis or endotoxaemia, or poor hepatic perfusion secondary to hypovolaemia and circulatory collapse. Biochemically, liver enzymes may be elevated, but in most instances, the problem is self-limiting and will resolve with adequate patient support following the removal of the uterus.

A non-regenerative, normocytic, normochromic anaemia often accompanies pyometra as a result of chronic inflammation causing erythropoietic bone marrow suppression, diapedesis into the uterus, blood loss intraoperatively and haemodilution caused by aggressive intravenous fluid therapy. Before treatment, the packed cell volume (PCV) and plasma total protein concentration may be used to assess the hydration status. Following aggressive fluid therapy, the results of PCV may be inaccurate in the assessment of anaemia. A blood transfusion should only be considered should the PCV fall below 20%.

Fortunately, disseminated intravascular coagulopathy (DIC) rarely occurs in animals with pyometra, but it should be considered a possible diagnosis in individuals showing unexpected haemorrhage, haemolysis, haemoglobinuria, delayed blood clotting and shock. The diagnosis and treatment of DIC is beyond a text of this size, and the reader is referred to a standard medical textbook for further information. The prognosis for this condition is poor.

Cardiac dysrhythmias are commonly encountered in animals with pyometra and often result from the toxic effects of the disease, shock, acid–base abnormalities, respiratory dysfunction and electrolyte imbalances. Ventricular premature complexes (VPCs) are most likely. Dysrhythmias should be assessed electrocardiographically and managed accordingly in conjunction with restoration of the physiological parameters discussed above.

The aim of surgical management is to remove the infected uterus without causing it to rupture and to ensure that intraoperative and post operative haemorrhage is kept to a minimum. Careful attention to anaesthetic detail is required and drugs causing cardiorespiratory depression should be avoided. Intraoperative haemorrhage should be guarded against with the judicious use of cautery and careful attention to surgical technique. *The procedure should not be considered as routine as an elective ovariohysterectomy for sterilization.*

The uterus should be exteriorized carefully, so as not to tear associated vessels or rupture the viscus. The structure should be handled gently, avoiding excessive manipulation. Once exteriorized it should be packed off from the remaining abdominal contents with numerous sterile saline-soaked swabs. Vessels in the broad ligament will often require ligation. The uterine pedicle should be transected through the cervix or cranial vagina in individuals with a 'closed' pyometra so as to avoid leakage of uterine contents. It is unnecessary to oversew and invert the uterine pedicle in either a 'closed' or 'open' pyometra. If the surgeon is concerned about contamination from this structure, omentum should be sutured over the stump prior to abdominal closure. An absorbable suture material is recommended rather than a permanent suture material, when sutures are to be placed through contaminated tissue. This will minimize the likelihood of the suture acting as a nidus for infection, which could result in abscess formation and fistulous tract development.

On occasion, the surgeon may be presented with an individual in which the uterus has ruptured preoperatively, or, in which iatrogenic intraoperative uterine rupture occurs. Remaining pus should be removed from the uterus with suction, to prevent further contamination, and the ovariohysterectomy should be completed. The peritoneal cavity should then be lavaged with copious amounts of body temperature sterile saline, then aspirated with suction. The flush/aspiration cycle should be continued until the aspirated fluid is clear of gross contamination. This procedure will prove very difficult to achieve effectively if surgical suction facilities are unavailable. The intra-abdominal instillation of antibiotics is unnecessary, but the animal must receive immediate broad-spectrum systemic antibiosis intravenously. Open peritoneal drainage may be necessary in severe cases; however, most animals can be treated effectively with abdominal lavage and prolonged systemic broad-spectrum antibiotic administration.

Postoperative complications of a pyometra are similar to those encountered following routine ovariohysterectomy. These are discussed under the section on ovariohysterectomy. The preoperative complications will continue into the post operative period and should be addressed as described above until they have resolved.

### Metritis

Acute inflammation of the uterus is most commonly observed in the early postpartum period, and may be associated with dystocia, obstetric fetal manipulation, fetal or placental retention.

Both medical and surgical treatment for the condition has been attempted. The treatment of choice, following patient stabilization, is ovariohysterectomy, although conservative medical treatment may be attempted in a valuable breeding bitch. This consists of systemic antibiotics and prostaglandins (described in Chapter 12) or surgical uterine drainage.

### Torsion

Uterine torsion is an uncommon condition in the bitch and queen. Reported cases are commonly associated with uterine distension, a pregnant gravid uterus or pyometra. The torsion may be uni- or bilateral. Clinical signs are very variable; the animal may be clinically normal or, at the other extreme, may present collapsed in shock. More commonly, signs of dystocia, abdominal pain, anorexia and vomiting are seen. Abdominal palpation and abdominal radiographs may be suggestive of a torsion, but diagnosis is often only confirmed during exploratory laparotomy. In the pregnant individual, torsion may develop following the delivery of a puppy or kitten. Viable fetuses may be present in the twisted horn and a Caesarean operation or ovariohysterectomy may allow delivery of live young. Reduction of a torsion may be possible but commonly treatment consists of patient stabilization followed by ovariohysterectomy.

### Prolapse

Prolapse of one or both uterine horns is rare, and follows uterine contraction. Parturition or abortion usually precede the prolapse. Single horn prolapse is termed partial and prolapse of both horns is termed complete. The prolapse process will commonly occlude or rupture the ovarian vessels resulting in haemoperitoneum. The prolapsed tissue can easily be palpated and/or observed at the vulva.

If replacement is to be attempted, it should be performed early following the prolapse, but it is often easier to treat the condition by ovariohysterectomy.

### Uterine rupture

Rupture of a gravid uterus can occur following abdominal trauma or spontaneously during parturition. It is a rare finding, but should be considered in near-term individuals that have received abdominal trauma. There are reports of intra-abdominally expelled fetuses maintaining circulation and living to term, but it is more likely that the rupture will result in peritonitis.

Complete ovariohysterectomy is indicated in most instances, although small ruptures may be sutured. There are also reports of unilateral ovariohysterectomy to preserve breeding potential.

Rupture of the uterus may also occur with pyometra. Commonly, the rupture occurs at the time of ovariohysterectomy for this disease and results in extensive peritoneal contamination with the uterine contents. These individuals should be managed aggressively to combat this life-threatening peritonitis (see Pyometra).

### Subinvolution of placental sites

Subinvolution of placental sites may follow a disturbance in the normal post-parturient placental degeneration and endometrial reconstruction processes. A persistent serosanguineous vaginal discharge occurs 7–12 weeks after parturition. It is reported to occur most commonly in bitches less than 2½ years of age, following their first or second whelping.

The disease is usually self-limiting, rarely requiring medical or surgical therapy. In individuals with severe anaemia and continuous bleeding, ovariohysterectomy is recommended.

### Uterine neoplasms

Uterine tumours in the bitch and queen are rare. In the bitch, reported tumours include the leiomyoma and the less frequent leiomyosarcoma. In the queen endometrial carcinoma has been most frequently reported.

Large masses may interfere with urinary function, causing dysuria. Many smaller tumours are found incidentally during ovariohysterectomy.

Surgical treatment is ovariohysterectomy.

## Ovaries

In both the bitch and queen, the ovaries are located at the caudal poles of the kidneys close to the abdominal wall. In the bitch, the ovary is within a pouch of peritoneum termed the ovarian bursa. This bursa is closed except for a small opening ventrally. In the queen the ovarian bursa is smaller, covering only the lateral surface of the ovary.

In both species, the ovary is attached to the abdominal wall by the mesovarium which is continuous cranially with the suspensory ligament and caudally with the mesometrium. The ovarian blood vessels are found within the mesovarium. The proper ligament of the ovary is a short continuation of the suspensory ligament connecting the caudal end of the ovary to the cranial end of the uterine horn.

### Congenital anomalies

Reported congenital anomalies of the ovaries include agenesis, hypoplasia and accessory ovaries. Ectopic supernumerary ovaries have also been reported. Congenital ovarian abnormalities are rare and are usually discovered incidentally during routine ovariohysterectomy.

Hermaphroditism refers to the presence of both male and female sex organs in the same individual. True hermaphroditism is unusual and may be bilateral, with male and female gonads on both sides, unilateral with male and female gonads on one side, or contralateral with a male gonad on one side and a female gonad on the other. In these individuals secondary sex organs may be male, female, or a combination of both.

Pseudohermaphroditism refers to a condition in which the external genitalia and some of the internal structures resemble the opposite sex to that represented by the gonads. For example, male pseudohermaphrodites may have testicular gonads but female-like external genitalia. The abnormalities are not always evident until the internal genitalia have been examined either grossly or microscopically.

### Cysts

A number of ovarian cystic structures occur in the bitch and queen.

The most common cystic structure occurs within the ovarian bursa, and may originate from either the mesonephric or paramesonephric tubules and ducts. They are more commonly encountered in dogs than cats, and are not endocrinologically active. The follicular cyst is the most common true ovarian cyst. These may be single or multiple and histologically they contain remnants of granulosa or theca cells. Clinical signs associated with follicular cysts include prolonged oestrus with a serosanginous vaginal discharge, cystic mammary hyperplasia and genital fibroleiomyomas.

Luteal cysts are usually multiple in the bitch and queen, and have been associated with prolonged metoestrus or anoestrus and with pyometra. They form from the corpus luteum following ovulation.

Most cystic structures of the ovary are encountered incidentally during ovariohysterectomy. Follicular cysts producing clinical signs have been treated successfully with luteinizing hormone, excision or rupture of the cyst. Attempts to manage luteal cysts medically with oestrogen and prostaglandins are rarely successful. Most cystic structures are managed surgically by ovariectomy during the ovariohysterectomy procedure.

#### Ovarian neoplasms

Ovarian tumours occur more frequently in older, nulliparous individuals, but they are rare in both dogs and cats.

Reported tumour types relate to the tissue types of the ovary. The granulosa–theca cell tumour is the commonest ovarian tumour in the bitch. The neoplastic cells of this neoplasm may differentiate into female-type cells, producing clinical signs in affected individuals associated with elevated levels of oestrogen, for example, vaginal hyperplasia.

The second most common sex cord–stromal tumour is the Sertoli–Leydig cell tumour. The cells of this neoplasm differentiate into male-type cells which are often hormonally active, producing cystic endometrial hyperplasia and pyometra.

Ovariohysterectomy is recommended as the surgical therapy for sex cord–stromal neoplasms. Local invasion is not uncommon, and may necessitate body wall resection and unilateral nephrectomy.

## SURGICAL CONDITIONS OF THE MALE

Genital tract surgical diseases of the tom are rare, and the majority of the following conditions are only observed in the dog.

### Penis and prepuce

#### Hypospadias

Hypospadias is the most common developmental anomaly of the male genitalia. The condition results from failure of fusion of the urogenital folds and incomplete formation of the penile urethra. The external urethral orifice can be located anywhere on the ventral aspect of the penis from the normal opening to the perineal region. Hypospadias may be glandular, penile, scrotal, perineal or anal, depending on the location of the external urethral orifice (Figure 15.7). Hypospadias is usually associated with other developmental abnormalities, such as failure of fusion of the prepuce and underdevelopment of the penis. The diagnosis of hypospadias is made by close examination of the penis and perineal region. Surgical correction is not usually attempted in the dog because the urethra cranial to the abnormal orifice is deficient. Severe defects of the penis may necessitate the formation of a scrotal urethrostomy, but in most instances the abnormal urethral orifice can be maintained.

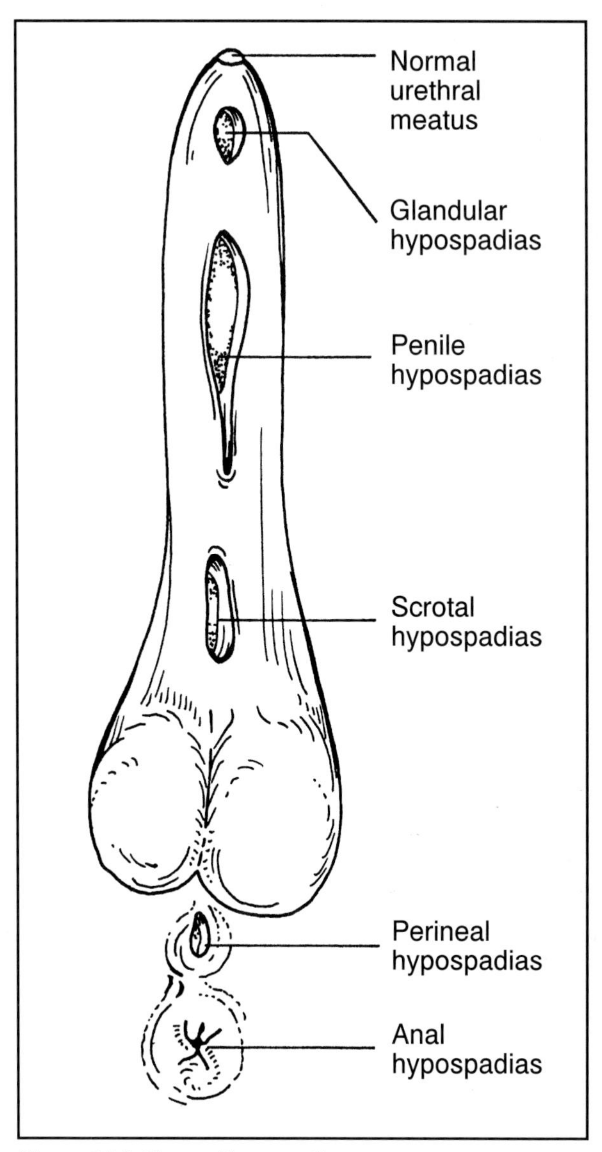


*Figure 15.7: Types of hypospadias.*

#### Persistent penile frenulum

Persistence of the penile frenulum is not uncommon in the dog. In some individuals there is an obvious deviation of the glans penis in a ventral direction. In others, the condition may only be recognized during periods of sexual excitement, when the ventral tethering of the glans may cause pain. Treatment is usually straightforward and involves the severing of the avascular connective tissue ventral to the glans.

### Phimosis

Phimosis refers to the inability to extend the penis from the prepuce because of a congenital or acquired stricture of the preputial orifice. It occurs in both the dog and cat. Usually, careful examination of the prepuce will confirm the diagnosis. A secondary balanoposthitis may also be evident. Treatment involves the resection of a small wedge of prepuce from the ventral preputial opening resulting in a significant widening of the orifice (Figure 15.8).

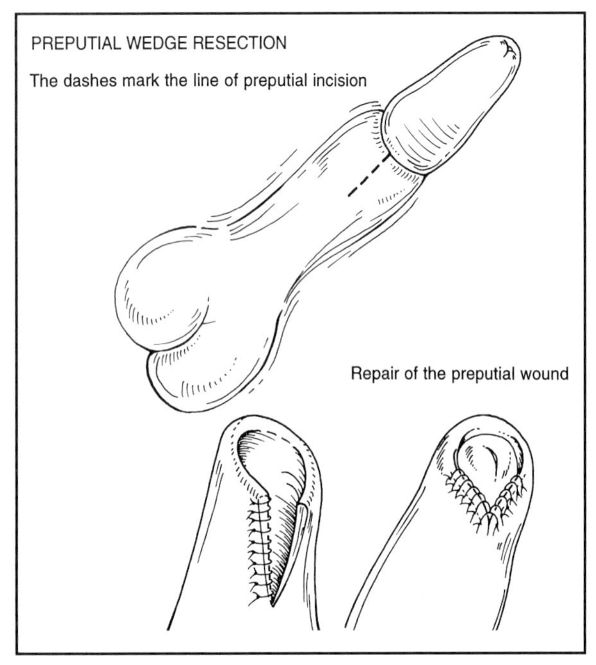


***Figure 15.8:*** *Preputial wedge resection.*

### *Technique for preputial wedge resection*

1. The prepuce and surrounding parapenile skin are clipped and cleaned.
2. The bladder is expressed manually or with the aid of urethral catheterization.
3. The dog is moved to the operating theatre, placed in dorsal recumbency and the surgical site prepared for surgery.
4. If the penis can be exposed, remove any secretions from its surface with moist swabs; try to prevent disinfectants or spirit entering the preputial cavity.
5. Insert the blunt tip of a pair of straight scissors into the preputial orifice, whilst stretching the prepuce forwards.
6. Cut caudally in the ventral midline for approximately 1 cm; the extent to which the orifice requires opening depends on its original size, and the size of the dog, but should be sufficient to allow free protrusion of the penis.
7. The edges of the cut mucosa and skin are co-apted on each side using a simple interrupted or simple continuous suture pattern.
8. An Elizabethan collar may be required to prevent self-trauma to the wound during healing.

### Paraphimosis

Paraphimosis refers to the inability to retract the penis once it is protruded from the prepuce. The condition is associated with some degree of phimosis. It usually occurs after mating or masturbation when the penis has been erect. The glans penis becomes trapped outside the prepuce because of a constricting band of prepuce which prevents the subsidence of the erection. The glans remains erect and becomes congested and engorged, thus further constricting the penis. The penis and prepuce should be examined for any other injuries and they should be cleansed with warm and cold saline in an attempt to reduce the erection. Lubricants may be used to help replace the penis within the prepuce. In long standing cases, a urethral catheter should be placed to allow the animal to urinate. Temporary or permanent enlargement of the preputial orifice may be necessary. Penile necrosis will require partial penile amputation. Recurrence of the paraphimosis is common. In cases where an inappropriately short prepuce is associated with the condition, it may be possible to extend the coverage of the penis by performing a preputial advancement.

### *Technique for partial amputation of the penis (Figure 15.9)*

1. The dog is placed in dorsal recumbency.
2. Catheterize the urethra with the largest urinary catheter that can be passed comfortably; the bladder may be expressed of urine.
3. Reflect the prepuce and tightly tie on a 2.5 cm bandage proximal to the bulbus glandis; this acts as a tourniquet and can be used to secure the penis vertically.
4. Make a stab incision into the midline of the dorsal surface of the penis and incise distally and laterally away from this point on both sides to produce a 'V'-shaped cut down to the os penis.
5. Gently dissect the urethra off the os penis with a blade, keeping the cutting surface towards the bone.
6. Free the urethra for about 1 cm distal to the base of incision and incise around its circumference at its distal attachment.
7. Cut the os penis as short as possible with bone-cutting forceps.
8. Push the amputated distal portion of the penis along the catheter and out of the operative field.
9. The two sides of the 'V'-shaped flap are sutured together using a 3-0 absorbable suture in a simple interrupted pattern.
10. The urethra is then slit longitudinally with a blade, on a lateral surface, to its point of emergence from the penis, and sutured back over the stump.
11. The tourniquet may then be removed and any haemorrhage assessed.
12. An Elizabethan collar may be required to prevent self-trauma to the wound during healing.

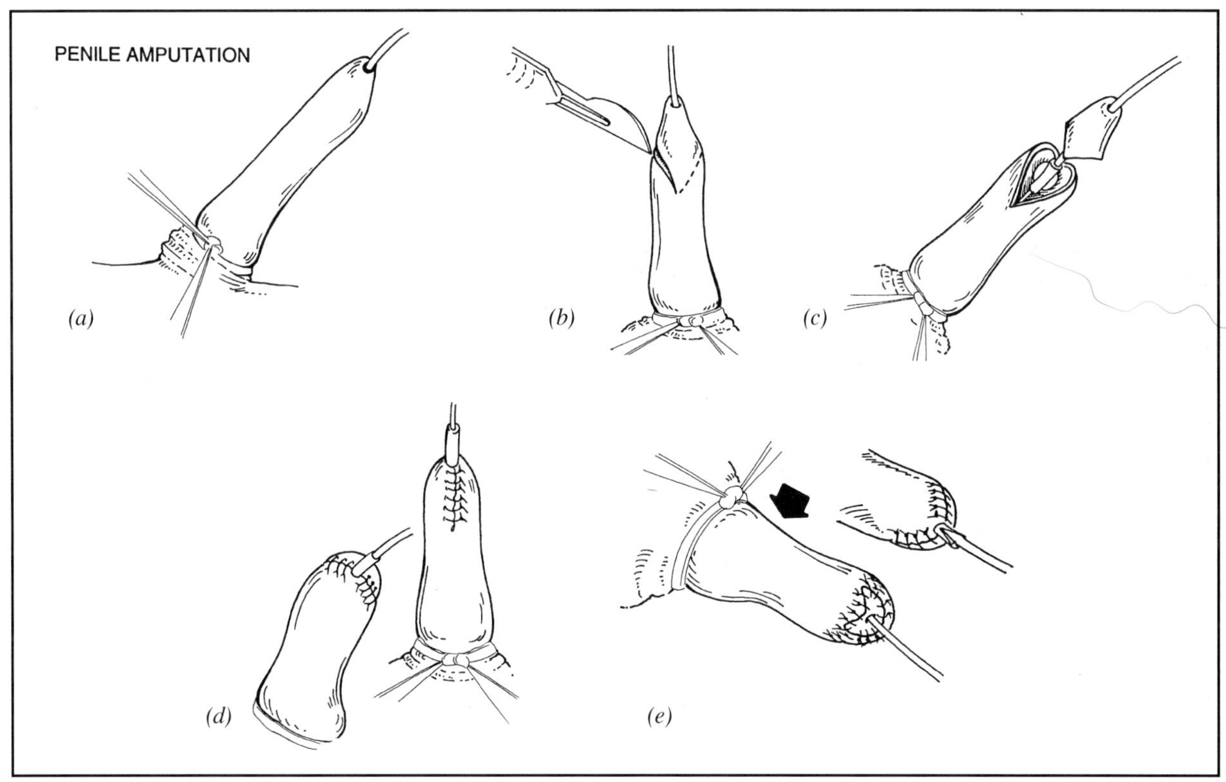


*Figure 15.9:* *Partial amputation of the penis. (a) The urethra is catheterized and a bandage used to reflect the prepuce and also to act as a tourniquet. (b) A V-shaped incision is made through the penis down to the os penis on the dorsal aspect, taking care not to cut the urethra on the ventral aspect. (c) The amputated portion of penis is removed along the catheter. (d) The incision is closed using simple interrupted sutures. (e) The urethra is sutured back over the penile stump.*

### *Technique for preputial advancement (Figure 15.10)*

1. The prepuce and surrounding parapenile and abdominal skin are clipped and cleaned.
2. Placement of a urethral catheter may aid in the manipulation of the penis during the surgical procedure.
3. The dog is moved to the operating theatre, placed in dorsal recumbency and the surgical site prepared for surgery.
4. An elliptical skin incision is made cranial to the prepuce passing laterally on both sides to a distance of 5–10 cm from the midline depending on the size of the dog.
5. A second elliptical skin incision is made cranial to the first, having a shorter radius of circumference.
6. On joining the lateral incisions of the ellipses on both sides, a crescent moon-shaped piece of skin is removed.
7. The bilaterally placed preputial muscles which extend from the area of the xiphoid cartilage to the dorsal wall of the prepuce are located and a 1–5 cm myectomy is performed on both sides, depending on the size of the dog and the degree of cranial advancement of the prepuce required.
8. The muscles are repaired with an absorbable suture in their new foreshortened position so that the prepuce has been advanced cranially to cover the penis.
9. The two elliptical skin incisions are co-apted with simple interrupted skin sutures.

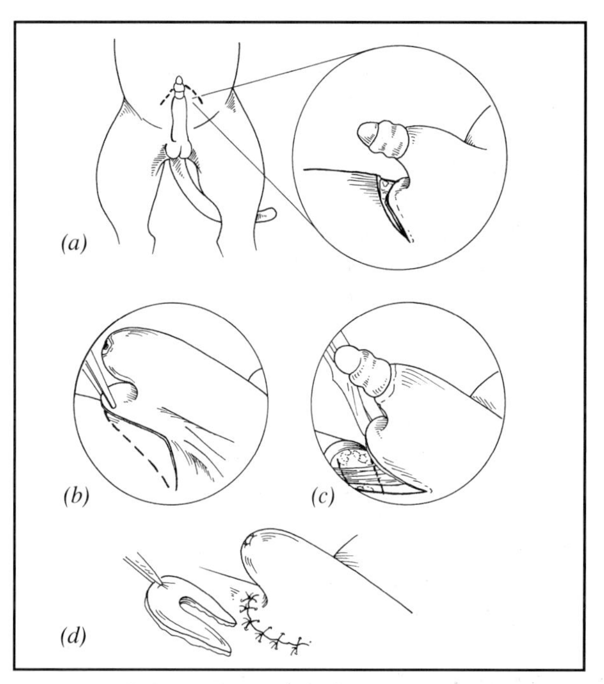


*Figure 15.10:* *Surgical preputial advancement. (a) An elliptical skin incision is made cranial to the prepuce, passing laterally on both sides to a distance of 5–12 cm from the midline. (b) The preputial skin is pulled cranially until the prepuce fully covers the penis. This gives an indication of the site for the second cranial elliptical skin incision (dotted line). (c) A bilateral myectomy and rejoining of the preputial muscles is performed. (d) The two elliptical skin incisions are co-apted with simple interrupted skin sutures. The crescent moon-shaped piece of skin that has been removed is also shown.*

### Deformity of the os penis
Rarely, the os penis of the dog may develop with an abnormal curvature. This results in an abnormal penile curvature, which may make full retraction of the glans into the prepuce impossible. The exposed penis is liable to infection. Straightening of the penis by osteotomy of the os penis is possible, but is rarely performed. Partial or complete penile amputation may be required in some cases.

### Fracture of the os penis
Fracture of the os penis is a rare occurrence. Most reported fractures are transverse in nature. Clinical signs depend on the degree of soft tissue damage and fracture displacement. Radiography is required to characterize the fracture fully. Conservative management of most fractures will be successful, although certain fractures will be amenable to internal fixation with small finger plates. Urinary catheterization may be required for several days in both conservatively and surgically managed cases.

### Preputial foreign bodies
Foreign bodies such as grass awns and stalks, and urinary calculi may occasionally become lodged within the prepuce. They will usually cause some irritation to the area, resulting in licking of the prepuce. Careful retraction of the prepuce and examination of the internal lining will reveal the cause of the irritation and allow its removal.

### Balanoposthitis
Infection of the penis and the prepuce is not uncommon in the dog. It is often accompanied by a copious yellow or blood-tinged preputial discharge. A slight creamy preputial discharge is normal in the mature dog. Treatment of balanoposthitis should be aimed at trying to establish an underlying cause for the infection (e.g. foreign body), removing the underlying cause, improving preputial hygiene (irrigation of the prepuce with warm saline) and in some instances the use of systemic antibiosis (consider culture and sensitivity of the discharge). The prostate should be assessed to ensure that disease of this structure is not the source of the discharge.

### Priapism
Priapism is persistent penile erection not associated with sexual excitement. It is very rare in small animals and should be differentiated from paraphimosis. With priapism, the penis can be manually replaced into the prepuce. Treatment may prove difficult, as the condition may result from a spinal cord injury. Spontaneous resolution of the condition has been reported.

### Prolapse of the urethra
Prolapse of the penile urethra is associated with excessive sexual excitement and genitourinary infection. The prolapse may be self-inflicted by a persistent penile irritation leading to rubbing of the structure along household furniture and carpets in an attempt to ease the discomfort. These activities may result in penile and preputial infection, and eventual prolapse of the urethra. The condition may be associated with intermittent haemorrhage from the penis. Diagnosis is confirmed by examination of the external urethral orifice on the end of the penis. The presence of a red pea-shaped mass at this site usually indicates a urethral prolapse. Conservative treatment by reducing the urethral prolapse is seldom effective in the long term, and resolution of the problem can be better achieved following the amputation of the prolapsed portion of the urethra and anastomosis of the urethral and penile mucosa.

### Trauma
Trauma to the penis is rare. Malicious application of a rubber band around the penis will lead to penile strangulation. The penis and prepuce should be carefully inspected for any evidence of such malicious acts. Chronic strangulation of the penis will necessitate partial or complete penile amputation. Accidental trauma to the external prepuce is more common and should be treated in a similar manner to any other skin wound.

### Neoplasms
Neoplasms of the cat penis are very rare, although they are encountered on occasion in the dog. Reported tumours include squamous cell carcinoma, papilloma and, in countries where it is enzootic, or occasionally in imported dogs in the United Kingdom, transmissible venereal tumour. Diagnosis should be confirmed by cytological and histological examination of aspirates and biopsies. Most penile tumours require aggressive surgical resection, which necessitates complete penile amputation and perineal urethrostomy in most instances. Transmissible venereal tumours may be treated with radiation therapy, chemotherapy and surgery.

## Prostate
Prostatic disease is common in the older entire dog. Most diseases that affect the prostate result in prostatomegaly, and dogs are generally presented with clinical signs associated with prostatic enlargement. Although the male cat possesses a prostate gland, prostatic diseases are only rarely described in this species.

The prostate gland is the only accessory sex gland in the male dog. It completely encircles the urethra from the bladder neck to the post prostatic membranous urethra. It is a bilobed structure, with a prominent dorsal longitudinal groove separating the prostate into left and right halves. The structure predominantly lies in the retroperitoneal space, having a bilateral neurovascular supply entering dorsolaterally on the left and right sides.

The vasa deferentia enter the craniodorsal surface of the prostate before coursing in a caudoventral direction to enter the urethra at the colliculus seminalis.

Physiologically, the prostate contributes the fraction of semen that is delivered in the third phase of ejaculation. There is some debate concerning the classification of prostatic diseases, but the following descriptions are generally accepted in the clinical situation. It should be remembered that there is often histological evidence of more than one disease process occurring in the prostatic tissue of the same individual.

In the intact male dog, the yearly check-up examination performed at booster vaccination should always include a per rectum palpation of the prostate to screen for the presence of early prostatic disease.

### Diagnostic approach to prostatic disease

Ultrasonography provides the safest and most informative screening test of the prostate and this imaging modality can be used to guide the operator when prostatic parenchyma biopsies are required. Diagnosis may involve the following:

- Complete history and thorough clinical examination
- Palpation of the prostate gland by concomitant rectal and abdominal palpation
- Routine haematology and serum biochemistries
- Urinalysis and urine culture
- Evaluation of prostatic fluid obtained by ejaculation or after prostatic massage
- Ultrasonography
- Survey and contrast radiography
- Prostatic aspiration
- Prostatic biopsy:
    - Percutaneous biopsy
    - Open surgical biopsy.

Fluid may be obtained from the prostate following prostatic massage. A urethral catheter is placed aseptically and passed into the urinary bladder. The bladder is emptied and a sample of urine saved for urinalysis and urine culture. The bladder is flushed several times with sterile saline to ensure that it is empty. The last flush of 5-10 ml is saved as the pre-prostatic massage sample. The catheter is then retracted to a position distal to the prostate. Rectal palpation will ensure the correct positioning of the catheter. The prostate is massaged both per rectum and via the abdominal wall for 1-2 minutes. Sterile saline (5-10 ml) is injected slowly through the catheter and the catheter is slowly advanced to the bladder with repeated aspiration across the prostatic urethra. Most of the fluid may not be aspirated until the bladder is reached. All samples should be cytologically examined and cultured.

In certain circumstances in which the prostatic urethra is involved in the disease process, it may be possible to achieve a cytological biopsy of the prostatic epithelial cells by applying negative pressure to the urethral catheter as it is moved back and forth across the prostatic urethra. Care should be taken not to induce serious iatrogenic damage to the urethra when performing this technique.

Percutaneous prostatic biopsy can be achieved either perirectally (Figure 15.11) or transabdominally. The biopsy needle is best guided by palpation in combination with ultrasonography. Parenchymal cystic structures may be aspirated rather than biopsied. Placement of a urethral catheter will help the operator avoid penetrating the prostatic urethra. This procedure should not be performed in dogs thought to be suffering from prostatic abscessation, since large numbers of bacteria may be seeded along the needle tract or a localized peritonitis may develop.

Open surgical prostatic biopsy is performed via a parapenile abdominal incision and will allow the whole of the prostate to be examined externally. Cysts or abscesses may be drained. A biopsy may be taken either using a skin punch or by resecting a wedge of prostatic parenchyma (Figure 15.12). Again, the posi-

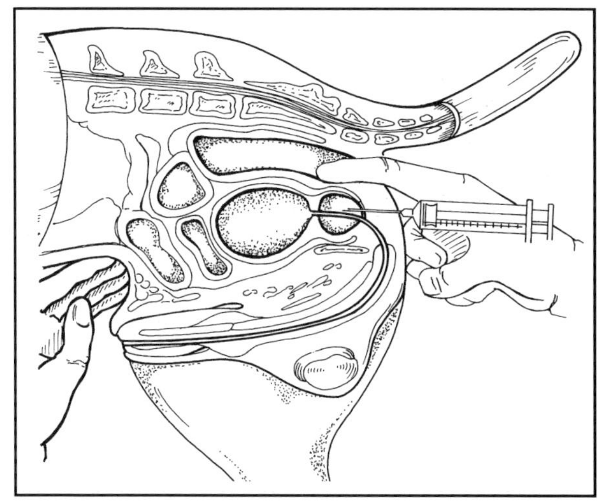

*Figure 15.11: Perirectal percutaneous aspiration/biopsy of the prostate. A catheter may be placed in the urethra to help avoid penetration of the intraprostatic urethra.*

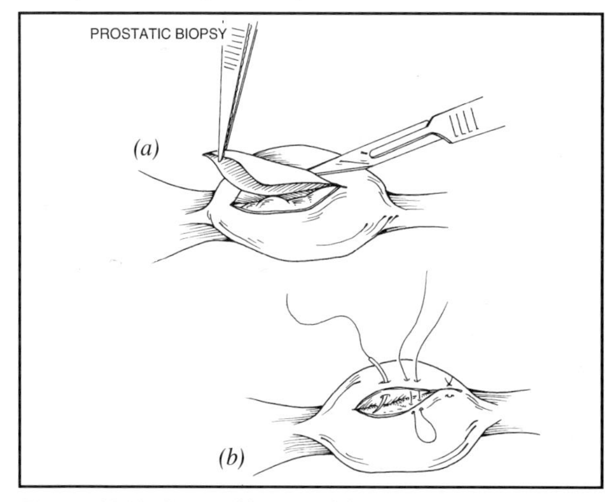


*Figure 15.12: Surgical biopsy of the prostate. (a) A wedge incision is made into the prostatic parenchyma. A catheter has been placed in the urethra to minimize the risk of cutting it. (b) The incision is closed using horizontal mattress sutures.*

tion of the prostatic urethra can be located and avoided if a urethral catheter is in place. The biopsy site is apposed with an absorbable suture material in a mattress pattern through the parenchyma.

**Prostatic metaplasia**

Excessive endogenous or exogenous oestrogens will induce prostatic epithelium to undergo squamous metaplasia. Oestrogens will also produce stromal hypertrophy of the gland. The major endogenous cause of hyperoestrogenism is a functional Sertoli cell tumour. The prostatomegaly associated with metaplasia is usually symmetrical although the process of metaplasia may lead to prostatic fluid stasis predisposing to infection with abscessation. In these individuals prostatic enlargement may be asymmetrical. In some individuals the secretory stasis may predispose to cyst formation.

The clinical signs associated with prostatomegaly and prostatic disease include faecal tenesmus, dysuria and urinary incontinence, urethral discharge, and in some instances systemic illness. It is important to remember that any prostatic disease, with the exception of acute prostatitis, can occur with no abnormal clinical signs evident to the owner. The treatment of squamous metaplasia requires removal of the oestrogen source. In the majority of cases, this means castration for the removal of a functional Sertoli cell tumour or the discontinuation of oestrogen-containing medication. Once the source of oestrogen has been removed the histological appearance of the prostatic tissue returns to normal.

**Benign prostatic hyperplasia**

Benign prostatic hyperplasia occurs in the older dog and requires the presence of the testes. It is associated with an age-related change in the androgen to oestrogen ratio within the affected individual. The process may begin in young individuals (2–3 years of age). It more commonly occurs in individuals above 4 years of age. The process is arguably a normal ageing change, although the physiological process of hyperplasia may become associated with pathological changes. Two types of benign prostatic hyperplasia have been described. These are: glandular hyperplasia, occurring in the younger population of dogs with a peak prevalence of 5–6 years of age; and complex or cystic hyperplasia, occurring in older dogs and affecting 70% of 8- to 9-year-old animals. It is suggested that intraprostatic cystic hyperplasia may be an extension of glandular hyperplasia.

Treatment of benign prostatic hyperplasia may only be required if the prostatomegaly is associated with clinical signs. The most effective form of treatment for the disease is castration. This procedure generally results in a 70% decrease in prostatic size. Sometimes, symptomatic treatment for the clinical signs associated with prostatomegaly may be required initially. Medical treatments for the condition include exogenous oestrogen, megoestrol acetate and the androgen receptor antagonist, flutamide. In general, these medical therapies are not recommended.

The diagnosis of benign prostatic hyperplasia is commonly made from clinical examination alone. Therefore, should castration fail to initiate a decrease in prostatic size, prompt investigation of other causes of prostatomegaly should be pursued.

**Bacterial prostatitis**

Bacterial infection of the prostate is not uncommon, and may result from ascending or descending urinary tract infection, haematogenous infection and infected semen. Ascent of bacteria up the urethra is considered the usual cause. Often, there are changes in prostatic architecture such as cysts, squamous metaplasia, or neoplasia associated with the bacterial prostatitis. A lower urinary tract infection is almost always seen in conjunction with bacterial prostatitis in the dog.

Clinical presentation may vary from signs of low grade infection to overt sepsis. Cultured organisms commonly include *Escherichia coli*, *Proteus*, *Staphylococcus*, *Streptococcus*, *Klebsiella* and *Pseudomonas* species. Micro-abscesses develop within the prostatic parenchyma, coalescing, leading to larger parenchymal abscesses. Eventually, a prostatic abscess may rupture, releasing bacteria and endotoxins into the retroperitoneal and caudal peritoneal cavities and possibly resulting in life-threatening septic peritonitis. Clinical signs may vary from a low grade urinary tract infection to septic shock and collapse. Often individuals are pyrexic, showing signs of caudal abdominal pain. They resent rectal palpation of the prostate. Diagnostic aids include radiography, ultrasonography, ultrasonographically guided aspiration and biopsy, with cytology of urethrally aspirated prostatic fluid following prostatic massage.

**Treatment of acute bacterial prostatitis**

During acute bacterial prostatitis, the blood prostatic fluid barrier may not be intact due to acute inflammation of the gland. This should allow good antibiotic penetration to the prostate. The choice of antibiotic should be based on results of urine culture and/or prostatic wash culture. The course of treatment should last for 21–28 days. Following therapy, the dog should be re-examined at regular intervals to ensure that the acute infection does not progress to a chronic infection. This re-examination should include urinalysis, urine culture and ideally examination of prostatic fluid.

**Treatment of chronic bacterial prostatitis**

In cases of chronic bacterial prostatitis, the blood-prostatic fluid barrier is likely to be intact. Factors such as lipid solubility, ionization, and protein binding affect the ability of antibiotics to cross this blood-prostatic fluid barrier. Current recommendations depend on whether the infectious agent is a Gram-positive or Gram-negative organism. If the causative organism is Gram positive, erythomycin, clindamycin, chloramphenicol or trimethoprim may be chosen, based on bacterial sensitivity. If the organism is Gram negative, chloramphenicol, trimethoprim or enrofloxacin may be indicated. Antibiotic therapy should be continued for at least 6 weeks and similarly to cases with acute bacterial prostatitis, continued surveillance of chronic bacterial prostatitis should be performed following resolution of clinical signs and discontinuance of antibiotic therapy.

In both acute and chronic bacterial prostatitis it is recommended that castration be performed as part of the treatment regimen. This procedure will reduce the quantity of prostatic tissue and may hasten the resolution of infection.

**Treatment of prostatic abscessation**

Prostatic abscessation is a severe form of chronic bacterial prostatitis. Prostatic micro-abscesses coalesce to produce larger areas of septic purulent exudate. *Escherichia coli* is the predominant organism cultured from such abscesses. The disease seems to occur most commonly in the older intact male dog. Many individuals present in septic shock; in some, the abscesses will have burst, leading to generalized peritonitis. Frequently, long-term antibiotic therapy in combination with castration will fail to resolve the condition of prostatic abscessation. Surgical therapy offers the best chance of resolution of signs. Various techniques have been described, but all rely on the surgical drainage of the abscess cavities.

In a dog showing signs of peritonitis on examination (painful abdomen, weak pulse, prolonged peripheral time, tachycardia, pale mucous membranes), abdominocentesis or a diagnostic peritoneal lavage should be performed to confirm the condition. Peritoneal lavage may be performed therapeutically to stabilize the dog prior to surgical exploration.

The prostate is exposed via a caudal midline coeliotomy and the abscesses are burst and explored in a controlled fashion with the aid of suction to remove the purulent material. All of the abscessations should be opened, usually with the aid of a finger. Care should be taken not to damage the prostatic urethra, which can often be more readily identified manually by previous placement of a urinary catheter. The gland should be flushed with copious amounts of sterile saline, while the remaining abdominal organs are packed off with laparotomy swabs. Penrose drains may be placed within the prostate parenchyma; these exit via separate ventral abdominal stab wounds. Considerable success has been achieved by the placement of omentum around the prostatic urethra following the breakdown of the abscess cavities. The prostatic omentalization procedure does not require the placement of Penrose drains. The reader is referred to a standard surgical text for the description of these procedures.

Following surgery, affected dogs must continue to have antibiotics administered, based on culture and sensitivity results. Similarly to chronic bacterial prostatitis, these animals should be monitored for further prostatic disease for a number of months following the resolution of the problem.

Prostatic abscessation could prove a difficult and expensive disease to manage, with many potential complications. Aggressive therapy is associated with the best long term outcome, but the prognosis should still be considered guarded.

**Prostatic neoplasia**

Reported primary prostatic neoplasms in the dog include adenocarcinoma, transitional cell carcinoma, squamous cell carcinoma and lymphoma. Of these, the commonest is the adenocarcinoma. All of the primary prostatic neoplasms are considered malignant.

Clinical signs are usually related to the prostatomegaly associated with the neoplastic condition. In addition, and unlike benign prostatic hypertrophy, prostatic carcinoma commonly causes urinary signs including dysuria and haematuria. Sometimes clinical signs may also be associated with metastatic spread to the lumbar vertebrae and bones of the pelvis. Radiographic and ultrasonographic studies should be used to investigate the disease and look for metastatic spread. Contrast radiography may be useful to demonstrate an irregular prostatic urethra. Unlike benign prostatic hypertrophy, palpation of the prostate per rectum often reveals an enlarged irregular, immobile, asymmetrical mass which may be painful.

Diagnosis is confirmed with needle biopsy or incisional biopsy of the prostate. Urethral washings may reveal malignant cells, but this method is unreliable.

**Treatment of prostatic neoplasia**

Total transurethral prostatectomy has been described as the treatment of early lesions without metastatic spread. However, this procedure is technically demanding and commonly results in urinary incontinence. In the author's opinion, there is rarely an instance where undertaking this procedure is justified.

Intraoperative external beam radiation therapy has been used in the treatment of prostatic carcinoma, but is rarely available.

The prognosis of prostatic carcinoma is poor, with local metastasis almost invariably present at the time a diagnosis is made. It should be remembered that clinical signs associated with prostatic disease relate primarily to prostatomegaly and, therefore, prostatic neoplasia should always be considered a differential diagnosis which requires ruling out.

**Paraprostatic and prostatic cysts**

The origin of most paraprostatic and prostatic cysts is usually unclear. It is suggested that the cysts may be remnants of the uterus masculinus or of prostatic origin. The terminology for these large cysts varies between different authorities, although most use the term paraprostatic cyst to describe a large cyst outside the prostatic parenchyma which does not have a patent connection to the prostatic gland. The term prostatic retention cyst is reserved for those cysts with a patent connection, especially if the structure seems to arise from the parenchyma of the prostate gland. Clinical signs are usually related to the size of the cyst and the effects of this space occupying lesion on the adjacent structures. Many may go unnoticed until they are diagnosed incidentally during investigation of an unrelated disorder.

### Treatment of paraprostatic and prostatic cysts

The recommended treatment for the condition is surgical drainage with excision or marsupialization of the cyst. Castration is also recommended. A prostatic biopsy should be taken to rule out any other prostatic disease, which may be the primary or secondary to the cyst formation. Prognosis depends on the underlying aetiology of the condition. If this is benign, then the prognosis is very favourable.

## Testes and epididymides

### Anorchism and monorchism

Congenital absence of both testes in small animals is rare. Monorchism is more commonly recognized, with the left testis usually being absent. A very careful intraoperative examination should be made in a cryptorchid individual before a diagnosis of monorchism is made. It is much more likely that the undescended testis has been missed rather than is absent, necessitating a thorough examination.

### Orchitis and epididymitis

Infection of the testis and epididymis is not unusual in the dog. Signs will often include testicular pain, tenderness and scrotal oedema. The animal may show systemic signs of illness and infection. Infection reaches the testis most commonly via the ductus deferens from the urinary bladder, urethra or the prostate. *Escherichia coli* is often isolated from the lesion. The affected testicle will often be inflamed and enlarged, although in chronic cases, the testicle may be small, irregular and firm. In most individuals the treatment of choice is orchidectomy. Breeding individuals may be medically managed with systemic antibiotics, but the condition is likely to return in the long term. It is important that possible underlying causes for the problem be excluded such as prostatic disease and chronic urinary tract infections.

### Testicular trauma

Testicular trauma is rare in dogs and cats. Malicious placement of rubber bands sometimes occurs leading to scrotal strangulation. Clinical signs associated with testicular trauma include local pain and swelling of the testis involved, hind limb lameness, scrotal swelling and oedema, and in severe cases, scrotal haematoma formation. Cases of minor testicular trauma may be treated conservatively, although regular re-examinations should be made to ensure the testicular swelling is reducing and that interscrotal haemorrhage is not occurring. In more severe cases, the treatment of choice is either unilateral or bilateral orchidectomy, possibly including a scrotal ablation.

### Testicular neoplasia

Testicular tumours occur relatively frequently, and represent 5–15% of all tumours in male dogs. Testicular tumours are very rare in the cat. There are three tumour types which are seen most commonly; the Sertoli cell tumour, the seminoma and the interstitial cell tumour. Cryptorchid dogs are at least 10 times more at risk of developing Sertoli cell tumour and seminoma than normal dogs. Metastatic spread is unusual with seminoma and interstitial cell tumour, although it occurs in about 10% of individuals suffering from Sertoli cell tumour. Regional lymph nodes, liver and lung are most likely to be affected. Sertoli cell tumour and seminoma often show the paraneoplastic syndrome of feminization due to their excretion of oestrogen-like substances. Feminization will usually present as gynaecomastia, pendulous prepuce, bilateral flank alopecia, prostatic metaplasia, atrophy of the contralateral testicle and attractiveness to other male dogs. Few clinical signs other than mild testicular enlargement are associated with interstitial cell tumour. Surgical castration will prove curative in

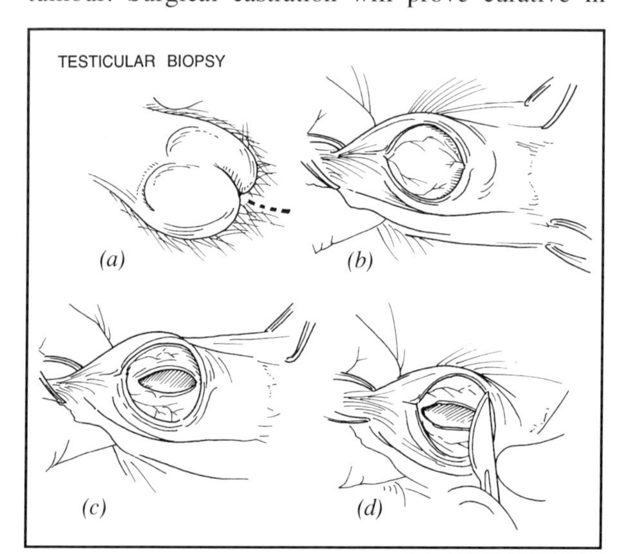


***Figure 15.13:*** *Biopsy of the testis. (a) The site of the skin incision (dotted line) is in the midline caudal to the scrotum. (b) Manual pressure on the testis in a caudal direction will position it under the skin incision. (c) The tunica albuginea is incised, revealing the testicular paraenchyma. (d) A blade is used to slice a biopsy specimen. The tunica albuginea is closed with an absorbable suture using a continuous pattern and the skin incision closure is routine.*

individuals with no metastatic disease. Owners of cryptorchid dogs should be advised to have their dogs castrated at an early age.

## INTERSEX

An intersex animal is one possessing the characteristics of both sexes. They are classified as either true hermaphrodites or pseudohermaphrodites and as either male or female on the basis of their gonads.

- Pseudohermaphrodites possess the gonads of one sex but the secondary sex characteristics and external genitalia of the opposite sex. Female pseudohermaphrodites have ovaries but are phenotypically male in appearance. Male pseudohermaphrodites have testes while possessing mixed or female genitalia
- True hermaphrodites possess both testicular and ovarian tissue in various combinations:
  - A testis may be found on one side and an ovary on the contralateral side
  - An ovatestis (both testicular and ovarian tissue found in a single gonad) only may be present
  - An ovatestis may be paired with a testis or ovary.

Intersex animals should not be confused with those animals with insufficient genital development. The latter animals are infertile males and females, typically possessing abnormal sex chromosome make-ups. The definitive diagnosis of intersexuality can only be made on histological examination of the gonads. Both gonads should be evaluated so that the possible combination of an ovary and testis will not be overlooked.

Pseudohermaphroditism is more commonly recognized than true hermaphroditism and male pseudohermaphroditism is recognized more commonly than the very rarely encountered female pseudohermaphroditism.

There are a number of reasons why an intersex animal may be presented to the veterinary surgeon and in many instances the reason for presentation is unrelated to the abnormal sexual development of the animal. Owners will often be unaware that any problem exists. The veterinary surgeon may be unaware of the condition, which is subsequently encountered at surgery for neutering when testes are found in place of the expected ovaries. Some of the more common reasons for presentation include:

- Difficulty in determining the animal's sex on examination of the external genitalia
- Males which are attractive to other male dogs and may show signs of oestrus
- Females which possess an enlarged clitoris protruding from the vulva
- Apparent male behaviour in a phenotypically female individual
- Individuals of either sex which assume a micturition posture opposite to that of their apparent sex.

There are a number of problems which can be encountered in both the intersex and normal individual and these include:

- Hypoplasia of the penis and prepuce
- Clitoral hypertrophy
- Cryptorchidism
- Pyometra
- Sertoli cell tumour development.

Anecdotally, many veterinary surgeons report the diagnosis of intersex at the time of presumed routine ovariohysterectomy. In individuals with a normal urinary system the management is often straightforward, with the removal of the internal genitalia in a similar manner to a routine ovariohysterectomy. The type of hermaphroditism may be determined by the histological evaluation of the resected gonadal tissue.

The management of diagnosed hermaphrodites and pseudohermaphrodites is determined by the degree of abnormality of the genitourinary system that exists in the affected individual. This determination may be difficult to ascertain on physical examination alone. Survey and contrast radiographic studies including a retrograde urethrogram and an intravenous urogram will be required to determine the anatomy of the animal's urogenital tract more fully.

The aims of management include:

- Removal of the internal genitalia
- Determination of the anatomy and function of the animal's urogenital tract
- Surgical reconstruction of the external genitalia to improve urinary function.

Suspected and confirmed intersex individuals should not be used for breeding.

### Associated problems

#### Hypospadias

Hypospadias is a congenital anomaly, which is occasionally seen in pseudohermaphrodite dogs, in which the penile urethra terminates along the ventral surface of the penis proximal to its normal opening. The condition and its management are more fully discussed in the section on penile conditions.

#### Os clitoris and os penis

This condition may be seen in phenotypically female individuals. Examination will reveal the presence of an os clitoris or os penis protruding from the vulva. An os clitoris may be differentiated from an os penis depending on whether a urethra is identified within the structure. If a urethra is present within the structure, it is classified as an os penis.

An os clitoris may cause no clinical problems in some dogs, and in these individuals no treatment for the condition is necessary. On the other hand, the structure may become an irritation to the animal, resulting in unacceptable licking of the area and self-trauma. In these individuals, amputation of the os clitoris is recommended. If significant clitoral hypertrophy is present, as is commonly the case, it is recommended that a total clitorectomy be performed. The procedure for clitorectomy is described in the section on clitoral conditions. The procedure of amputation of the os clitoris involves the resection of the os from within the clitoris. Care should be taken to ensure that the urethral opening into the vestibule is preserved. In individuals with an os clitoris, the junction between the cranial limit of the clitoris and the urethral opening may be very short. It is imperative that the urethral opening is recognized and a urinary catheter placed. The os clitoris can be removed via a ventral incision of the clitoral mucosa following manual eversion of the clitoris from its fossa. The os is separated and removed from the surrounding soft tissue with sharp and blunt dissection. The mucosal defect can then be closed with an absorbable suture using a continuous suture pattern. An Elizabethan collar may be required in the postoperative period to prevent interference with the surgical site.

Resection of an os penis is technically demanding, since the urethral opening is abnormally placed at some point along the length of the structure. Resection requires the dissection of the distal urethra from the dorsal aspect of the os penis and the repositioning of the urethral opening on the ventral wall of the vestibule. Similarly to all cases of intersex, a careful assessment of the remaining urogenital tract should be made prior to embarking on the such surgical repair.

### Persistent penile frenulum

Persistent penile frenulum is seen occasionally in male pseudohermaphrodites. The condition and its management is discussed in the section on penile conditions.

### Urinary incontinence

It is not uncommon for individuals with hermaphroditism and pseudohermaphroditism to suffer from urinary incontinence. Concurrent urinary incontinence with intersex conditions is often of non-neurogenic origin associated with developmental anomalies of the vagina, the urethra, or both. Developmental anomalies reported include urethrovaginal fistula, vaginal membrane obstructing the normal outflow of urine, double urethra and a persistent urogenital sinus. Surgical management of these conditions may improve or resolve the urinary incontinence. Urinary incontinence due to urethral sphincter mechanism incompetence is also a possible differential diagnosis which may require investigation.

### Testicular neoplasia

Similarly to the normal male with a retained testicle, the presence of a retained testicle in a male pseudohermaphrodite predisposes to malignant testicular changes (commonly a Sertoli cell tumour). A Sertoli cell tumour in the male pseudohermaphrodite will behave similarly to that in the normal male. Clinical signs of hyperoestrogenism, including attraction to other male dogs, gynaecomastia and bilaterally symmetrical alopecia, may be present. If the male pseudohermaphrodite possesses a uterus it may be enlarged as a result of the endometrial response to the hyperoestrogenism.

Management consists of removal of the neoplastic testicle and the other internal genitalia. Removal of the accessory sexual organs such as the uterus should be performed at the same time.

### Pyometra

The uterus of a pseudohermaphrodite can undergo pathological changes similar to those seen in the normal female. Documented conditions include cystic endometrial hyperplasia, mucometra and pyometra. The diagnosis of pyometra in a phenotypically male dog can be very challenging. These unusual, but possible findings make a full examination and diagnostic investigation essential in such cases. Surgical treatment is recommended and is similar to that for pyometra in the normal female.

## REFERENCES AND FURTHER READING


Howard PE and Bjorling DE (1989) The intersexual animal. *Problems in Veterinary Medicine* **1**, 74–84

Gregory SP (1996) Management of urinary incontinence. In: *Manual of Canine and Feline Nephrology and Urology*, ed. J Bainbridge and J Elliot, pp. 161–173. BSAVA, Cheltenham

Kyles AE, Aronsohn M and Stone EA (1996) Urogenital surgery. In: *Complications in Small Animal Surgery*, ed. AJ Lipowitz *et al.*, pp. 455–525. Williams & Wilkins, Baltimore

Manfra Marretta S, Matthiesen DT and Nichols R (1989) Pyometra and its complications. *Problems in Veterinary Medicine* **1**, 50–62

Pearson H (1973) The complications of ovariohysterectomy in the bitch. *Journal of Small Animal Practice* **14**, 257–266

Pearson H (1996) Genital surgery in the bitch and cat. In: *Veterinary Reproduction and Obstetrics, 7th edn*, ed. GH Arthur *et al.*, pp. 332–341. WB Saunders, London

Slatter D (ed.) (1993) *Textbook of Small Animal Surgery, 2nd edn*. WB Saunders, Philadelphia

Wykes PM and Olson PN (1985) Surgical management of dystocia. In: *Textbook of Small Animal Surgery*, ed. D Slatter, pp. 1689–1691. WB Saunders, Philadelphia.

CHAPTER SIXTEEN

# Pharmacological Control of Reproduction in the Dog and Bitch

*Gary C.W. England*

## INTRODUCTION

Recently there has been a significant increase in our knowledge of reproductive biology in dogs and bitches. Combining this knowledge with the increased availability of new hormonal preparations has allowed treatment options to be improved for many clinical disease conditions.

This chapter aims to review the main categories of drugs that may be used to control reproductive activity, to highlight the availability of drugs within each group and to discuss the specific clinical applications of the available agents.

## REPRODUCTIVE ENDOCRINOLOGY

Female and male reproductive endocrinology are fully discussed in Chapters 1, 2 and 6. For the present chapter, however, a simplistic overview is suitable to allow an understanding of the normal biology and the mechanisms of action of various pharmacological preparations.

### Bitch

Figure 16.1 demonstrates the changes in selected hormone concentrations during the pregnant cycle of the bitch.

- During late anoestrus, the pulse frequency and mean concentration of luteinizing hormone (LH) and the mean concentration of follicle stimulating hormone (FSH) are increased
- Follicular development is initiated and oestrogen concentrations are elevated
- Oestrogen has a negative feedback effect on the hypothalamus-pituitary, and LH and FSH concentration is reduced
- Oestrogen concentrations peak and then decline, removing the negative feedback effect which potentiates a surge release of LH and FSH
- Progesterone concentration increases from the time of the LH surge and is dramatically raised after ovulation
- Progesterone concentrations are similar in pregnancy and non-pregnancy
- Prolactin is elevated during the late luteal phase and is a principal luteotrophic factor
- After the luteal phase the bitch enters anoestrus where gonadotrophin and steroid hormone release are basal.

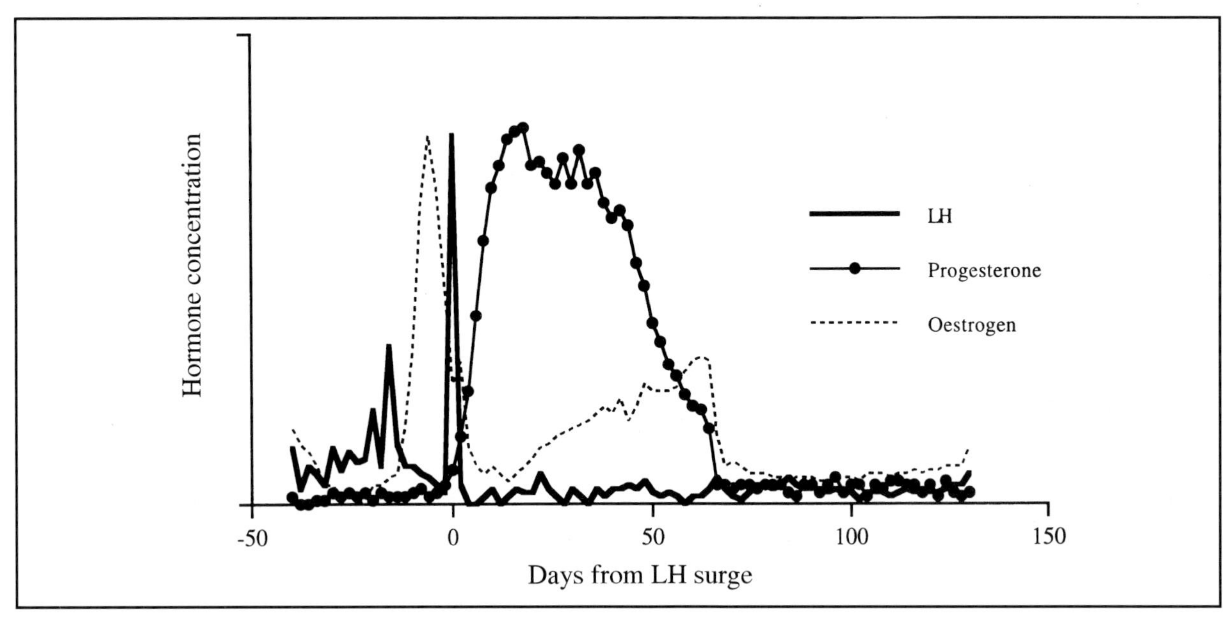


***Figure 16.1:*** *Representation of the hormonal changes during the oestrous cycle of a pregnant bitch.*

The feedback control mechanisms and site of action of exogenous drugs are shown in Figure 16.2.

## Dog

- Gonadotrophin control of testicular activity is provided by LH (also called interstitial cell stimulating hormone), and FSH.
- LH stimulates the interstitial (Leydig) cells to produce testosterone and small quantities of oestradiol.
- LH secretion is regulated by a feedback mechanism involving testosterone and possibly oestradiol.
- FSH stimulates spermatogenesis indirectly by an action upon the Sertoli cells.
- Androgen binding protein is secreted by the Sertoli cells and is involved in regulating the concentration and transport of testosterone within the epididymides.
- Inhibin is secreted by the Sertoli cells and acts upon the pituitary to regulate the secretion of FSH.

The feedback control mechanisms and sites of action of exogenous drugs are shown in Figure 16.3.

# PRODUCTS USED TO CONTROL THE REPRODUCTIVE SYSTEM

A variety of physiological processes and disease conditions may be influenced by the administration of pharmacological preparations. In Tables 16.1 and 16.2, common conditions are listed, together with agents that may be used for their control or treatment. The agents, their effects and adverse effects, and clinical uses are described later in the chapter.

Throughout the text, reference is given to pharmaceutical products that may be useful for the control of reproduction. Not all of these agents have a licensed indication in the dog, and some are as yet unavailable in the UK. For reference the UK-licensed products are given in Table 16.3.

# PROGESTOGENS

Progesterone is produced by the corpora lutea of the bitch, and is not naturally produced by the dog. Endogenous progesterone suppresses spontaneous myo-

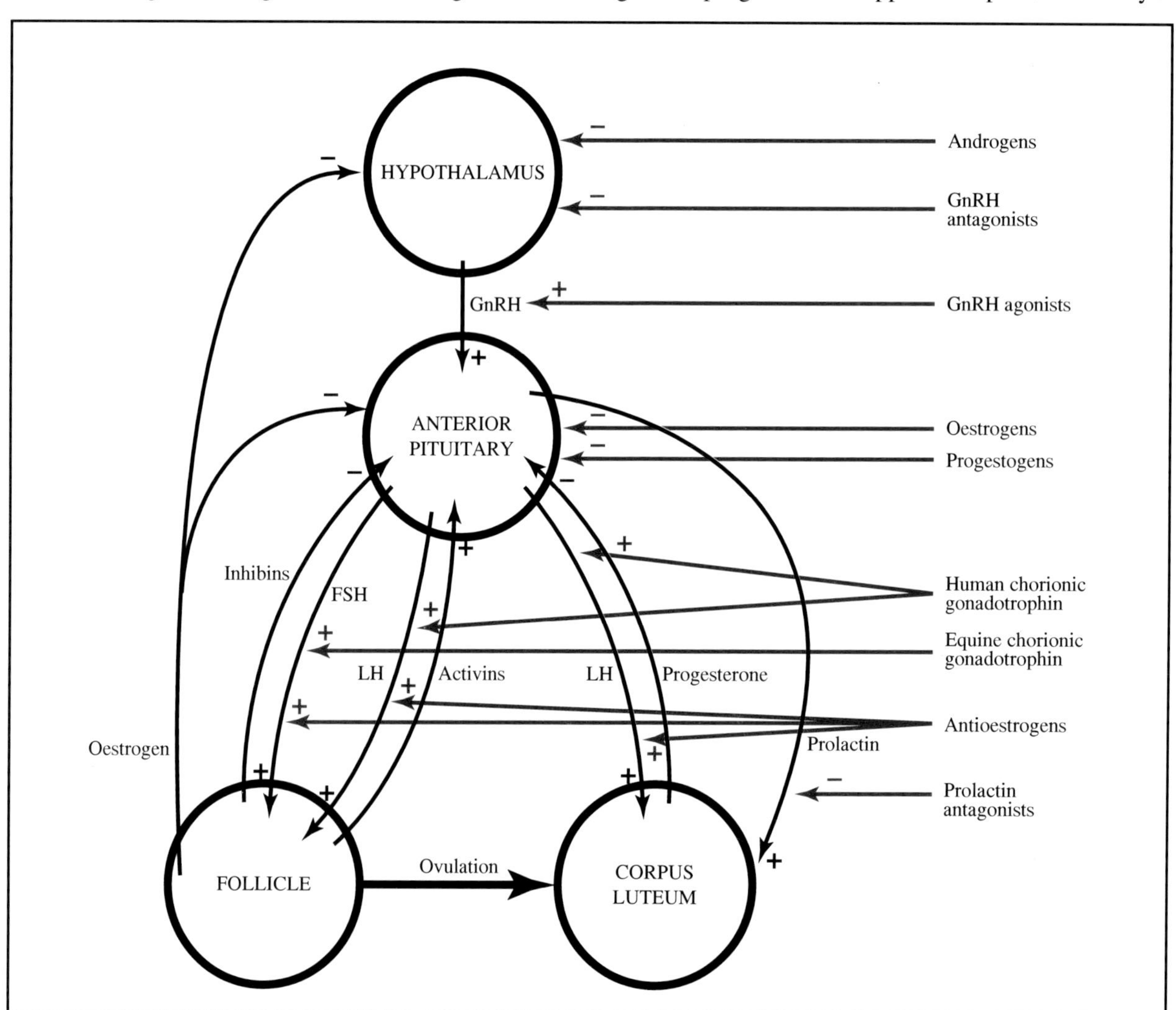


*Figure 16.2: Schematic representation of hormonal action and feedback control. The red arrows indicate the actions of exogenous drugs in the bitch. + = stimulation; – = inhibition.*

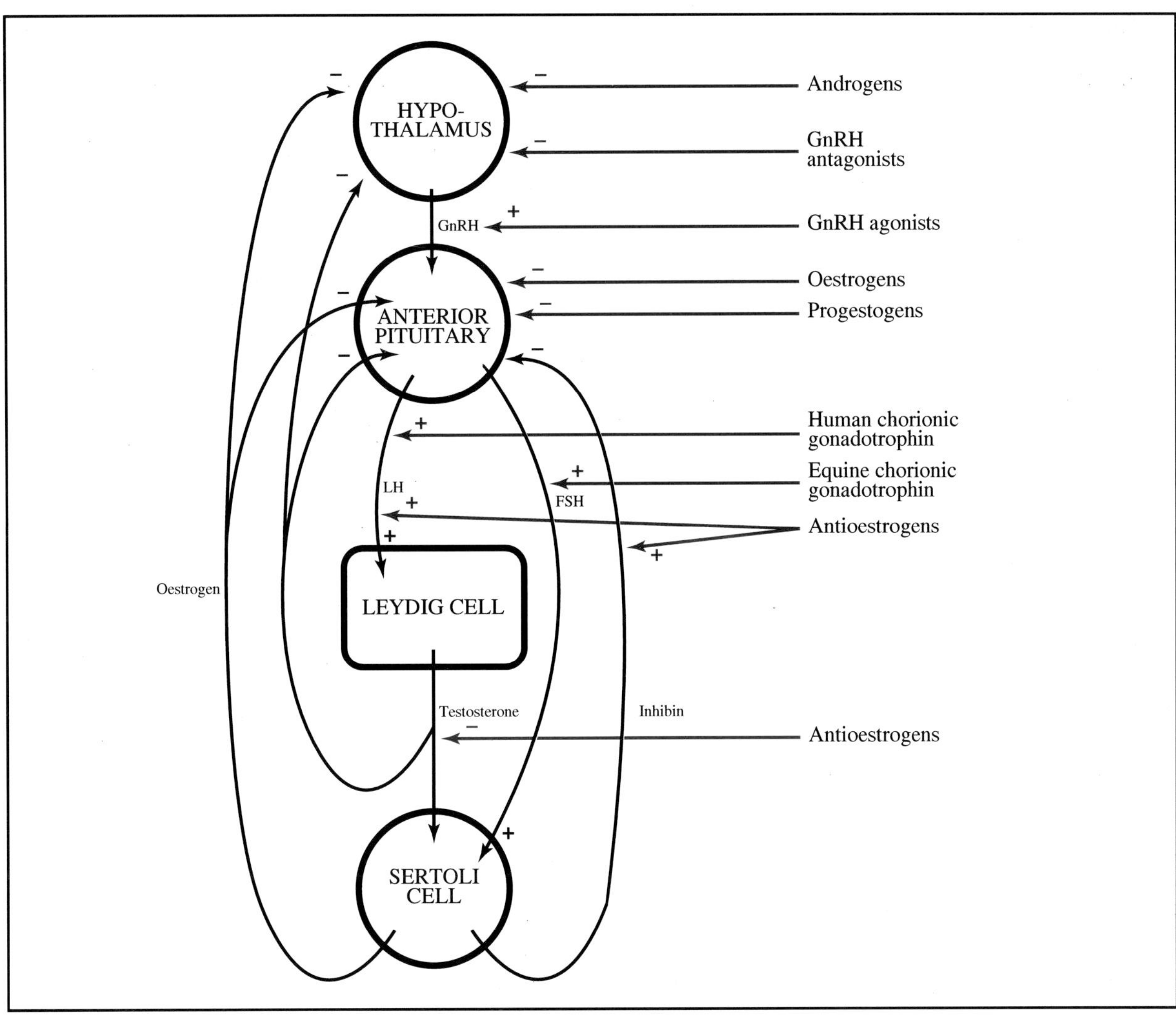


***Figure 16.3:*** *Schematic representation of hormonal action and feedback control. The red arrows indicate the actions of exogenous drugs in the male dog. + = stimulation; – = inhibition.*

metrial activity and stimulates endometrial growth; it is also responsible for mammary gland development during the luteal phase. Progesterone has a negative feedback effect upon the hypothalamus and pituitary gland (Figure 16.3). Progestogens are compounds with progesterone-like activity, and therefore they exhibit the actions above. It is, however, their negative feedback effect upon the hypothalamus and pituitary gland that is the principal reason why they are so widely used for the control of reproduction.

It was initially thought that the negative feedback was directly against FSH and LH; however, recent evidence suggests that whilst there is a reduction in circulating FSH concentration, there is no change in plasma LH concentration, only a decreased responsiveness to gonadotrophin-releasing hormone (GnRH). Progestogens therefore appear to exert their action by preventing a rise in gonadotrophin secretion, which is the trigger for the transition from one phase of the oestrous cycle to another. Progestogens do however have a negative feedback effect upon the release of prolactin, and can reduce, to some extent, circulating concentrations of testosterone and oestrogen by their actions upon gonadotrophin secretion.

A further effect demonstrated by high doses of progesterone and progestogens is that they are centrally sedative in action.

Progesterone and progestogens are commercially available in a variety of formulations including oral therapy that must be given daily, and oily suspensions and implants that provide a slow release over several weeks or months (Table 16.4).

## Adverse effects of progestogens

### General

Many transient effects may follow the administration of progestogens. These include increased appetite and weight gain, lethargy, mammary enlargement and occasional lactation (following the withdrawal of the agent), hair and coat changes, and temperament variations. These effects vary in their incidence between the different progestogens, although in general they are less frequent with the more recently developed ones.

The subcutaneous administration of some progestogens (especially the depot preparations) may

| Classification | Condition | Potential treatments |
|---|---|---|
| Prepubertal | Juvenile vaginitis | Oestrogens |
| | Delayed puberty | Gonadotrophins<br>Oestrogens<br>Prolactin antagonists |
| Mature | Control of oestrus | Progestogens<br>Androgens<br>Prolonged GnRH agonists and antagonists<br>(ovariectomy / ovariohysterectomy) |
| | Prolonged anoestrus | Gonadotrophins<br>Oestrogens<br>Prolactin antagonists |
| | Prolonged pro-oestrus | hCG<br>GnRH agonists |
| | Unwanted mating | Oestrogens<br>Prolactin antagonists<br>Prostaglandins<br>Prolactin antagonists + prostaglandins<br>Progesterone antagonists<br>GnRH agonists and antagonists<br>Synthetic antioestrogens<br>(ovariohysterectomy in early pregnancy) |
| | Pseudopregnancy | Progestogens<br>Androgens<br>Androgens + oestrogens<br>Prolactin antagonists |
| | Pyometra | Prostaglandins<br>Progesterone antagonists (ovariohysterectomy) |
| Pregnant | Termination of pregnancy | See unwanted mating (above) |
| | Habitual abortion | ? Progestogens |
| | Uterine inertia | Oxytocin<br>Ergot preparations |
| | Dystocia | Oxytocin<br>Ergot preparations |
| Post-partum | Fetal and/or placental retention | Oxytocin (hysterotomy) |
| | Post-partum haemorrhage | Ergot preparations<br>Oxytocin |
| | Post-partum metritis | Oxytocin<br>Prostaglandins<br>Ergot preparations |
| | Subinvolution of placental sites | ? Oxytocin as prevention |
| | Agalactia | Oxytocin |
| Ovariectomized | Urinary incontinence | Oestrogens (phenylpropanolamine) |
| | Alopecia | Oestrogens |
| | Vaginitis | Oestrogens |
| | Identification of remnant ovarian tissue | hCG<br>GnRH (laparotomy) |
| Ageing | Mammary tumours | Progestogens<br>Androgens<br>Synthetic antioestrogens<br>(surgery + ovariectomy or ovariohysterectomy) |
| | Debility | Androgens |

***Table 16.1:*** *Common clinical conditions in bitches (classified according to approximate age of occurrence) and possible pharmacological treatments. Treatments are listed with the most suitable agent first, related to product availability within the UK. Further details of the treatment rationale and therapeutic regimens are given in the text.*

| Classification | Condition | Potential treatments |
|---|---|---|
| Prepubertal | Cryptorchidism | None<br>(bilateral castration) |
| | Testicular hypoplasia | None |
| | Identification of testicular tissue | hCG |
| Adult | Antisocial and behavioural problems | Progestogens<br>Oestrogens<br>(behaviour modification therapy)<br>(castration)<br>(euthanasia) |
| | Contraception | Progestogens<br>Androgens<br>Progestogens + androgens<br>Prolonged GnRH agonists and antagonists<br>(castration) |
| | Poor libido | Gonadotrophins<br>hCG |
| | Poor semen quality | Synthetic androgens<br>Synthetic antioestrogens |
| | Epilepsy | Progestogens |
| Ageing | Benign prostatic hyperplasia | Progestogens<br>Synthetic antiandrogens<br>Oestrogens<br>GnRH agonists<br>(castration) |
| | Prostatic neoplasia | Progestogens<br>Synthetic antiandrogens<br>Oestrogens<br>GnRH agonists |
| | Circum-anal ademomata | Progestogens<br>Oestrogens<br>(castration) |
| | Feminization caused by testicular tumours | Androgens<br>(castration) |
| | Debility | Androgens |

***Table 16.2:*** *Common clinical conditions in male dogs (classified according to approximate age of occurrence) and possible pharmacological treatments. Treatments are listed with the most suitable agent first, related to product availability within the UK. Further details of the treatment rationale and therapeutic regimens are given in the text.*

produce hair discoloration and local alopecia at the site of injection. It is therefore recommended that administration should be performed in an inconspicuous site, although this is usually impractical.

All progestogens may potentially induce the production of growth hormone. Chronic over-secretion of growth hormone may result in the clinical signs of acromegaly and peripheral insulin antagonism which may result in diabetes mellitus.

Progestogen therapy may also produce adrenocortical suppression.

### Specific adverse effects in the bitch

Progestogens may result in the development of cystic endometrial hyperplasia and pyometra. The risk is related to the particular progestogen used, the amount administered, and the duration of treatment. This action appears to be potentiated by oestrogen, and for this reason certain depot progestogens (medroxyprogesterone acetate) are not licensed for use when the bitch is in pro-oestrus (when oestrogen concentrations are elevated); they are therefore not advised for the suppression of oestrus. Other depot progestogens (proligestone and delmadinone acetate) have been shown to be safe when administered at practically any stage of the oestrous cycle, although delmadinone acetate is only licensed for use in the male dog. Oral therapy with megoestrol acetate and medroxyprogesterone acetate has been shown to produce only a low incidence of adverse effects.

There are no preparations recommended for use on the first oestrous period or in prepubertal bitches.

Benign mammary nodules can be induced by progestogen therapy, and it has been suggested that

| Drug | Trade name | Formulation | Dose | Distributor |
|---|---|---|---|---|
| **PROGESTOGENS** | | | | |
| Progesterone | Progesterone Injection | 25 mg/ml for injection | 1-3 mg/kg | Intervet UK Ltd |
| Medroxyprogesterone acetate | Perlutex for Injection | 25 mg/ml for injection | 2.5-3.0 mg/kg at intervals of 5 months | Leo Laboratories Ltd |
| | Perlutex tablets | 5 mg tablets | 10 mg/bitch/day for 4 days then 5 mg/bitch/day for 12 days (double the dose if weighs more than 15kg) | Leo Laboratories Ltd |
| | Promone-E | 50 mg/ml for injection | 50 mg/bitch at intervals of 6 months | Pharmacia and Upjohn Ltd |
| Megoestrol acetate | Ovarid | 5 mg and 20 mg tablets | 0.5 mg/kg/day for up to 40 days commencing in anoestrus or, 2.0 mg/kg/day for 8 days commencing in pro-oestrus or, 2.0 mg/kg/day for 4 days then 0.5 mg/kg/day for 16 days commencing in pro-oestrus | Mallinckrodt Veterinary Ltd |
| Proligestone | Delvosteron | 100 mg/ml for injection | 10-33 mg/kg in pro-oestrus or, a series at 3,4, and then 5 month intervals | Intervet UK Ltd |
| Delmadinone acetate | Tardak | 10 mg/ml for injection | 1.0-2.0 mg/kg | Pfizer Ltd |
| **OESTROGENS** | | | | |
| Oestradiol benzoate | Mesalin | 0.2 mg/ml for injection | 0.01 mg/kg on day 3 and 5 post mating (a 3rd dose may be given on day 7) | Intervet UK Ltd |
| Ethinyl oestradiol (+ methyltestosterone) | Sesoral | 0.005 mg tablets (+ 4 mg) | 0.0008 mg/kg for 5 to 10 days (+ 0.7 mg/kg) | Intervet UK Ltd |
| **ANDROGENS** | | | | |
| Methyltestosterone | Orandrone | 5 mg tablets | 0.5 mg/kg daily | Intervet UK Ltd |
| Methyltestosterone (+ ethinyl oestradiol) | Sesoral | 4 mg tablets (+ 0.005 mg) | 0.7 mg/kg for 5 to 10 days (+ 0.0008 mg/kg) | Intervet UK Ltd |
| Testosterone phenylpropionate | Androject | 10 mg/ml for injection | 0.5 - 1.0 mg/kg every 7-10 days | Intervet UK Ltd |
| Testosterone propionate + testosterone phenylpropionate + testosterone isocaproate + testosterone decanoate | Durateston | 50 mg/ml for injection | 2.5 - 5.0 mg/kg every 6 months | Intervet UK Ltd |
| **GONADOTROPHINS** | | | | |
| Serum gonadotrophin (FSH-like) | Folligon | 1000 IU crystal + solvent | 20 IU/kg daily for 10 days | Intervet UK Ltd |
| | PMSG-Intervet | 5000 IU crystal + solvent | 20 IU/kg daily for 10 days | Intervet UK Ltd |
| Chorionic gonadotrophin (LH-like) | Chorulon | 1500 IU crystal + solvent | 100-500 IU/animal | Intervet UK Ltd |
| **ECBOLIC AGENTS AND DRUGS PROMOTING MILK LETDOWN** | | | | |
| Oxytocin | Oxytocin-S Injection | 10 IU/ml for injection | 2-10 IU/bitch | Intervet UK Ltd |
| Pituitary extract | Hyposton | 10 IU/ml for injection | 2.5-20 IU/bitch | Pharmacia and Upjohn Ltd |
| Pituitary extract (synthetic) | Pituitary Extract (Synthetic) | 10 IU/ml for injection | 2-10 IU/bitch | Animalcare Ltd |

***Table 16.3:*** *Commercially available products licensed in the UK that may be useful for the manipulation of reproduction in the dog and bitch. Drug data sheets should be consulted for full dose regimens.*

| Drug | Trade name | Formulation | Distributor |
|---|---|---|---|
| Progesterone | Progesterone Injection | 25 mg/ml for injection | Intervet UK Ltd |
| Medroxyprogesterone acetate | Perlutex for Injection | 25 mg/ml for injection | Leo Laboratories Ltd |
| | Perlutex tablets | 5 mg tablets | Leo Laboratories Ltd |
| | Promone-E | 50 mg/ml for injection | Upjohn Ltd |
| Megoestrol acetate | Ovarid | 5 mg and 20 mg tablets | Mallickrodt Veterinary Ltd |
| Proligestone | Delvosteron | 100 mg/ml for injection | Intervet UK Ltd |
| Delmadinone acetate | Tardak | 10 mg/ml for injection | Pfizer Ltd |
| Altrenogest | * Regumate Equine | 2.2 mg/ml solution for oral admin. | Hoechst Rousell Vet Ltd |
| | * Regumate Porcine | 4.0 mg/ml solution for oral admin. | Hoechst Rousell Vet Ltd |
| Norethisterone acetate | †Primolut N | 5 mg tablets | Schering Health Care |
| | †Micronor | 350 μg tablets | Ortho Division of Cilag Ltd |

***Table 16.4:*** *Commercially available progestogens that may be useful for the manipulation of reproduction in the dog and bitch.*
** Veterinary products available in the UK but not licensed for use in this species.*
*† Human medical preparations.*

progestogens may induce mammary neoplasia, although this does not appear to be the case with proligestone.

Progestogen therapy during pregnancy may delay or prevent parturition, and may produce masculinized female and cryptorchid male fetuses.

### Specific adverse effects in the dog

High doses of progestogens cause marked changes in semen quality, the majority of which are related to a direct effect upon the epididymides. Changes in semen quality may result in infertility. It is, however, surprising that lower doses administered for short periods of time produce no measurable effect on semen quality or fertility.

## Clinical use of progestogens

### Bitch

***Control of oestrus:*** Administration of progestogens to bitches in anoestrus prevents gonadotrophin secretion increasing above basal values, and therefore prevents a return to cyclical activity. If given when there is follicular activity (pro-oestrus or early oestrus), ovulation is inhibited and the bitch returns relatively quickly to anoestrus. This second action of progestogens is probably mediated by inhibition of increasing concentration of gonadotrophins, although in other species there is evidence that progestogens may have a direct action upon the ovary.

When progestogens are given daily or as a depot preparation, they mimic a luteal phase, which is followed by a state of anoestrus. Therefore cyclical activity is inhibited for longer than the duration of progestogen treatment.

Progestogens are generally administered in one of four regimens to control oestrus in the bitch. A large number of agents and formulations are available and the reader should consult specific data sheet recommendations for doses and treatment regimens.

***Subcutaneous administration of depot preparations during anoestrus:*** Depot progestogens (medroxyprogesterone acetate, delmadinone acetate, and proligestone) administered subcutaneously during anoestrus will prevent a subsequent oestrus, and regular repeated dosing (at 4- to 6-month intervals) can be used to prevent oestrus on a long-term basis. Medroxyprogesterone acetate and proligestone are licensed for this purpose; delmadinone acetate is unlicensed and has to be administered more frequently because of its shorter duration of action.

It may be inadvisable to prevent oestrus for more than 2 years, since this may increase the risk of cystic endometrial hyperplasia and pyometra. However, providing that a breakthrough cycle does not occur, prolonged prevention may have few adverse effects, especially when the more recently developed progestogens are used. Indeed it has been reported that the incidence of pyometra and mammary tumours are reduced compared with untreated females. When therapy is ceased, most females cycle normally, although progestogens may induce cystic endometrial hyperplasia and therefore reduce fertility. The time to return of oestrus is variable.

***Oral administration of progestogens during anoestrus:*** Low doses of orally active progestogens (megoestrol acetate, medroxyprogesterone acetate, altrenogest, norethisterone acetate) can be used to prevent oestrus for as long as administration continues. Megoestrol acetate and medroxyprogesterone acetate are licensed for this purpose in the bitch. The drugs are best given during late anoestrus before the anticipated oestrus. However, should the animal enter pro-oestrus during the first few days of treatment, the dosage can be increased. A period of anoestrus usually follows therapy so the animal does not return to oestrus immediately after cessation of treatment.

***Oral administration of progestogens during pro-oestrus:***

High doses of orally active progestogens (megoestrol acetate, medroxyprogesterone acetate, altrenogest, norethisterone acetate) may be given during pro-oestrus to suppress the signs of oestrus. Megoestrol acetate and medroxyprogesterone acetate are licensed for this purpose in the bitch. Usually the signs of pro-oestrus/oestrus disappear within approximately 5 days. Treatment during late pro-oestrus may not prevent ovulation in the bitch, although conception is unlikely to occur. However, treatment too early may lead to a return to oestrus soon after dosing. A reducing dose regimen administered from the first signs of pro-oestrus and continued for up to 16 days is usually efficacious. This regimen is frequently followed by a variable period of anoestrus, and oestrus will return between 4 and 6 months after the end of medication.

***Subcutaneous administration of depot preparations during pro-oestrus:*** Administration of the new generation depot progestogens (proligestone) to bitches in early pro-oestrus may be used to suppress the signs of that oestrus. Older generation progestogens (medroxyprogesterone acetate) which have potent effects upon the uterus are not recommended, since they increase the risk of uterine disease, especially pyometra. The signs of pro-oestrus/oestrus disappear within approximately 5 days. Following the depot progestogen, there is a variable anoestrus, and oestrus returns approximately 6 months later (range 3–9 months).

Other agents that may be used for the control of oestrous cyclicity include long-term administration of androgens commencing in anoestrus.

***Treatment of pseudopregnancy:*** Administration of progestogens to bitches with clinical signs of pseudopregnancy produces a suppression of prolactin secretion, and clinical signs rapidly disappear. Prolactin concentration is reduced for the duration of progestogen administration, but may increase again if progestogen therapy is rapidly withdrawn. In this situation there may be a return of the clinical signs.

Progestogens may be administered either orally daily (megoestrol acetate 2 mg/kg/day), or by depot injection (proligestone 20 mg/kg or delmadinone acetate 1.0 mg/kg), although some of these products are not licensed for this purpose. First-generation progestogens such as medroxyprogesterone acetate are not recommended, since they have marked effects upon the uterus and might potentiate the development of a pyometra.

In general, depot progestogen therapy works well, since the progestogen concentration gradually reduces in the circulation. Oral therapy is often associated with relapse of the clinical signs if therapy is terminated too quickly; this can usually be prevented by gradually reducing the dose over a period of approximately 7 or 10 days.

Care must be taken to ensure that the bitch is not pregnant, especially if depot progestogens are to be used, since these can delay or prevent parturition.

It is likely that the return to oestrus will be delayed following the administration of progestogens.

Other options for the treatment of pseudopregnancy include the administration of androgens, oestrogens, or combinations of these, or the use of prolactin antagonists such as bromocriptine or cabergoline.

***Treatment of mammary tumours:*** Both oestrogen and progesterone receptors have been identified in canine mammary tumours. Until recently, some clinicians have used progestogen therapy (megoestrol acetate) for the control of oestrogen-dependent mammary tumours. This therapy is empirical, and may not be efficacious since many advance tumours possess few steroid receptors and appear to have an autonomous growth pattern. Care should be taken if progestogen use is contemplated, since these agents may promote mammary tumourigenesis via their induction of growth hormone overproduction. Other agents that may be used to control mammary tumours include androgens, and the anti-oestrogens like tamoxifen.

***Treatment of habitual abortion:*** There is only anecdotal evidence that habitual abortion occurs in the bitch; in many alleged cases pregnancy has never been confirmed by a reliable method, and most are probably non-pregnant bitches that are repeatedly mated at an inappropriate time. An abnormal uterine environment (such as cystic endometrial hyperplasia) may result in repeated pregnancy failure; however, there is no evidence that a low progesterone concentration is the cause of repeated pregnancy loss.

It is therefore not appropriate to supplement bitches with progestogens to prevent habitual abortion unless a persistently low plasma progesterone concentration has been documented. Progestogen supplementation during pregnancy may produce masculinized female puppies and cryptorchid male puppies, and may possibly impair or delay parturition, resulting in fetal death. Progestogen use should be restricted to those cases in which a true luteal insufficiency has been diagnosed; in these cases oral therapy is most appropriate.

### Dog

***Antisocial behaviour and other behavioural problems:*** Problems such as aggression, roaming, territory marking, copulatory activity, destruction and excitability exhibited by both entire and castrated dogs may be controlled in some cases by progestogen administration.

The action of progestogens in these cases relates both to their anti-androgenic effect and their central sedative action. Depot progestogen therapy may need to be repeated every month for the shorter acting preparations (delmadinone acetate 1–2 mg/kg) up to every 6 months for the longer acting preparations (medroxyprogesterone acetate 3.0 mg/kg and proligestone 20 mg/kg). Oral therapy has also been shown to be effective and has the advantage that the dose may be adjusted to the effect. Commonly oral therapy commences with an initial high dose for 2 weeks and a reduction over several months.

Behaviour modification training is an essential adjunct to progestogen therapy, and entire dogs should be considered for castration.

***Benign prostatic hyperplasia:*** Prostatic enlargement may encroach upon the structures within the pelvic cavity and produce clinical signs of dysuria or faecal tenesmus. These signs are often preceded by haemospermia, which progresses to haematuria.

Progestogens are both anti-androgenic and have a direct effect upon the prostate gland (see anti-androgens later), and these actions are responsible for the rapid reduction in the clinical signs associated with regression of the prostate gland. Depot therapy usually causes remission of clinical signs within 4 days, although a second treatment may be necessary after a short time in some individuals. In stud dogs delmadinone acetate (1-2 mg/kg) is commonly used because of its short period of action and its relative sparing effects upon sperm production. However, the use of other depot or orally active progestogens is equally efficacious, and high doses or long treatment periods are necessary to interfere with fertility. In dogs not required for breeding castration in the treatment of choice.

Other agents that may be used for the treatment of benign prostatic hyperplasia include oestrogens and anti-androgens such as flutamide and finasteride.

***Prostatic neoplasia:*** Some clinical improvement may be seen in dogs with prostatic neoplasia following the administration of progestogens. The effect is, however, usually only short term, and progestogens offer little relief in metastatic disease. Other agents that may be used include oestrogens, anti-androgens such as flutamide and finasteride, and GnRH analogues such as buserelin.

***Circum-anal adenomata:*** The anti-androgenic effect of progestogens given as a depot (delmadinone acetate 1-2 mg/kg) or oral preparation is useful in causing temporary reduction of the size of these benign tumours. Tumours with central necrosis may not respond well, and the remission time is related to the duration of action of the progestogen used. In dogs not required for breeding, castration is the treatment of choice. Other agents that may be used include oestrogens.

***Contraception:*** Long-term administration of high doses of progestogens may be useful to induce spermatozoal abnormalities and possible suppression of spermatogenesis. The concurrent administration of depot androgens appears to be more efficacious and allows the progestogen dose to be reduced. Castration is the method of choice in dogs that are not required for breeding.

***Epilepsy:*** Castration and/or progestogen therapy has been used for the control of some epileptiform convulsions. The actions of progestogens may be related to their central sedative effect. Only delmadinone acetate (1-2 mg/kg) is licensed for this purpose, although any depot or oral progestogen may be useful.

## OESTROGENS

Oestrogen is produced by ovarian follicles of the bitch, and the Leydig cells of the dog. Endogenous oestrogens (Table 16.5) are responsible for the development of the female sexual characteristics, including uterine growth, the production of pheromones, swelling of the vulva, thickening of the vaginal mucosa, secretion from the cervical glands, and mammary development. Oestrogens are necessary for normal secretion and functioning of the uterine tubes, and changes in oestrogen concentration regulate the gonadotrophin surge which stimulates ovulation. In late pregnancy, oestrogen concentrations are elevated and may be involved in the initiation of parturition. In the male, oestrogens are involved in the feedback control of Leydig cell function.

In both the male and female, high doses of exogenous oestrogens cause a negative feedback at the hypothalamic-pituitary axis (Figures 16.2 and 16.3) and subsequent suppression of gonadotrophin secretion, but at low doses oestrogens enhance the release of FSH.

| Drug | Trade name | Formulation | Distributor |
|---|---|---|---|
| Oestradiol benzoate | *Oestradiol benzoate | 5 mg/ml for injection | Intervet UK Ltd |
| Oestradiol benzoate | Mesalin | 0.2 mg/ml for injection | Intervet UK Ltd |
| Ethinyl oestradiol (+ methyltestosterone) | Sesoral | 0.005 mg (+ 4 mg) tablets | Intervet UK Ltd |
| Ethinyl oestradiol | †Ethinyloestradiol | 10 µg, 50 µg and 1 mg tablets | Evans Medical Ltd |
| Stilboestrol | †Apstil Tablets | 1 mg and 5 mg tablets | Approved Prescription Services Ltd |
| Oestradiol valerate | †Climaval | 1 mg and 2 mg tablets | Novartis Pharmaceuticals UK Ltd |
| Oestradiol | †Oestradiol Implants | 25 mg, 50 mg and 100 mg implants | Organon Laboratories Ltd |
| Dienoestrol | †Ortho-dienoestrol | 0.01% cream for topical application | Janssen-Cilag Ltd |

***Table 16.5:*** *Commercially available oestrogens that may be useful for the manipulation of reproduction in the dog and bitch.*
** Veterinary products available in th UK but not licensed for use in this species.*
*† Human medical preparations.*

Oestrogens increase osteoblastic activity and cause the retention of calcium and phosphorus, produce an increase in total body protein and metabolic rate, and affect skin texture and vascularity.

## Adverse effects of oestrogens

### General

Oestrogens have been shown to produce a dose-related bone marrow suppression which results in a severe and possibly fatal anaemia and thrombocytopenia. There is considerable individual variation in the toxic dose, which for some animals may lie within the manufacturer's normal recommended dose range. Toxic effects are however generally dose-related and toxicity is less likely if low doses are given over a long period of time. Oral therapy should be prescribed whenever possible, so that administration can cease if adverse effects are noted.

Prolonged oestrogen therapy may produce a non-pruritic bilaterally symmetrical alopecia and skin hyperpigmentation.

### Specific adverse effects in the bitch

Whilst oestrogens do not stimulate cystic endometrial hyperplasia or pyometra *per se* they potentiate the stimulatory effect of progesterone on the uterus. Additionally, they cause cervical relaxation and allow vaginal bacteria to enter the uterus. For these reasons, oestrogen administration may result in the development of cystic endometrial hyperplasia and pyometra.

Large doses of oestrogens may stimulate signs of oestrus in both entire and ovariohysterectomized bitches, whilst administration during oestrus may cause the signs of that oestrus to be lengthened.

If oestrogens are administered during pregnancy, they may cause cervical relaxation and abortion, possibly as a result of the inhibition of LH, which is a major luteotrophic factor. Oestrogen administration during pregnancy may cause congenital defects in the developing fetuses including abnormalities of phenotypic sex.

### Specific adverse effects in the dog

The anti-androgenic effects of oestrogen administration may, after prolonged therapy, result in abnormalities of semen quality and a resultant reduction in fertility. Prostatic size initially decreases during oestrogen therapy, although subsequently reversible prostatic metaplasia may result, causing the gland to increase in size.

Prolonged oestrogen administration may result in gynaecomastia.

## Clinical use of oestrogens

### Bitch

***Unwanted mating:*** Oestrogens alter zygote transport time and impair implantation when administered soon after mating. Oestrogens may also produce a short luteal phase by interfering with LH support of the corpora lutea.

In clinical practice, diethylstilboestrol, oestradiol cypionate and mestranol have been widely used to treat unwanted matings. Only oestradiol benzoate is licensed for the treatment of unwanted mating in the UK. This is administered at a low dose (0.01 mg/kg), 3 and 5 (and possibly also 7) days after mating. The low dose regimen has been advocated in an attempt to reduce the possibility of adverse effects.

Other treatment options are to induce resorption or abortion by the use of prostaglandins, prolactin antagonists or combinations of these after pregnancy has been diagnosed (Chapter 11).

***Pseudopregnancy:*** Oestrogens may be used to produce an inhibition of prolactin secretion in bitches with pseudopregnancy. Both parenteral and oral diethylstilboestrol and parenteral oestradiol benzoate have been used, but neither is specifically licensed for this purpose. A preparation of ethinyl oestradiol (0.8 μg/kg/day) combined with methyltestosterone (0.7 mg/kg/day) is available, and produces a good clinical response, although the dose suggested seems arbitrary. Recurrence of the clinical signs may follow abrupt termination of treatment, and a reducing dose regimen may be necessary especially in bitches with recurrent or persistent clinical signs.

Other treatment options include the administration of progestogens, androgens or specific prolactin antagonists such as bromocriptine and cabergoline.

***Juvenile vaginitis:*** A mucoid vulval discharge, which may become purulent, is commonly seen in bitches from 8 weeks of age onwards. Signs may also include frequent licking and attractiveness to male dogs. The condition usually regresses after the first oestrus and often does not warrant treatment; however, if the clinical signs are severe some control may be effected using low doses of oral oestrogens daily for 5 days. Oestrogens are not licensed for this purpose and should not be given to bitches required for breeding. An alternative therapy is the topical application of oestrogen-containing creams (Vagifem: NovoNordisk Pharmaceuticals Ltd).

***Oestrus induction:*** Low doses of oestrogen enable FSH to stimulate the formation of LH receptors on granulosa cells, and therefore increase the responsiveness to basal concentrations of LH. The result is follicular growth and production of oestrogen; gonadotrophin concentrations are low due to the negative feedback effect. As follicles mature, their walls luteinize and progesterone is produced. The decline in the oestrogen:progesterone ratio facilitates the pre-ovulatory LH surge.

The induction of oestrus with diethylstilboestrol

has been used clinically for some time. Good results can be achieved using 0.3 mg/kg/day for up to 10 days (treatment is stopped 1 day after the onset of a sanguineous vulval discharge).

Other treatment regimens include the administration of equine chorionic gonadotrophin combined with human chorionic gonadotrophin, or the use of prolactin antagonists such as bromocriptine and cabergoline.

***Attractiveness in ovariectomized bitches:*** Some bitches may become attractive to male dogs several years after ovariohysterectomy. These bitches do not have an ovarian remnant (although this should be eliminated as a differential diagnosis), and appear to have a low grade vaginitis. Bacterial examination usually reveals commensal organisms. A good clinical response may be achieved using low doses of diethylstilboestrol (0.06 mg/kg/day) or other orally active oestrogens such as ethinyl oestradiol (0.06 mg/kg/day) for up to 7 days, or the topical application of oestrogen-containing creams (Vagifem: NovoNordisk Pharmaceuticals Ltd).

***Alopecia in ovariectomized bitches:*** Some ovariectomized bitches develop poor hair growth and alopecia. Unsuitable terminology has been used to describe these cases, which are often referred to as having ovarian imbalance. Oestrogen-responsive alopecia has a poorly understood aetiology and the clinical response to exogenous oestrogen administration probably relates to the epitheliotrophic action of oestrogen, the primary effects of which are to promote keratinization and to suppress sebum production.

***Urinary incontinence:*** Urinary incontinence is frequently described as a complication of ovariectomy or ovariohysterectomy. The aetiology of the condition is uncertain, but it may relate to changes in the thickness of the urethral mucosa in the absence of oestrogen. Incontinence may occur between months and many years after the surgery. Some cases respond to oestrogen therapy, whilst in others the response is short lived or absent.

Parenteral oestradiol benzoate has been recommended daily for 3 days, with subsequent injections every third day. The use of oral diethylstilboestrol (0.03 mg/kg/day) and ethinyl oestradiol (0.03 mg/kg/day) is also efficacious. Daily therapy is administered for up to 3 weeks, with a response being observed after a few days; the length of treatment is kept to a minimum required to produce a clinical effect, and treatment is repeated as necessary.

Other agents that may be used include the drugs that act directly upon the urethral muscle such as phenylpropanolamine (Propalin Syrup).

### Dog

***Antisocial behaviour and other behavioural problems:*** The anti-androgenic actions of oestrogens may be useful in dogs with antisocial behaviour. However, due to the potential adverse effects of long-term oestrogen therapy, and the superior action of progestogens, the latter are more commonly used in clinical practice.

***Benign prostatic hyperplasia:*** Repeated administration of oestrogens will result in a reduction in prostatic size and the amelioration of clinical signs. The effect of oestrogens upon semen quality and fertility have not been documented, and the treatment of choice is probably low dose progestogens which have a minimal effect on fertility, or castration in the animal that is not required for breeding. Prolonged oestrogen therapy may result in prostatic metaplasia and a resultant increase in the size of the prostate gland. Other treatment options include the use of anti-androgens and GnRH agonists.

***Prostatic neoplasia:*** Clinical signs of prostatic neoplasia may be controlled in the short term by the administration of oestrogens (diethylstilboestrol, 0.03 mg/kg/day). Progestogens may also be useful in the short term, whilst in man anti-androgens and GnRH analogues are widely used.

***Circum-anal adenomata:*** The anti-androgenic effect of oestrogens may be useful to cause a decrease in the size of these tumours. Other options are the administration of progestogens, although the treatment of choice is castration in dogs that are not required for breeding.

## ANDROGENS

The naturally occurring androgens testosterone and dihydrotestosterone are produced by the interstitial cells of the testes. In the male, androgens mediate the development and maintenance of primary and secondary sexual characteristics and normal sexual behaviour and potency, as well as playing an important role in the initiation and maintenance of spermatogenesis (Table 16.6).

Androgens are generally not present in significant concentrations in the non-pregnant female; however, it is clear that androgen receptors are present in oestrogen target tissues and that androgen administration can antagonize oestrogenic function.

Androgens also have a negative feedback effect upon the hypothalamic–pituitary axis (Figures 16.2 and 16.3), and will influence, among other things, the release of the gonadotrophins and prolactin.

Several synthetic androgens are available, and the duration of their activity is related to the nature of the synthetic ester. Androgens may be considered to have either primarily virilizing actions, or primarily anabolic actions. The virilizing effects include the development of the secondary sexual characteristics, including physical changes and the promotion of

| Drug | Trade name | Formulation | Distributor |
|---|---|---|---|
| Methyltestosterone | Orandrone | 5 mg tablets | Intervet UK Ltd |
| Methyltestosterone (+ ethinyl oestradiol) | Sesoral | 4 mg (+ 0.005 mg) tablets | Intervet UK Ltd |
| Testosterone phenylpropionate | Androject | 10 mg/ml for injection | Intervet UK Ltd |
| Testosterone propionate + testosterone phenylpropionate + testosterone isocaproate + testosterone decanoate | Durateston | 50 mg/ml for injection | Intervet UK Ltd |
| Mibolerone | # Cheque Drops | 1.0 mg/ml for oral admin. | Upjohn Company Ltd, USA |
| Mesterolone | † Pro-Viron | 25 mg tablets | Schering Health Care Ltd |

***Table 16.6:*** *Commercially available androgens that may be useful for the manipulation of reproduction in the dog and bitch.* # *Products not available in the UK.* † *Human medical preparations.*

libido and spermatogenesis. The anabolic effects include the stimulation of appetite, promotion of protein synthesis and muscle deposition, and the retention of certain elements including nitrogen, potassium, phosphorus and calcium.

## Adverse effects of androgens

### General

Androgen therapy may produce virilizing effects such as aggression, and their use may be contraindicated in dogs with existing behavioural problems. In prepubertal animals, premature epiphyseal growth plate closure may occur.

Androgens are contraindicated in nephrotic conditions, since the anabolic component causes both sodium and water retention. Hepatic dysfunction has also been reported following androgen administration.

### Specific adverse effects in the bitch

Prolonged androgen administration in bitches may result in clitoral hypertrophy, and very rarely the development of an os clitoris. Repeated or prolonged androgen therapy may result in a persistent vaginitis, and severe urogenital abnormalities may develop in female fetuses if androgens are administered to bitches during pregnancy.

### Specific adverse effects in the dog

High doses of androgen produces severe changes in spermatozoal morphology and fertility. This effect is probably mediated by the negative feedback mechanism described above.

## Clinical use of androgens

### Bitch

***Control of oestrus:*** Androgens may be administered to bitches in anoestrus to prevent a return to cyclical activity. The exact mechanism of action has not been determined, although it is likely to be similar to that of progestogens in preventing the increase in gonadotrophin secretion above basal values. Androgens are not useful for inhibiting oestrus in females with follicular activity (pro-oestrus or early oestrus), and administration must commence in late anoestrus, more than 30 days before the onset of pro-oestrus. When androgens are used in this manner, they do not mimic the luteal phase, and there is no subsequent anoestrus. Therefore cyclical activity returns rapidly after the cessation of treatment.

In the UK, the most common regimen is the administration of depot androgens during anoestrus either as a prolonged release implant, or depot injection of mixed testosterone esters. It is not uncommon for depot therapy to be supplemented by daily oral therapy. For example, mixed testosterone esters (25 mg/kg) may be given intramuscularly every 4–6 weeks, and methyltestosterone (0.25–0.5 mg/kg) is given orally daily.

In other countries including the USA, an orally active synthetic androgen, mibolerone, is available. This has been shown to be effective for the long-term prevention of oestrus in bitches. The compound has adverse effects typical of other androgens, including clitoral hypertrophy, vaginitis, and behavioural changes. In addition, anal gland abnormalities, obnoxious body odour and obesity have also been recorded.

Other agents that may be used to control oestrus include the progestogens.

***Pseudopregnancy***: Relatively high doses of androgens may be used to produce an inhibition of prolactin secretion in bitches with pseudopregnancy. This results in a rapid resolution of the clinical signs. Androgens may be more useful than either progestogens or oestrogens, since they do not have any adverse effects on the uterus. Oral methyltestosterone or parenteral testosterone esters are efficacious; however, an orally active licensed preparation of methyltestosterone (0.7 mg/kg/day) combined with ethinyl oestradiol (0.8 μg/kg/day) is available which produces a good clinical response.

Other treatment options for pseudopregnancy include the administration of progestogens, oestrogens alone, or specific prolactin antagonists such as bromocriptine and cabergoline.

***Mammary tumours:*** Androgens may be useful for the control of certain mammary tumours. Their action is related either to a negative feedback upon gonadotrophin release or to a direct local anti-oestrogenic effect.

Androgen therapy is empirical, and may not be efficacious, since many advanced tumours possess few steroid receptors and appear to have an autonomous growth pattern.

Other agents which may be used to control mammary tumours include the anti-oestrogens like tamoxifen.

***Anabolic therapy:*** Various androgens both as depot or repeated oral therapy may be used for their anabolic effect in aged or debilitated dogs. Parenteral therapy (mixed testosterone esters, 25 mg/kg) is usually given every 1 or 2 weeks, whilst oral therapy (methyl testosterone 0.25–0.5 mg/kg) is recommended daily.

### Dog

***Poor libido:*** There is no evidence in the dog that poor libido is caused by low circulating androgen concentrations. The condition is more likely to have a psychological background or to be the result of musculoskeletal pain during copulation. Although both oral and depot androgen administration have been advocated for the treatment of poor libido, androgens should be used with care since they produce significant changes in spermatozoal morphology and fertility via the negative feedback mechanism (Figure 16.3).

Other agents that may be useful in dogs with poor libido include human chorionic gonadotrophin, which produces an increase in endogenous testosterone concentration. This may be more physiological, although the efficacy of this regimen has not been proven.

***Poor semen quality:*** There are many causes of poor semen quality in the dog. Those related to inadequate endocrine support have not been fully elucidated in males of any species. It is not appropriate to administer androgens without a definitive diagnosis; nevertheless, androgen supplementation is commonly used, and this may produce disastrous results because the androgens suppress spermatogenesis via the negative feedback mechanism.

An analogue of dihydrotestosterone, mesterolone, is available in the UK for human use and is unusual in that it is not aromatized to oestradiol and therefore, at therapeutic doses, does not significantly suppress the release of pituitary gonadotrophins. It has been used in dogs (1.5 mg/kg/day) but its effects have not, as yet, been fully evaluated.

***Contraception:*** Long-term administration of high doses of androgens may be useful for reducing semen quality by a negative feedback suppression of spermatogenesis. The concurrent administration of progestogens appears to be more efficacious, although castration is the method of choice in dogs that are not required for breeding.

***Cryptorchidism:*** Cryptorchidism means hidden testicle and refers to the failure of one or both testes to descend into the scrotum. It is believed to be a sex-limited, autosomal recessive trait with a definite breed predisposition. Cryptorchid dogs and their parents should not be used for breeding because of the hereditary nature of the disorder; however, this is not always practical for the dog breeder. Castration of affected animals should be advised because of the increased risk of testicular neoplasia and testicular torsion, and to prevent breeding. Medical therapy with androgens has not been shown to be efficacious and is unethical.

***Feminization:*** Testicular tumours may be endocrinologically active and produce oestrogen. This may produce a variety of clinical signs including feminization. The feminizing effects may be suppressed by the anti-oestrogenic properties of various androgens, although the treatment of choice is castration. Mixed testosterone esters have been used to reverse the oestrogenic effects after castration, although the clinical signs should subside naturally provided that there are no metastases.

***Anabolic therapy:*** Various androgens, as depot or repeated oral therapy, may be used for their anabolic effect in aged or debilitated dogs. Parenteral therapy is usually given every 1 or 2 weeks, whilst oral therapy is recommended daily.

## GONADOTROPHINS

The gonadotrophins FSH and LH are secreted by the anterior pituitary gland. In the bitch, FSH stimulates initial follicular development and its surge release is associated with the process of ovulation. LH also stimulates follicular growth and is the trigger for ovulation. LH is one of the principal luteotrophic agents in the bitch (Figure 16.2) and is responsible for prolonging the lifespan of the corpora lutea in the late luteal phase of pregnancy and non-pregnancy. In the dog FSH stimulates spermatogenesis indirectly by an action upon the Sertoli cells, and LH stimulates the Leydig cells to produce testosterone (Figure 16.3).

Neither FSH nor LH is available for use in the dog; however, equine chorionic gonadotrophin (eCG) and human chorionic gonadotrophin (hCG) are commercially available (Table 16.7). eCG is produced in the mare during pregnancy and is mainly FSH-like in activity, but it does have some LH activity. hCG is extracted from the urine of pregnant women and is primarily LH like in effect.

| Drug | Trade name | Formulation | Distributor |
|---|---|---|---|
| Serum gonadotrophin (PMSG) | PMSG-Intervet | 5000 IU vials + solvent | Intervet UK Ltd |
| | Folligon | 1000 IU vials + solvent | Intervet UK Ltd |
| | * Fostim | 6000 IU vials + solvent | Pharmacia & Upjohn Ltd |
| Chorionic gonadotrophin (hCG) | Chorulon | 1500 IU vials + solvent | Intervet UK Ltd |

***Table 16.7:*** *Commercially available gonadotrophins that may be useful for the manipulation of reproduction in the dog and bitch.*
** Veterinary products available in the UK but not licensed for use in this species.*

## Adverse effects of gonadotrophins

### General

There is a risk of inducing anaphylactoid reactions and antibody formation following the injection of these protein preparations.

### Specific adverse effects in the bitch

The administration of a single dose of either gonadotrophin to anoestrous bitches is unlikely to have any effect. However, eCG if given to oestrous bitches may cause luteinization of follicles and interfere with ovulation. Hyperstimulation of the ovary by the administration of high doses or the use of repeated therapy may result in non-ovulating follicles and persistent oestrous behaviour. A persistent elevation in plasma oestrogen concentration may lead to adverse effects typical of oestrogen toxicity.

### Specific adverse effects in the dog

hCG administration causes an endogenous rise of plasma testosterone. This effect is short lived, however in a small number of cases temporary changes in temperament may occur.

## Clinical use of gonadotrophins

### Bitch

***Induction of oestrus:*** When repeated doses of eCG (20 IU/kg for 5 days) are given to anoestrous bitches, follicular growth and the production of oestrogen are stimulated. Ovulation may occur spontaneously, following an endogenous surge of LH, or may be induced using hCG (25 IU/kg on day 5). These regimens are more effective in late anoestrous bitches compared with those in early anoestrus. Care must be observed when using eCG and hCG combinations, since endogenous hyperoestrogenism may occur and result in inhibition of implantation, bone marrow suppression and death. Low doses of gonadotrophins are suggested to produce more physiological plasma oestrogen profiles.

A relatively low plasma progesterone concentration or short luteal phase following gonadotrophin-induced oestrus is common. This effect might be mediated by the high endogenous or exogenous oestrogen concentrations inhibiting release of the luteotrophic gonadotrophin LH.

Other agents which may be useful for the induction of oestrus include low dose oestrogens, or prolactin antagonists such as bromocriptine and cabergoline.

***Delayed puberty:*** Some bitches, especially those of the larger breeds, do not reach the first oestrus at the anticipated age. Puberty may normally occur as late as 2.5 years, and therefore treatment is not warranted until after this age. Therapy to induce oestrus may then be attempted using an eCG / hCG regimen.

***Hastening of ovulation:*** Bitches that repeatedly fail to conceive are sometimes given hCG at the time of mating on the assumption that ovulation has not occurred or that early development of the corpora lutea is inadequate. There is no evidence that this is the case, and such 'blind' therapy cannot be recommended. However, some bitches may have prolonged pro-oestrus or oestrus and the administration of hCG (25 IU/kg) may possibly hasten ovulation. In these cases, hCG is administered when more than 90% of vaginal epithelial cells are anuclear, or when a slight rise in plasma progesterone concentration has been detected. Premature administration of hCG may result in follicular luteinization and failure of ovulation. GnRH may be used in a similar manner to hasten ovulation.

***Identification of ovarian tissue:*** It may be difficult to determine clinically whether a bitch has been ovariohysterectomized (or ovariectomized - which is common in Europe). The administration of hCG (25 IU/kg) to bitches with ovaries results in an increase in plasma oestrogen concentration; in the absence of ovaries, this does not occur. This method may therefore be clinically useful for detecting the presence of ovaries, although it has not been fully investigated. Similarly, the administration of GnRH may be clinically useful for the identification of spayed bitches.

### Dog

***Diagnosis of testicular tissue:*** In male dogs with no scrotal testes, the presence of testicular tissue can be confirmed by performing an hCG stimulation test. Plasma testosterone concentration is measured before and 60 minutes after the i.v. administration of 50 IU/kg hCG; a significant rise in testosterone concentration is diagnostic of testicular tissue.

Dogs with functional testicular tissue, whether intra-abdominal or extra-abdominal generally have a higher resting plasma testosterone concentration than castrated dogs. However, there are large fluctuations in plasma testosterone throughout the day and therefore the hCG stimulation test is the preferred method of diagnosis.

***Cryptorchidism:*** Cryptorchid dogs do not respond to the administration of gonadotrophins. The condition is likely to be inherited (see above) and medical treatment is unethical; these dogs should be bilaterally castrated to remove them from the breeding programme and to prevent the development of testicular tumours and the risk of testicular torsion.

***Hypogonadism:*** In some dogs, the abnormal development of the germinal epithelium causes the formation of degenerate spermatocytes resulting in oligospermia and azoospermia. The condition may be congenital and possibly hereditary. Treatment with gonadotrophins is not successful.

***Poor libido:*** A transient increase in libido may follow the administration of hCG to dogs. This effect is the result of increased endogenous testosterone concentrations.

## GONADOTROPHIN RELEASING HORMONE AGONISTS AND ANTAGONISTS

Gonadotrophin releasing hormone (GnRH) controls gonadotrophin synthesis via a neuroendocrine mechanism. GnRH is released by hypothalamic nerve endings into the portal capillaries and is delivered to the anterior pituitary gland by the hypophyseal portal veins. The GnRH neuronal network is extremely complex and is influenced by several different areas of the brain. Both the frequency and the amplitude of GnRH release are important features of its function in stimulating the release of FSH and LH.

Several synthetic GnRH agonists are available (Table 16.8) although none is licensed for use in the dog. The administration of these agents causes an increase in the production of FSH and LH (Figures 16.2 and 16.3), and in general it is the action of these gonadotrophins which produces any clinical effect.

An alternate method of use of GnRH agonists is their repeated administration resulting in a down-regulation of the receptors in the pituitary, preventing the release of FSH and LH.

GnRH antagonists produce a reduction in circulating gonadotrophins within a few hours. The most commonly investigated agent, detirelix, is not available in the UK.

### Adverse effects of GnRH agonists and antagonists

#### General

No general adverse effects have been reported with these agents in dogs.

#### Specific adverse effects in the bitch

Repeated administration of both GnRH agonists and antagonists may result in absence of oestrus.

#### Specific adverse effects in the dog

Repeated administration of both GnRH agonists and antagonists may cause a reduction in libido, and long-term administration may cause reduced spermatozoal morphology and fertility.

### Clinical use of GnRH agonists and antagonists

#### Bitch

***Identification of ovarian tissue:*** The i.v. administration of GnRH (0.01 μg/kg) to intact bitches produces an increase in LH and subsequent rise in plasma

| Drug | Trade name | Formulation | Distributor |
|---|---|---|---|
| **GnRH AGONISTS** | | | |
| Buserelin | * Receptal | 0.004 mg/ml for injection | Hoechst Rousell Vet Ltd |
| Fertirelin | * Ovalyse | 50 μg/ml for injection | Pharmacia & Upjohn Ltd |
| Gonadorelin | * Fertagyl | 0.1 mg/ml for injection | Intervet UK Ltd |
| Buserelin | † Suprefact | 1.0 mg/ml for injection | Hoechst UK Ltd |
| Goserelin | † Zoladex | 3.6 mg depot implant for 1 month therapy | Zeneca Pharmaceuticals |
| | † Zoladex LA | 10.8 mg depot implant for 3 months therapy | Zeneca Pharmaceuticals |
| Nafarelin | † Synarel | nasal spray - no suitable veterinary formulation | Searle DG and Co. Ltd |
| **GnRH ANTAGONISTS** | | | |
| Detirelix | Not available | Not available | |

***Table 16.8:*** *Commercially available GnRH agonists and antagonists that may be useful for the manipulation of reproduction in the dog and bitch.*
** Veterinary products available in the UK but not licensed for use in this species. † Human medical preparations.*

oestrogen concentration. This test may be useful for the detection of bitches that have been ovariectomized or ovariohysterectomized where no rise in oestrogen will occur. An alternative method is to administer hCG and monitor plasma oestrogen concentration, which increases in bitches with ovaries but not those that have been ovariectomized.

***Hastening of ovulation:*** Single i.v. doses of GnRH (0.01 μg/kg) may be useful to hasten ovulation in bitches in oestrus. The use of hCG, however, may be more efficacious, although neither drug has been adequately evaluated for this purpose.

***Control of oestrus:*** Analogues of GnRH, such as nafarelin, may be used to prevent reproductive cyclicity, since they cause initial stimulation and then receptor down-regulation. Prevention of oestrus may be achieved using s.c. implanted sialastic devices providing a sustained release of GnRH for up to 1 year; however, suitable preparations of these agents are not commercially available.

GnRH antagonists may be used to prevent cyclical activity when administered during pro-oestrus, because they produce a reduction in circulating gonadotrophins within a few hours. Bitches therefore rapidly return to anoestrus.

Other agents that may be used to control oestrus include progestogens and androgens.

***Unwanted mating:*** The down-regulation effects of continuous GnRH administration may be used in attempts to withdraw gonadotrophin support of the corpus luteum. Similarly, the daily administration of GnRH antagonists successfully suppresses luteal function. A single dose of detirelix produces resorption or abortion depending upon the stage of pregnancy. Administration very early in the pregnancy is however not efficacious even when the drug dose is increased.

Other agents that may be used to induce resorption or abortion include oestrogens, prostaglandins, prolactin antagonists and progesterone antagonists.

### Dog

***Contraception:*** Both GnRH agonists and antagonists may be useful for the suppression of LH and FSH, causing reduced libido and infertility. In particular, depot preparations appear to be useful as reversible methods of contraception. Since these agents are not widely commercially available, other products such as progestogens, androgens or progestogen–androgen combinations may be considered, although castration is the method of choice in the animal that is not required for breeding.

***Prostatic disease:*** Repeated or depot administration of GnRH agonists may be useful in dogs which have benign prostatic hyperplasia and prostatic neoplasia, although no clinical trials with these agents have been reported.

## PROSTAGLANDINS

The reproductive prostaglandins are synthesized in the endometrium and are luteolytic and spasmogenic in nature. In the bitch, prostaglandin release is involved in the termination of the luteal phase and the initiation of parturition.

There are no naturally occurring or synthetic prostaglandin analogues licensed for use in the bitch. The use of the naturally occurring prostaglandin $F_2\alpha$, dinoprost, and several synthetic prostaglandin analogues, especially cloprostenol and luprostiol, have, however, been widely described (Table 16.9).

When administered to bitches, they cause lysis of the corpora lutea and the termination of the pregnant and non-pregnant luteal phase. However, bitch corpora lutea are not very responsive to the action of prostaglandin, and repeated therapy (often twice daily) may be necessary. Prostaglandin administration will also cause uterine contractions via its direct spasmogenic effect.

### Adverse effects of prostaglandins

#### General

Prostaglandin administration may be followed by restlessness, hypersalivation, vomiting, abdominal pain, pyrexia, tachycardia, ataxia and diarrhoea; high doses may be lethal. These effects may develop quickly after administration and usually persist for up to 60 minutes.

| Drug | Trade name | Formulation | Distributor |
|---|---|---|---|
| Cloprostenol | * Estrumate | 265 μg/ml for injection | Mallickrodt Veterinary Ltd |
| | * Planate | 87.5 μg/ml for injection | Mallickrodt Veterinary Ltd |
| Dinoprost | * Lutalyse | 5 mg/ml for injection | Pharmacia & Upjohn Ltd |
| Tiaprost | * Iliren | 0.15 mg/ml for injection | Hoechst UK Ltd |
| Luprostiol | * Prosolvin | 7.5 mg/ml for injection | Intervet UK Ltd |

***Table 16.9:*** *Commercially available prostaglandins that may be useful for the manipulation of reproduction in the bitch.*
** Veterinary products available in the UK but not licensed for use in this species.*

**Specific adverse effects in the bitch**
Prostaglandin administration to pregnant bitches may cause resorption or abortion. Cases of closed-cervix pyometra given prostaglandin may develop uterine rupture.

**Specific adverse effects in the dog**
Prostaglandins are not used in male dogs.

## Clinical use of prostaglandins

### Bitch

***Termination of pregnancy:*** Low doses of dinoprost (150 µg/kg) or cloprostenol (0.025 mg/kg) given daily, or twice daily, for several days can be used to produce luteal regression. Repeated low doses cause lysis of the corpora lutea and induce abortion, especially when given later than 23 days after ovulation. Earlier treatment may not be efficacious, since developing corpora lutea are more resistant to the effects of prostaglandin. Recent studies have, however, shown that daily treatment commencing 5 days after the onset of metoestrus (dioestrus) using high doses (250 µg/kg) of dinoprost may be efficacious, although adverse effects are common.

Cloprostenol can be used to produce abortion 23 days after ovulation when administered as a prolonged release intravaginal device, however, such a product is not commercially available.

A better therapeutic option is to use a combination of prostaglandin (dinoprost, 5.0 µg/kg every other day for 10 days) and a prolactin antagonist such as cabergoline (5.0 µg/kg/day for 10 days), although the latter agents may also be used alone.

***Treatment of pyometra:*** Cystic endometrial hyperplasia and pyometra occur during the luteal phase of the oestrous cycle. Progesterone-induced cystic endometrial hyperplasia usually precedes the development of pyometra which is associated with bacterial invasion of the uterus during oestrus. These cases may be described as either open-cervix or closed-cervix pyometras. In the former, there is usually a copious vulval discharge, and following its diagnosis, prostaglandins may be used to induce uterine emptying (the result of their spasmogenic activity) and to remove the stimulatory effects of progesterone (the result of their luteolytic action). Dinoprost administered at low doses (150 µg/kg) twice daily for 5 days, combined with appropriate antibiotic and fluid therapy may be used as a successful treatment. Following treatment approximately 20% of bitches return to fertility. The long-term complications include recurrence of pyometra, anoestrus, failure to conceive and abortion.

Other agents that may be used to treat cases of pyometra include progesterone receptor antagonists such as RU486 and RU46534.

***Post-partum metritis:*** Post-partum metritis in the bitch may be treated by using prostaglandins to induce uterine evacuation. Antimicrobial preparations are mandatory; however, in these cases there is no underlying hormonal component (plasma progesterone concentration is low) and therefore a rapid resolution should be expected. A regimen similar to that used for the treatment of pyometra may be employed. Other treatment options include the administration of oxytocin or ergometrine.

***Induction of oestrus:*** Whilst prostaglandins may be used to induce luteolysis, the luteal phase is followed by a variable period of anoestrus. Therefore shortening of the luteal phase does not produce a rapid return to oestrus. Methods of oestrus induction include the use of gonadotrophins, oestrogens and prolactin antagonists.

### Dog
There are no therapeutic roles for reproductive prostaglandins in the dog.

## OXYTOCIN AND ERGOT PREPARATIONS

Oxytocin is synthesized within the supraoptic nucleus and is transported axonally to storage areas in the posterior pituitary gland. Oxytocin release is stimulated by mechanical-neural reflexes. Oxytocin receptors develop on the myoepithelial cells of the mammary gland shortly before parturition. Oxytocin release stimulated by sucking results in contraction of the myoepithelial cells and causes active milk ejection. Uterine oxytocin receptors also increase in number close to parturition, and oxytocin release stimulates contraction of the oestrogen-primed uterus. Oxytocin

| Drug | Trade name | Formulation | Distributor |
|---|---|---|---|
| Oxytocin | Oxytocin-S Injection | 10 IU/ml for injection | Intervet UK Ltd |
| | Oxytocin Leo | 10 IU/ml for injection | Leo Laboratories Ltd |
| Pituitary extract | Hyposton | 10 IU/ml for injection | Pharmacia & Upjohn Ltd |
| Pituitary extract (synthetic) | Pituitary Extract (Synthetic) | 10 IU/ml for injection | Animalcare Ltd |
| Ergometrine (+ oxytocin) | † Syntometrine | 500 µg/ml (+ 5 IU/ml) for injection | Novartis Pharmaceuticals UK Ltd |

***Table 16.10:*** *Commercially available oxytocin and ergot preparations that may be useful for the manipulation of reproduction in the bitch.*
*† Human medical preparations.*

is also released in some species at coitus and is involved in the transportation of gametes within the tubular genital tract.

Several formulations of oxytocin and posterior pituitary extract are available commercially, and can be used to stimulate the smooth muscle of the mammary and genital tract (Table 16.10).

Ergotamines are alkaloid extracts of a rye fungus. They are ecbolic agents like oxytocin, however, they produce a prolonged spasm of uterine muscle. Relaxation occurs only after 1 or 2 hours when the uterus then contracts rhythmically in a similar manner to that produced by oxytocin. There are no preparations licensed for use in the dog, although parenteral preparations of ergometrine maleate combined with oxytocin are available.

## Adverse effects of oxytocin

### General

There is a low incidence of skin sloughing or abscess formation recorded by some manufacturers following the subcutaneous administration of oxytocin. The agent is therefore best given by the i.m. route.

Ergot alkaloids may produce emesis and slight stimulation of the central nervous system.

### Specific adverse effects in the bitch

Oxytocin administration to a bitch in late pregnancy may initiate premature parturition.

### Specific adverse effects in the dog

Oxytocin and ergot preparations have no clinical indications in the dog.

## Clinical use of oxytocin

### Bitch

***Uterine inertia:*** Cases of primary uterine inertia or secondary uterine inertia (following correction of the dystocia) may respond to parenteral oxytocin administration (0.5 IU/kg). Oxytocin has a short half-life and therefore repeated administration (every 15–30 minutes) may be necessary to maintain a clinical effect. The use of a continuous low dose intravenous infusion does not appear to have been investigated in bitches, although it is a common treatment regimen in humans.

Oxytocin is contraindicated in obstructive dystocia, since uterine rupture may result.

Ergot preparations may also be used in cases of inertia; however, they produce prolonged uterine contractions and are less suitable.

***Placental retention or fetal retention:*** Retention of a puppy or a placenta is rare in the bitch. A presumptive diagnosis may be made by the persistence of a green-coloured vulval discharge after parturition. Accurate diagnosis requires ultrasound examination. Parenteral administration of oxytocin (0.5 IU/kg) is usually effective at producing uterine contraction and expulsion of the puppy and/or placenta. The response to oxytocin rapidly decreases after parturition, and in cases that are presented late, a hysterotomy may be necessary. Ergot preparations may be used in these cases but are less efficacious because of the prolonged contractions that they produce.

***Post-partum haemorrhage:*** Physical damage to the uterine wall may occur during parturition. Bleeding may therefore continue after parturition but usually rapidly decreases as the uterus involutes. Persistent haemorrhage may be controlled by oxytocin administration. Ergometrine may be a more useful agent in these cases because of the initial prolonged uterine contractions. Excessive haemorrhage, although very rare, may necessitate a hysterotomy to identify and correct the cause.

***Post-partum metritis:*** Uterine evacuation may be induced by oxytocin administration in cases of post-partum metritis. In these cases, appropriate antibiotic therapy is mandatory. Other treatment options include the administration of prostaglandins or ergometrine.

***Subinvolution of placental sites:*** Partial uterine non-involution may occur in the bitch after parturition. The aetiology of the condition is not known, but cases are characterized by an intermittent haemorrhagic discharge from the vulva for 3–6 months after parturition. Most cases regress spontaneously after the bitch has returned to oestrus. Oxytocin has little effect once the clinical signs are established, but some anecdotal reports suggest that treatment at the time of parturition may prevent the condition from developing.

***Agalactia:*** Lack of milk production is rare in the bitch. However, an absence of milk letdown is relatively common. The former condition has no treatment whilst the latter can be treated by the administration of oxytocin. Oxytocin does not increase milk production.

### Dog

There are no clinical applications for oxytocin therapy in male dogs.

# PROLACTIN ANTAGONISTS

Prolactin is the primary luteotrophic factor of the pregnant and non-pregnant luteal phase. The corpora lutea initially appear to be autonomous and prolactin concentrations increase from day 20 onwards, and are then maintained for the duration of the luteal phase.

The use of prolactin antagonists (Table 16.11) causes a rapid decrease in prolactin concentration

| Drug | Trade name | Formulation | Distributor |
|---|---|---|---|
| Cabergoline | Galastop<br>† Dostinex | 50 μg/ml for oral admin.<br>0.5 mg tablets | Boehringer Ingelheim Ltd<br>Pharmacia |
| Bromocriptine | † Parlodel | 1 mg and 2.5 mg tablets | Novartis Pharmaceuticals UK Ltd |

***Table 16.11:*** *Commercially available prolactin antagonists that may be useful for the manipulation of reproduction in the bitch. † Human medical preparations.*

(Figure 16.2) and a subsequent decline in plasma progesterone concentration. Prolonged prolactin suppression results in a termination of the luteal phase, and the bitch enters anoestrus. Continual administration of prolactin antagonists during anoestrus causes a return to oestrus sooner than anticipated. The mechanism of this oestrus-inducing action is unknown.

Bromocriptine and cabergoline are ergot derivatives which inhibit prolactin release by direct stimulation of dopaminergic receptors of prolactin-releasing cells in the anterior pituitary gland. Cabergoline appears to be a more specific agent that has fewer side effects attributable to dopaminergic stimulation of the central nervous system. Neither bromocriptine nor cabergoline is licensed for use in dogs in the UK. Cabergoline is licensed for this species in Europe, and both agents are available as human preparations in the UK.

## Adverse effects of prolactin antagonists

### General

Dopaminergic stimulation may result in nausea and vomiting and this is the most common adverse effect reported following bromocriptine administration. Lethargy and occasional constipation may also be noticed. These effects are uncommon following the administration of cabergoline. The adverse effects of bromocriptine can be reduced by using the minimal effective dose and mixing the drug with food. Specific anti-emetics such as metoclopramide may prevent vomiting, but whilst they are clinically useful, their administration does not make pharmacological sense because they also work via dopamine receptors.

### Specific adverse effects in the bitch

Administration of prolactin antagonists during pregnancy may cause abortion.

Repeated administration will result in a shortening of anoestrus and a return to oestrus.

### Specific adverse effects in the dog

There are no clinical applications of prolactin antagonists in male dogs.

## Clinical use of prolactin antagonists

### Bitch

***Pseudopregnancy:*** Suppression of prolactin causes a rapid resolution of the clinical signs of pseudopregnancy.

Bromocriptine (20 μg/kg/day) has been used for this purpose for some time; however, its effect of causing vomiting is a common reason for owner non-compliance and cessation of treatment. Vomiting can be reduced by starting with a very low dose and increasing this gradually over several days until a clinical effect occurs. Mixing the drug with food, or the prior administration of metoclopramide may also be useful as discussed above.

Cabergoline (5.0 μg/kg/day) is widely available in Europe and has a higher activity, longer duration of action, has better tolerance, and produces less vomiting than bromocriptine. The human medical preparation of cabergoline available in the UK is expensive.

Other agents that may be used for the treatment of pseudopregnancy include progestogens, androgens, oestrogens and androgen/oestrogen combinations.

***Termination of pregnancy:*** Repeated administration of the prolactin antagonists produces a reduction in plasma progesterone concentration and resorption or abortion. Both agents are more effective when given in the second half of pregnancy, since the role of prolactin is more important at this time. Efficacy can be increased, and treatment may be given earlier, when prolactin antagonists and prostaglandins are given simultaneously (see prostaglandins above). One regimen is to use a combination of prostaglandin (dinoprost, 5.0 μg/kg) every other day for 10 days and a prolactin antagonist such as cabergoline (5.0 μg/kg) daily for 10 days.

Other agents that may be used to terminate pregnancy include prostaglandins or progesterone antagonists.

***Induction of oestrus:*** Apparently normal and fertile oestrous periods may be induced in anoestrous bitches following continual administration of bromocriptine (20 μg/kg/day) or cabergoline (5.0 μg/kg/day). The time to the onset of oestrus seems to relate to the stage of anoestrus; bitches in late anoestrus respond more quickly than those in early anoestrus. Treatment starting in the luteal phase may also be effective but the treatment period is longer.

The induced oestrus appears to be physiological, and whilst the mechanism of action is uncertain it may be the result of inhibition of remnant progesterone production by the corpora lutea.

Other agents that may be used to induce oestrus

| Drug | Trade name | Formulation | Distributor |
|---|---|---|---|
| Mifepristone (RU486) | † Mifegyne Tablets | 200 mg tablets | Roussel Laboratories Ltd |
| Aglepristone (RU46534) | # Alzin | 30 mg/ml for injection | Hoechst Roussel Vet, France |

***Table 16.12:*** *Commercially available progesterone antagonists that may be useful for the manipulation of reproduction in the dog and bitch.*
*# Products not available in the UK. † Human medical preparations.*

include low dose oestrogens or gonadotrophin combinations.

### Dog

There are no therapeutic roles for prolactin antagonists in the male dog.

## PROGESTERONE ANTAGONISTS

Specific progesterone receptor antagonists (Table 16.12) have antiprogestogenic effects in several species.

The availability of these agents varies from one country to another, and mifepristone (RU486) appears to be most widely obtainable. This product is available as an orally active product, and sale is limited in some countries because of the potential human misuse. A similar progesterone antagonist (RU46534) has recently been marketed for use in bitches in France, and is available as an injectable solution.

Administration of these agents produces no initial change in plasma progesterone concentration; however, progesterone action is blocked.

### Adverse effects of progesterone antagonists

#### General

Adverse effects following the administration of these products have not been reported.

#### Specific adverse effects in the bitch

Administration of progesterone antagonists to pregnant bitches will result in resorption, abortion or premature parturition depending upon the time of administration.

#### Specific adverse effects in the dog

There are no therapeutic uses of progesterone antagonists in male dogs.

### Clinical use of progesterone antagonists

#### Bitch

***Pregnancy termination:*** The administration of mifepristone to pregnant bitches at 28 days after ovulation produces fetal death and resorption within 5 days of treatment. Earlier administration requires a longer treatment period. In later pregnancy, mifepristone administration results in abortion, or premature parturition. RU46534 (4.0 mg/kg) may be used to terminate pregnancy at any stage; it may even be used to prevent conception by administration immediately after ovulation. Mifepristone administered orally and RU46534 administered by injection are not associated with any adverse clinical effects.

***Pyometra:*** RU46534 has been investigated in bitches with pyometra. Administration on days 1, 3, 5, 8 and 16 after presentation was successful for the treatment of both open- and closed-cervix pyometra in the majority of bitches. The regimen produced no adverse effects, and emptying of uterine fluid occurred quickly producing resolution of the clinical signs. Some of the bitches returned to normal fertility after treatment.

#### Dog

There are no clinical indications for the use of progesterone antagonists in male dogs.

## SYNTHETIC ANTI-OESTROGENS

The pharmacology of the anti-oestrogens, clomiphene citrate and tamoxifen, may superficially appear confusing. These agents (Table 16.13) are antagonists at the hypothalamic receptor level; they displace oestrogen from the receptor and lead to a decreased negative feedback of the oestrogen. As a result, gonadotrophin

| Drug | Trade name | Formulation | Distributor |
|---|---|---|---|
| Tamoxifen | † Tamofen | 10, 20 and 40 mg tablets | Pharmacia |
| | † Nolvadex | 20 mg and 40 mg tablets | Zeneca Pharmaceuticals |
| Clomiphene citrate | † Serophene | 50 mg tablets | Serono Laboratories UK Ltd |
| | † Clomid | 50 mg tablets | Merrel Dow Ltd |

***Table 16.13:*** *Commercially available antioestrogens that may be useful for the manipulation of reproduction in the dog and bitch.*
*† Human medical preparations.*

release is increased, and in humans this may result in follicular growth and ovulation. Peripheral receptor antagonism also occurs and this may be useful in conditions that are stimulated by oestrogen.

These agents also have important action on the uterus; they may interfere with normal endometrial maturation, and therefore prevent implantation. Anti-oestrogens also inhibit progesterone production from the corpora lutea and therefore interfere with the luteal phase.

In males, the enhanced secretion of LH and FSH may stimulate spermatogenesis and Leydig cell function. However, clomiphene has significant intrinsic oestrogenic activity which may directly impair spermatogenesis. Tamoxifen has no such oestrogenic activity and may be more suitable in the male.

These agents are not licensed for use in the dog, and suitable studies into their actions have not been performed.

## Adverse effects of synthetic anti-oestrogens

### General

The anti-oestrogens have been reported to produce gastrointestinal disturbances in man.

### Specific adverse effects in the bitch

A high incidence of pathological changes of the bitch's reproductive tract are induced by tamoxifen, including ovarian cysts and endometritis. This drug cannot be recommended for use in animals required for future breeding.

Tamoxifen may produce some oestrogenic adverse effects including vulval swelling and discharge and attractiveness to male dogs. These effects may be mediated by an oestrogenic metabolite of tamoxifen.

### Specific adverse effects in the dog

No male-specific abnormalities have been reported.

## Clinical use of synthetic anti-oestrogens

### Bitch

***Termination of pregnancy:*** Tamoxifen has been shown to prevent or terminate pregnancy when given at relatively high doses twice daily during pro-oestrus, oestrus, or early metoestrus (dioestrus). The mode of action during pro-oestrus and oestrus may relate to changes in the rate of zygote transport, or there may be a direct effect upon the uterus preventing implantation. In the luteal phase the action may be related to the effect upon progesterone production, although this has not been evaluated.

Other agents that may be used for the termination of pregnancy include oestrogens, prostaglandins, prolactin antagonists and combinations of these. The majority of these regimens are more suitable than tamoxifen because of its high incidence of pathological changes induced in the reproductive tract.

***Mammary neoplasia:*** There is little evidence of clinical response of canine mammary tumours to therapy with anti-oestrogens. However, tamoxifen is used by some clinicians, and anecdotally may control progression of the tumour. The high incidence of oestrogenic adverse effects (see above) results in therapy being withdrawn in a large number of cases.

### Dog

***Poor semen quality:*** There are many causes of poor semen quality in dogs, the majority of which have not been adequately investigated. In humans, both clomiphene and tamoxifen have been reported to improve semen quality in cases of oligozoospermia. The intrinsic oestrogenic property of clomiphene (which may directly impair spermatogenesis) is the reason why tamoxifen is more widely used. Neither agent has, however, been investigated for use in the dog. Other products which may be used in these cases include synthetic androgens such as mesterolone.

# SYNTHETIC ANTI-ANDROGENS

Anti-androgens (Table 16.14) may have several mechanisms of action. Cyproterone acetate is a progestogen which is principally anti-androgenic. Flutamide inhibits androgen uptake and/or nuclear binding of androgen. Finasteride is a specific 5-alpha reductase inhibitor which prevents the conversion of testosterone into dihydrotestosterone. Formestane is an aromatase inhibitor which inhibits the conversion of androgen to oestrogen in peripheral tissue. There has recently been considerable interest in finasteride in dogs since it has no intrinsic androgenic, oestrogenic or progestogenic properties and it has no affinity for the androgen receptor. However, whilst all of these anti-androgens are commercially available, none is licensed for use in the dog.

## Adverse effects of synthetic anti-androgens

### General

Adverse effects of these agents have not been reported in dogs, but in humans include nausea, gastrointestinal upsets and gynaecomastia.

### Specific adverse effects in the bitch

These agents have no clinical indication in the bitch.

### Specific adverse effects in the dog

Prolonged anti-androgen administration may result in a reduction in semen quality and poor fertility.

## Clinical use of synthetic anti-androgens

### Bitch

There are no clinical indications for the use of these products in the bitch.

| Drug | Trade name | Formulation | Distributor |
|---|---|---|---|
| **PROGESTOGENS** | | | |
| Cyproterone acetate | † Androcur | 50 mg tablets | Schering Health Care Ltd |
| | † Cyptostat | 50 mg and 100 mg tablets | Schering Health Care Ltd |
| see also Table 16.3 | | | |
| **INHIBITORS OF ANDROGEN UPTAKE** | | | |
| Flutamide | † Drogenil | 250 mg tablets | Schering-Plough Ltd |
| **REDUCTASE INHIBITORS** | | | |
| Finasteride | # Proscar | 5.0 mg tablets | MSD, Belgium |
| **AROMATASE INHIBITORS** | | | |
| Formestane | † Lentaron | 250 mg vial + solvent | Ciba Laboratories |

***Table 16.14:*** *Commercially available anti-androgens that may be useful for the manipulation of reproduction in the dog and bitch. # Products not available in the UK. † Human medical preparations.*

### Dog

***Benign prostatic hyperplasia:*** Dogs with prostatic hyperplasia may respond to the administration of anti-androgens by a reduction in prostatic size and disappearance of the clinical signs. The concern with the use of these products is that there may also be a reduction in fertility. However, flutamide has been shown to reduce prostatic size with no change in libido or sperm output. More recently, finasteride (0.3 mg/kg/day) has been shown to produce a dramatic decrease in prostatic size and a reduction in the secretion of prostatic fluid. However, total spermatozoal output did not change and fertility remained unaltered. Formestane has not been evaluated for use in dogs.

Other preparations that may be used to treat benign prostatic hyperplasia include progestogens and oestrogens.

***Prostatic neoplasia:*** Anti-androgens have not been reported for use in cases of prostatic neoplasia; however, flutamide, finasteride and formestane have all been shown to be useful for the control of clinical signs in man.

## CONCLUSIONS

Recent studies have improved our understanding of reproductive physiology in dogs and the mechanism of action of specific hormones. It is clear that, with the increased availability of specific pharmaceutical preparations, there is considerable potential for the manipulation of reproductive function, to prevent reproduction, to treat reproductive disease and possibly to increase reproductive performance.

## REFERENCES AND FURTHER READING

England GCW (1994) Hormonal manipulation of breeding in the bitch. *The Veterinary Annual* **34**, 189-200

England GCW (1994) Reproductive endocrinology in the dog and bitch. In practice. *Journal of Postgraduate Clinical Study* **16**, 275-279

England GCW (1996) Reproductive biology in the dog. *The Veterinary Annual* **36**, 187-201

England GCW and Allen WE (1991) The effect of the synthetic androgen mesterolone upon seminal characteristics of dogs. *Journal of Small Animal Practice* **32**, 271-274

CHAPTER SEVENTEEN

# Pharmacological Control of Reproduction in the Cat

*John P. Verstegen*

## INTRODUCTION

Specific knowledge of therapeutic agents which may be used in cats is limited, particularly in the field of reproduction. Many of the drug regimens used are extrapolated from the dog. The dramatic increase in our knowledge of reproductive biology in dogs has proved beneficial for cats too, for example in the use of prostaglandins or antiprolactin agents. The aim of this chapter is to review the main clinical problems encountered in cats and to refer to categories of drugs that may be used in treatment (Table 17.1).

## THE QUEEN

### Oestrus induction, infertility, superovulation

Induction of follicular growth, oestrous behaviour and ovulation can easily be achieved in queens. Reasons for oestrus induction are delayed puberty, prolonged anoestrus or for synchronization purposes. These conditions must be fully investigated before deciding to induce oestrus.

#### Delayed puberty

It is important not to diagnose this too early, since the onset of puberty may vary greatly from one animal to another and from one breed to another. It is not exceptional to have a pedigree longhair cat reaching puberty at 18 or 24 months of age. Thus a clear history and physical examination has to be performed before making this diagnosis and deciding on treatment.

#### Prolonged anoestrus

Many cases of prolonged anoestrus are either related to poor oestrus detection and management, or secondary to progestogen administration.

It is important to recognize that cats are highly responsive to oestrus induction treatment and overstimulation leading to superovulation or the production of many anovulatory follicles with a cystic appearance is not rare. This is particularly true in prepubertal or young adult animals.

Many treatments have been proposed to induce oestrus in cats and these are summarized in Table 17.2. This author's preferred treatment is: 100 IU pregnant mare serum gonadotrophin (PMSG; also called equine chorionic gonadotrophin, eCG) in a single bolus administered to anoestrous cats, followed 5–7 days later by an injection of hCG (50 IU). This treatment produces ovulation and pregnancy rates comparable to natural matings. However, it is important to note that repeated injections of exogenous gonadotrophins may lead to the production of antibodies against PMSG which then results in a decreased response to stimulation. It is highly advisable not to repeat this kind of treatment.

In bitches with prolonged anoestrus, induction of oestrus by the use of buserelin implants or antiprolactin drugs has been successful. The possibility of using these approaches in cats should be investigated. Recent data from our laboratory suggest that cabergoline, a dopamine agonist, is efficient at 5 μg/kg per day to induce a fertile oestrous in cats as in dogs. GnRH has been recommended at 1 μg/kg s.c. until signs of oestrous behaviour are noted or for a maximum of 10 days.

### Oestrus prevention and/or suppression

Progestogens are routinely used to control or suppress oestrus in the queen. The side-effects in cats are similar to those in the dog and include: uterine disease, increased body weight, mammary tumours, hypoadrenocorticism and diabetus mellitus.

In cats, oestrus prevention is commonly achieved using orally administered megoestrol acetate at a dose of 2.5 mg/cats once weekly or 5 mg/cat every 2 weeks for up to several months.

Oestrus suppression is only advisable if treatment commences a maximum of 1–3 days after the beginning of pro-oestrus. Experience has indeed demonstrated that successful mating can occur if treatment commences late. If it is essential to suppress an oestrus that has begun, injectable agents cannot be used. If treatment is begun too late, queens may still be mated, and become pregnant. Since the injectable progestogen acts for more than 3 months, the progesterone concentration will still be elevated at the expected time of parturition, and therefore kittening will not occur.

| Main hormone | Administration | Clinical use | Dosage | Analogues |
|---|---|---|---|---|
| **Male hormones** | | | | |
| Testosterone (T) | i.m., s.c., oral, implants | Infertility, hypogonadism, poor libido, skin diseases, testicular descent (?) | Highly variable: low dose stimulates, high dose inhibits | Testosterone phenyl propionate, methyl testosterone |
| **Female hormones** | | | | |
| Oestrogen | i.m., s.c., oral | No longer used | 10–100 μg/kg depending on the used drug | Oestradiol cypionate, diethylstilboestrol, oestrodiol, benzoate |
| Progesterone | i.m., s.c., oral, implants | Hypoluteism, contraception, satyriasis, hyperactivity, urination | | Medroxyprogesterone acetate, proligestone, delmadinone, megoestrol acetate |
| Prostaglandin | s.c. | Abortion, metritis, pyometra | Natural: 100–500 μg/kg Synthetic: 2–5 μg/kg | Dinoprost, cloprostenol |
| Oxytocin | i.v., s.c., i.m. | Myocontraction, uterine bleeding, placental retention | 0.5–1 IU/kg every 20–30 minutes | |
| Ergot derivatives | Oral, s.c. | Abortion, lactation inhibition | Variable | Bromocriptine 10–25 μg/kg, metergoline, cabergoline 5 μg/kg |
| **Pituitary hormones** | | | | |
| Prolactin | Not clinically available | Stimulation of lactation, control of fertility in both genders | Unknown | Dopamine antagonist: stimulates prolactin: metoclopramide (5 mg/kg ) Dopamine agonists: inhibit prolactin secretion: cabergoline, metergoline, bromocriptine |
| LH | i.v., s.c., i.m. | Oestrus induction, ovulation induction testicular descent | Unknown for natural product | hCG (50–100 IU Chorulon) |
| FSH | i.v., s.c., i.m. | Oestrus induction | Unknown for natural product | PMSG (100–500 IU Folligon) |
| **Hypothalamic/ brain hormones** | | | | |
| GnRH | i.v., i.m., s.c. | Oestrus induction, infertility, testicular descent, ovulation | | Gonadorelin, buserelin (0.5–5 μg) |
| Melatonin | Oral patches, i.m., s.c., implants | Oestrus control Skin diseases | NA | NA |

***Table 17.1:*** *Main hormones and analogues used in cats.*

| Protocols | Dosages | Results | Comments |
|---|---|---|---|
| PMSG alone | 100–1000 IU 5-7 days | Poor, large number of unovulated follicles | Cline *et al.*, 1980 |
| PMSG + hCG | 100–150 IU once day 1 + 50–100 IU once day 5 -7 | Good, similar to natural oestrus | Cline *et al.*, 1980; Donoghue *et al.*, 1997 |
| FSH + hCG | 2 mg/day 4-5 days + 50–250 IU | Good | Dresser *et al.*, 1987; Goodrowe *et al.*, 1988 |
| Prolactin inhibitors | Cabergoline 5 μg/kg per day up to oestrous behaviour maximum 15 days | Good in terms of oestrous behaviour. Not enough data in terms of pregnancy | Verstegen, unpublished |

***Table 17.2:*** *Described treatments to induce oestrus in cats.*

For suppression of oestrus, oral progestogens can be given: 2.5 mg/cat per day for 2 weeks. If the administration is then stopped, the cat usually returns to oestrus a few days or sometimes several weeks later. If the treatment has to be continued, 2.5 mg/cat every week or 5 mg/cat every other week is recommended.

## Ovulation induction, nymphomania, ovarian cysts

During oestrus, ovulation can be induced mechanically by vaginal stimulation. Exogenous hormones can also be used: GnRH (5–25 μg/cat) or hCG at 50–250 IU may be administered i.m. or i.v. to induce ovulation and therefore terminate oestrus. Ovulation occurs 24–36 hours after the injection which will then be followed by 30–45 days of pseudopregnancy. Care has to be taken to ensure that the queen is not mated in the few days following injection, otherwise pregnancy may result.

Nymphomania is a complex syndrome in queens and can be related to frequent overlapping follicular waves (often observed in Siamese cats), the presence of ovarian cysts (old animals or after oestrus induction and superovulation), or may be of central (pituitary or hypothalamic) origin. Ovulation induction using GnRH or hCG can temporarily block the symptoms when they are related to ovarian activity. Central origin nymphomania is a rare condition sometimes observed in sterilized animals (particularly Siamese cats) continuously showing oestrous behaviour. The exact cause of this condition is still unknown; however, it may sometimes be associated with pituitary tumours. When central in origin, the administration of progestogens may be useful.

## Abortion induction

Veterinary surgeons are less frequently requested to induce abortion in cats than in dogs. However, when it is necessary to induce an abortion, this can be achieved in the same way as for the bitch. Oestrogens can be used before implantation (before day 10 after mating). Oestrogen treatment induces congestion, oedema and alteration of the uterine milk, therefore preventing normal embryo migration and embryo implantation. In cats, fewer side-effects are observed after oestrogen treatment than in dogs. Bone marrow aplasia and uterine disease is rarely observed. Nevertheless, the use of oestrogen is considered inadvisable, as new treatments with fewer side-effects exist. Prostaglandins can be used. Doses of up to 500 μg/kg of natural analogues 2–3 times a day for 5 days may be given after pregnancy has been diagnosed. Side-effects are less severe than in dogs, and these generally increase with the dose administered but decrease with the repetition of treatment. Among the side-effects are: emesis and muscle jerking. No data exist concerning the use of synthetic analogues (e.g. cloprostenol) alone in this species.

Prolactin inhibitors have been demonstrated to be potent abortion induction agents in cats as in dogs. Recently, we have demonstrated that after day 25 of pregnancy, abortion can be induced in cats using three injections of cloprostenol (5 μg/kg) at 48-hour intervals associated with every day oral treatment

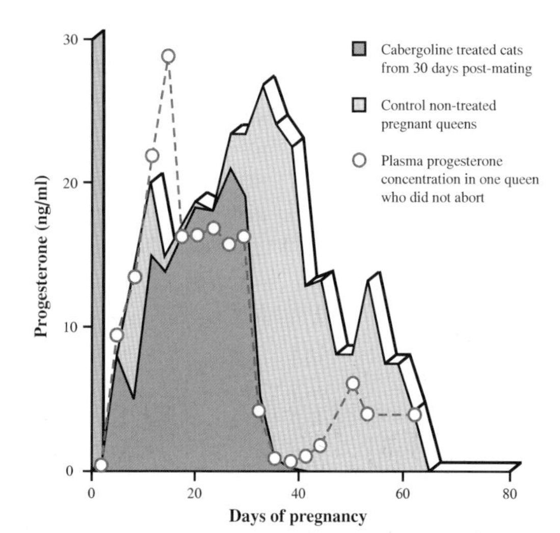


***Figure 17.1:*** *Progesterone evolution in queens during anti-prolactin and prostaglandin abortion induction treatment. Abortion is observed when progesterone plasma concentration decreases below 1 ng/ml (3 nmol/l).*

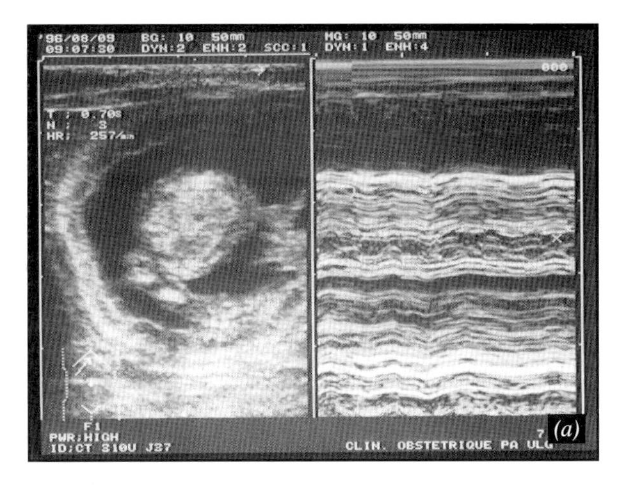


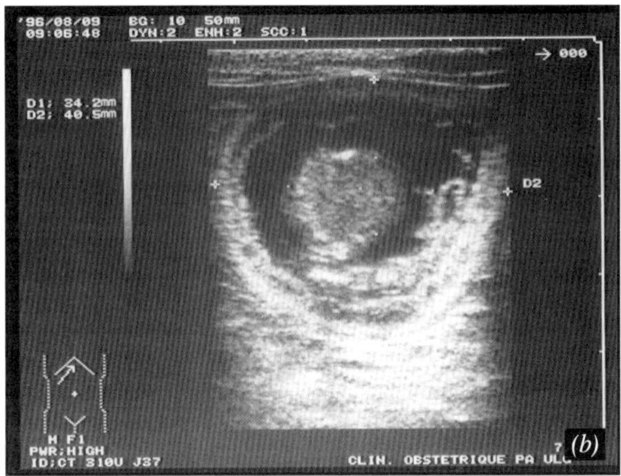


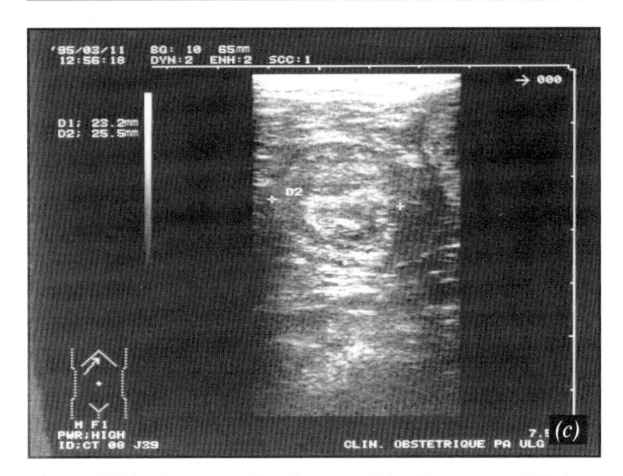


***Figure 17.2****: Sonographic pictures of (a, b) a normal 38 day post-mating pregnancy and (c) a uterus with fetal resorption induced by cabergoline associated with prostaglandin.*

with the prolactin inhibitor cabergoline (5 μg/kg once a day for 8 days). Abortion is observed along with a drop of progesterone below the threshold required to maintain pregnancy in this species (Figure 17.1). The main benefit of this treatment is that it is not associated with any side-effects. If abortion is induced before day 40, fetal resorption occurs with no expulsion of fetuses. If administered later, expulsion of perfectly normal kittens may occur (Figure 17.2). Table 17.3 summarizes the different abortion induction treatment available for cats.

## Pregnancy maintenance

As in dogs, there is no clear evidence that habitual abortion due to hypoluteism exists in cats. However, at around day 30–45 of pregnancy some significant changes occur in the function of the corpora lutea in cats. The pregnant corpus luteum differs from the pseudopregnant one, and it is possible that poor luteal support and transition may lead to an abortion. Oral progestogen (megoestrol acetate, 2.5 mg/cat every other day up to day 55 of pregnancy) may be used as a precaution, but should be restricted to cases in which true luteal insufficiency has clearly been diagnosed (see below). Progesterone during pregnancy may cause developmental defects (feminization of male kittens, cryptorchidsim) and parturition may be impaired (delayed parturition). The minimum progesterone requirement in cats during pregnancy is 3 nmol/l (1 ng/ml) (less than in dogs). The side-effects of high progestogen administration during pregnancy are the same as in dogs.

## Uterine inertia

Cases of primary or secondary uterine inertia may respond to administration of oxytocin. A dosage of 0.5–1 IU/kg may be given and repeated a maximum of 3 times at 20–30 minute intervals. The cause of the secondary inertia has to be carefully determined before administering oxytocin. Cases of obstructive dystocia are a clear contra-indication, since uterine rupture may result.

## Lactation: stimulation or inhibition

Stimulation of lactation can be obtained by the administration of dopamine antagonists such as metoclopramide (5 mg/kg for 3–5 days). They can be given once or twice daily to stimulate lactation after parturition and particularly after Caesarian operation. Dopamine antagonists stimulate prolactin release. Milk

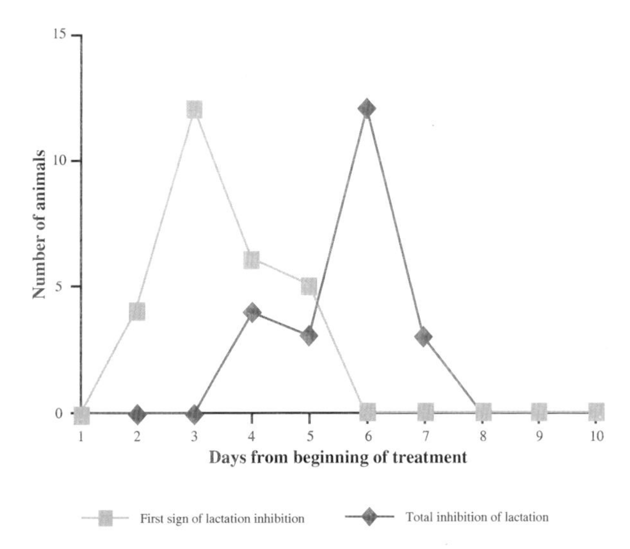


***Figure 17.3:*** *Effects of an anti-prolactinic drug on post-partum lactation and mammary gland hyperplasia. Queens were treated from day 10 post-kittening.*

| Protocols | Dosages | Results | Comments |
|---|---|---|---|
| Natural prostaglandin alone: dinoprost (Lutalyse, Dinolityc) | 250–500 µg/kg 1–3 times a day for 5–7 days from pregnancy diagnosis | Good. Ideally this treatment has to be used before fetal ossification to ensure resorption. Not to be used after day 50 | Some side-effects can be observed: prostration, emesis, salivation, muscle jerking |
| Synthetic prostaglandin alone | 2.5–5 µg/kg 1–3 times a day for 5–7 days from pregnancy diagnosis | Good. Ideally this treatment has to be used before fetal ossification to ensure resorption. Not to be used after day 50 | Fewer side-effects are observed than with natural PGF2α: prostration, emesis of short duration |
| Oestrogen: oestradiol benzoate or cypionate | 2–3 injections of 0.01–0.1 mg/kg every 2 days from mating. Not later than day 10 | Good but no longer advisable | Oestradiol benzoate is less toxic than cypionate. Side-effects are less frequent in cats than in dogs. Have to be given before a pregnancy diagnosis is made |
| Prolactin inhibitors: cabergoline, bromocriptine, metergoline | Cabergoline 5 µg/kg per day for 7 days from pregnancy diagnosis. Bromocriptine 25–50 µg/kg 3 times/day for 1 week | Contradictory results exist concerning the efficacy of this treatment before day 40. Seem to be efficacious after that period but unfortunately abortion is then associated with fetal expulsion. Not to be given after day 55 because will then induce premature parturition instead of abortion | Cabergoline seems to be more efficacious than the two other agents due to longer duration of action and a more important inhibitory effect on prolactin secretion. Side-effects are weak or absent with cabergoline compared to bromocriptine |
| Combined cabergoline (prolactin inhibitor) and cloprostenol (synthetic prostaglandin) | 5 µg/kg per day for 7–10 days orally + 2.5 µg/kg s.c. at days 1, 3 and 5. This treatment can be administered as soon as pregnancy diagnosis can be performed: approximately day 25 | Good. When the treatment is applied before fetal ossification (day 40–45) abortion occurs by resorption | No side-effects are observed. This is the author's recommended treatment: efficacious, early in pregnancy, no side-effects |

***Table 17.3:*** *Described treatments to induce abortion in cats.*

letdown but not production can be stimulated by the use of oxytocin post-partum.

Lactation can be inhibited by inhibiting prolactin release directly at the level of the pituitary by giving dopamine agonists (bromocriptine 10–25 µg/kg or cabergoline 5 µg/kg for 5 days; Figure 17.3) or indirectly by modifying dopamine release at the hypothalamic level by giving serotonin antagonists (metergoline). Progestogens, androgens and androgen–oestrogen combinations will also create a negative feedback and reduce prolactin secretion and lactation. However, they are non-specific agents, associated with effects not desired in the lactation inhibition. They are therefore no longer recommended, particularly since specific therapies devoid of any side-effects exist.

## Metritis/pyometra, placental retention

Natural prostaglandins such as dinoprost may be used to induce stimulation of the uterine smooth muscle and to cause luteolysis in order to treat cases of pyometra. The mechanism of action is to induce a functional arrest and a luteolysis of the corpora lutea, promote the opening of the cervix, and induce uterine contractions. Natural prostaglandins may also be used in cases of metritis and retained placenta, although here the action is solely uterine contraction. Doses of 20–50 µg/kg administered 3–5 times a day or

200–500 μg/kg once or twice a day for 5–7 days associated with the appropriate antibiotic, and fluid therapy is the most successful treatment. Fertility appears to remain normal after resolution of the condition. However, if fertility is not required, ovario-hysterectomy is the preferred option. No significant data are available concerning the use of progesterone antagonists such as RU486 or RU46534 in cats.

Placental retention can also be treated using the administration of oxytocin if diagnosed early after kittening, as uterine sensitivity to oxytocin rapidly decreases after parturition.

### Post-partum haemorrhage

Post-partum haemorrhage can be treated using oxytocin and/or ergot derivatives such as ergometrine which induce uterine contraction and vasoconstriction. However, the cause of the uterine haemorrhage has to be clearly ascertained, as such treatment is contraindicated in uterine rupture.

## THE TOM

Table 17.4 gives conditions, dosage and effects of several agents which may be used in tomcats.

### Urine spraying, aggression and excitability, overt sexual behaviour

Urination in tomcats is used as a communication tool to mark a territory. This behaviour is observed equally in neutered and entire male cats. Whilst accepted as normal in feral cats, in the house cat urination is considered unpleasant by the owner. It may be related to the presence of other cats or animals, or possibly to social, environmental or emotional changes. In entire males, castration decreases plasma testosterone concentrations and can diminish or abolish the behaviour and reduce the odour of the urine. The specific male odour is due both to testosterone and to retrograde ejaculation which is normal in the cat. Both are prevented by castration.

Administration of progestogens can be useful in decreasing this behaviour. Medroxyprogesterone acetate, megoestrol acetate, injectable delmadinone acetate, MAP (100 mg s.c.) or proligestone, act by their central hormonal feedback and central nervous relaxing effects, and may be effective. Repeated high dosages need to be given to gain significant effects; however, there may be decreasing effectiveness following repeated administration. Prolonged administration of such agents cannot be recommended because of their metabolic and hormonal side-effects at the high dosages used (diabetus mellitus, hypoadrenocorticism, increase in body weight). Sedative or psychoactive drugs (diazepam, carbamazepine) can be used to control these behavioural changes temporarily. Behavioural training should always be associated with medical treatment.

### Cryptorchidism

No real treatment exists to stimulate testicular descent in cats. Furthermore, such treatments are probably unethical because of the probable inherited nature of the condition. In other species, successful treatment has, however, been claimed using GnRH, LH or testosterone, although the results must be considered equivocal. Testosterone cannot be recommended due to its general androgenic and anabolic effects. Indeed, testosterone administration in prepubertal animals will not only have an irreversible negative feedback on the hypothalamus but will also have anabolic effects and block the development of the growth plates. Bilateral castration should be the treatment of choice.

### Stimulation of libido

Poor libido related to gonadal abnormalities, immaturity or behavioural inhibition can be treated using repeated low doses of testosterone, GnRH or exogenous gonadotrophins (hCG). It is, however, important to note that there is no evidence that poor libido is related to low plasma levels of testosterone. For each of the above therapies, no real dose-titration studies have been performed. As androgens may have negative effects on spermatozoal morphology or continuous negative feedback effects on the hypothalamus, the use of GnRH and hCG is preferred by the author. Doses of 1–2 μg of GnRH or 50–100 IU of hCG (Chorulon), or 0.1–1 mg methyl testosterone or testosterone propionate every 48–72 hours on three to five occasions are suggested. As reproductive behaviour in cats is essentially an acquired trained behaviour, often a temporary stimulation of libido is sufficient to permanently correct the problem.

### Other conditions in tomcats

#### Epilepsy

As described in dogs, castration and progestogen therapy may be useful as main or adjunct therapy to control some types of epileptic behaviour. The progestogen action is related to its central sedative effects.

#### Contraception

Drugs of the GnRH family, both analogues or antagonists, can be used to block FSH and LH production. The antagonists directly block the receptor, whilst agonists produce down-regulation after an initial stimulation. Both agents therefore reduce spermatogenesis and androgen production. Significant results have been obtained in other species but have not yet been obtained or clearly demonstrated in cats. Furthermore, such drugs are presently unavailable for clinical use.

Immunosterilization is another way to control reproduction; however, to date, results have been disappointing, mainly due to the need for an adjuvant

| Drug | Indication | Dosage | Side-effects | Trade name/ Comments |
|---|---|---|---|---|
| Testosterone phenylpropionate | Infertility, reduced libido | 0.1–1mg every 48–72 h, 3–5 injections i.m. or s.c. | High doses can affect fertility<br>Early growth plates closure | Androject |
| Testosterone esters | Infertility, reduced libido | 0.1–1 mg every 48–72 h, 3–5 injections i.m. or s.c. | High doses can affect fertility<br>Early growth plates closure | Durateston |
| hCG, chorionic gonadotrophin | Infertility, reduced libido, testis descent, hormonal testing | 50–100 IU repeated if necessary | None | Chorulon |
| Medroxy-progesterone acetate | Overt male behaviour, epilepsy, contraception at high doses, psychogenic alopecia | 25–100 mg to be repeated every 2 weeks, then every month, then every 2 months at half the initial dose. s.c. injection.<br>For contraception use more than 50 mg with repeated dosing<br>Urine spraying :100 mg s.c. | Mammary development and benign tumours after long-term use<br>Hypoadrenocorticism<br>Diabetus mellitus<br>Growth of white hair or disappearance of the hair at injection side | Depo-promone<br>Depo-Provera<br><br>Consistency of the contraceptive effects in male cats is poor |
| Megoestrol acetate | Overt male behaviour<br>Skin diseases (psychogenic alopecia, eosinophilic granuloma complex) | 2.5 mg/cat every day for 2 weeks, then to effect<br>Doses for dermatological problems are empirical | Mammary development and benign tumours after long-term use<br>Hypoadrenocorticism<br>Diabetus mellitus | Ovarid<br>Megecat<br>Ovaban |

***Table 17.4:*** *Main treatment, indications and dosage for reproductive problems in tomcats.*

to stimulate immunity. More recently, a new approach to cat neutralization (destruction of spermatogenesis and partial reduction of androgen production) and/or sterilization (definite functional arrest and fibrosis of the testes) using sclerosing agents has been proposed, but with unsatisfactory results. Castration remains the method of choice in animals not required for breeding.

## REFERENCES AND FURTHER READING


Allen D, Pringle J, Smith D and Conlon P (1993) *Handbook of Veterinary Drugs*, p. 678. Lippincott, Philadelphia

Chastain C, Graham C and Nichols C (1981) Adrenocortical suppression in cats given megestrol acetate. *American Journal of Veterinary Research* **42**, 2029–2035

Cline E, Jennings L and Sojka N (1980) Breeding laboratory cats during artificially induced estrus. *Laboratory Animal Science*, 1003–1005

Davidson A, Feldman E and Nelson R (1992) Treatment of feline pyometra in cats using prostaglandines F2 alpha: 21 cases. *Journal of the American Veterinary Medical Association* **200**, 825–832

Donoghue A, Johnston L, Munson L, Brown J and Wildt D (1992) Influence of gonadotropin treatment interval on follicular maturiation, in vitro fertilisation, circulating steroid concentrations and subsequent luteal function in the domestic cat. *Biology of Reproduction* **46**, 972–980

Dresser B, Sehlhorst C, Wachs K, Keller G, Gelwicks E and Turner J (1987) Hormonal stimulation and embryo collection in the domestic cat. *Theriogenology* **28**, 915–927

Feldman E and Nelson R (1989) Diagnosis and treatment alternatives for pyometra in dogs and cats. In: *Current Veterinary Therapy X*, ed. R Kirk. WB Saunders, Philadelphia

Goodrowe KL, Howard J and Wildt D (1988) Comparison of embryo recovery, embryo quality, oestradiol, and progesterone profiles in domestic cats at natural or induced oestrus. *Journal of Reproduction and Fertility* **82**, 553–561

Goodrowe KL, Howard J, Schmidt P and Wildt D (1989) Reproductive biology of the domestic cat with special reference to endocrinology, sperm function and in-vitro fertilisation. *Journal of Reproduction and Fertility* **39**, 73–90

Henik R, Olson P and Rosychuck R (1989) Progesterone therapy in cats. *Compendium on Continuing Education for the Practising Veteri-*

*narian* **7**, 132–144

Johnston S (1989) Premature gonadal failure in female dogs and cats. *Journal of Reproduction and Fertility* **39**, 65–72

Lein D (1989) Male reproduction. In: *The Cat: Disease and Clinical Management*, ed. R Sherding, p. 1475. Churchill Livingstone, New York

Lein D and Concannon P (1983) Infertility and infertility treatments and management in the queen and tomcat. In: *Current Veterinary Therapy VIII*, ed. R Kirk, pp. 936–942. WB Saunders, Philadelphia

Schwartz S (1994) Carbamazepine in the control of aggressive behaviour in cats. *Journal of the American Animal Hospital Association* **30**, 515–519

# Index

Numbers in italics represent figures